Praise for *The Gourmet Butcher's Guide to Meat*

"Is there a bible of meat? There is one now. Cole Ward's book demystifies the whole process of how animals are raised, slaughtered, and eventually make it to your plate. From learning about breeds to cutting up your own side of beef, you will be a more empowered meat eater once you read this book."

—**Rebecca Thistlethwaite**, author, *Farms with a Future*

"After nearly forty years of concentrating, industrializing, and deskilling the livestock and meat industries, a few global meat companies have separated the eater from the farmer, land, communities, and animals that we depend on for food. Cole Ward helps restore the lost craftsmanship of meat production by sharing critical knowledge about where meat comes from and how it's produced, processed, and marketed. His book takes the reader on an important journey from animal husbandry through the fading art of butchery to recipes for preparing a healthy meal, all interwoven with explanatory pictures, notes, and interesting trivia."

—**Mike Callicrate**, owner, Callicrate Cattle Company
and Ranch Foods Direct

"This comprehensive book is far more than a guide to cutting meat—it's for anyone who wants a better understanding of meat (and we all should). Engaging, informative, and, yes, fun!"

—**Nicolette Hahn Niman**, rancher and author,
Righteous Porkchop: Finding a Life and Good Food Beyond Factory Farms

"Cole Ward has done an extraordinary job of balancing the widely diverse components of meat production, marketing, and quality in this comprehensive and uniquely informative book. The author has taken every effort to present even the most contentious issues surrounding meat production from a balanced and accurate perspective. His thorough treatment of these issues provides the reader the opportunity to make a well-informed decision as a matter of personal choice, unencumbered by emotion or innuendo.

However, the real value of the book is in the articulate way Ward connects the reader to both the science and the artisanship of gourmet butchering. His comfortable style and incomparable knowledge of gourmet butchering make this a valuable resource for quality meat aficionados and a must read for chefs, butchers, and meat lovers everywhere."

—**Mark Boggess**, PhD, animal scientist and meat-industry expert

To Mike
from Greylaine Farm
Kate Jarmin's
Best

The GOURMET BUTCHER'S GUIDE to MEAT

Pasture-raised in Vermont

Greylaine Farm

MEATS & EGGS

The GOURMET BUTCHER'S GUIDE *to* MEAT

How to SOURCE IT ETHICALLY,
CUT IT PROFESSIONALLY,
and PREPARE IT PROPERLY

COLE WARD *with* KAREN COSHOF

Chelsea Green Publishing
White River Junction, Vermont

Acquiring Editor: Makenna Goodman
Developmental Editor: Fern Marshall Bradley
Copy Editor: Laura Jorstad
Proofreader: Eric Raetz
Indexer: Shana Milkie
Designer: Melissa Jacobson

Printed in the United States of America.
First printing January, 2014.
10 9 8 7 6 5 4 3 2 1   14 15 16 17 18

Our Commitment to Green Publishing

Chelsea Green sees publishing as a tool for cultural change and ecological stewardship. We strive to align our book manufacturing practices with our editorial mission and to reduce the impact of our business enterprise in the environment. We print our books and catalogs on chlorine-free recycled paper, using vegetable-based inks whenever possible. This book may cost slightly more because it was printed on paper that contains recycled fiber, and we hope you'll agree that it's worth it. Chelsea Green is a member of the Green Press Initiative (www.greenpressinitiative.org), a nonprofit coalition of publishers, manufacturers, and authors working to protect the world's endangered forests and conserve natural resources. With the exception of the cover *The Gourmet Butcher's Guide to Meat* was printed on FSC®-certified paper supplied by QuadGraphics that contains at least 10% postconsumer recycled fiber.

Library of Congress Cataloging-in-Publication Data

Ward, Cole.
  The gourmet butcher's guide to meat : how to source it ethically, cut it professionally, and prepare it properly / Cole Ward with Karen Coshof.
    pages cm
  Includes bibliographical references and index.
  ISBN 978-1-60358-468-5 (hardback) — ISBN 978-1-60358-469-2 (ebook)
1. Meat industry and trade. 2. Meat cuts. 3. Slaughtering and slaughter-houses. I. Coshof, Karen. II. Title.

  TS1970.W37 2014
  664'.9029—dc23
                        2013035609

Chelsea Green Publishing
85 North Main Street, Suite 120
White River Junction, VT 05001
(802) 295-6300
www.chelseagreen.com

To my sons, Todd and Christopher, and to my grandsons, Benjamin and Nicholas. Todd is an awesome butcher and Christopher a financial wizard; both have encouraged me in everything I do. Ben and Nick are the apples of my eye; Ben is my old soul buddy and Nick, my musical pal. I'm proud of them both. I must also include my sisters, Jeannette, Lillie, and Priscilla, for always believing in me. I am blessed to have such a caring family.

A painted trillium in the Vermont woods.

CONTENTS

ACKNOWLEDGMENTS

Without Karen Coshof this book wouldn't be possible. Thank you for your hard work, research, and dedication—and for pushing me to do what I thought impossible.

Thanks to Mike Taylor for his encouragement and for laughing at all my off-color jokes. Thanks to Joe and Lorraine Padulo for being my friends, for constant encouragement, and for loaning me their beautiful kitchen to shoot *The Gourmet Butcher*'s DVD.

Thanks to my wonderful cousin Judy Butler for her support and many hours spent typing up my unreadable handwriting.

Thanks to Ryan Clapper for his feedback and opinions, and to Todd Keworth and Jon Wilson (former meat department and store managers) for giving me free rein on the butcher blocks I ran for them, as well as for their encouragement, friendship, and support.

Thanks to Joe Monthey for being a great apprentice and for uncomplainingly hauling my equipment around.

Thanks to my former employer Anthony J. La-Frieda for teaching me the finer aspects of the meat business and for all of his praise during the time I worked for him.

Thank you to my best friend, Karen Warren, for supporting me, believing in me, and loving me. She has gotten me over a lot of rough spots.

Thanks to Sam Fuller of NOFA Vermont for offering me the opportunity to teach so many of NOFA's workshops, believing I was the best, and introducing me to so many wonderful locavores. Thanks to Jenn Campus and Roberto, and to my "other Jenn"—Jenn Colby—who asked me to do workshops at the Vermont Grassfed Conference.

Thanks to Ben Hartwell of the Maine Grassfed Conference as well as to the staff at Sterling College in Craftsbury, Vermont, for inviting me on so many occasions to present a meat-cutting workshop.

Thanks to Walter and Holly Jeffries and their kids for supporting me, for getting the word out on all of our events, for being great farmers, and most of all for just being the Jeffries—great friends at Sugar Mountain Farm, West Thompson, Vermont.

Thank you to Mike Bowen at North Hollow Farms of Rochester, Vermont, for just being Mike. Mike is one of the most honest farmers I know. Many, many thanks to him for supplying the animals processed in this book.

Thanks to Paul List for his contributions to the book and who as a farmer is doing it right.

Thanks to the farmers and butchers who shared their time and stories with us (including their many unprintable jokes).

Thanks to my customers past and present who have followed me for years. Their feedback and loyalty have been invaluable.

Many thanks to Chelsea Green Publishing for giving me the opportunity to write this book, with special thanks to Makenna Goodman and the wonderful support staff.

Thanks to the Essex Center for Technology for inviting me for so many years to teach a workshop for their culinary students.

Thanks to Chef Courtney Contos, friend, chef, and co-star of *The Gourmet Butcher*'s DVD who gives way too much praise. Her cooking tips are the best.

Finally, a from-the-heart thank-you to Fern Marshall Bradley, whose gentle editing and dead-on suggestions have made this a much better book.

HOW TO USE THIS BOOK AND CD

When Karen Coshof and I began thinking about this book, it occurred to us that simply explaining how to cut meat just wouldn't . . . cut it. If you love something, you want to know more about it, and the world of meat is, believe me, a very complicated place. For a substance that's virtually everywhere—particularly inside our bodies—you'd think people would know more about it.

This book will take you into the world of meat: how it came into our lives, how butchers developed and thrived, how meat animals were domesticated and evolved into the umpteen breeds we see in pastures today, how farmers raise them, and so much more, including:

- The sources of meat.
- The terminology of the meat sector (to help you be a smart shopper).
- Meat safety issues.
- The extremely confusing world of meat marketing claims ("organic!" "natural!" "free-range!" "grass-fed!" "pasture-raised!" and so on).

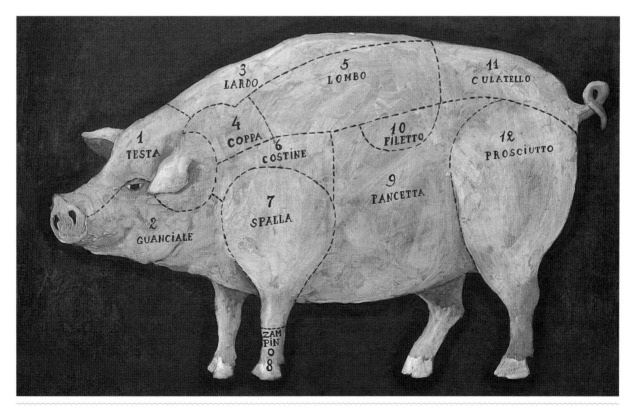

Courtesy Macelleria Falorni, Italy.

- Which cuts come from what place on a carcass.
- How-to instructions for cutting up a side of beef, lamb, and pork, plus a whole chicken.
- Butchering tools and equipment.

WHAT'S YOUR PLEASURE?

So . . . this is *not* just a cutting guide. It's as close to an *everything-you-need-to-know-about-the-meat-you-eat* guide as we can write. And it's not just a book, either. As we developed the manuscript and gathered photos, we realized that we could never provide all the visual guidance we wanted to offer on the pages of a book. So the project evolved into a book and CD set, and that's what you're holding in your hands. In the back cover of this book is a CD that contains an entire butchery course—the same course I teach to culinary students. So when you're ready to cut up your own beef, pork, and lamb, or discover the butcher's method of cutting up a chicken, just pull out the CD, slip it into your computer, and away you go. Over 800 step-by-step photos of the entire process, in simple English that even I understand.

How you use this book will depend on what you want to do and learn. Don't be cowed by the book's hefty size—you don't have to start at page 1 and read the whole thing. Instead, jump right into the chapter that interests you most.

SO IF YOU'RE . . .

- Fascinated by history: You'll like chapter 2, the history of butchery.
- Interested in animal breeds: Take a look at the descriptions and photos of livestock breeds in chapters 7, 8, and 9 (and you thought you knew what a cow looked like).
- Confused about how beef is produced in North America: See chapter 7.
- Ditto pork and lamb: See chapters 8 and 9.

Ten Myths About Meat

Just to get you thinking, ponder the following statements and ask yourself—are they true or false? And we've given you a hint about where to look to double-check your answers.

1. All butchers and meat-cutters are the same. (chapter 1)
2. Eating meat has nothing to do with being human. (chapter 2)
3. The more I pay for a cut of meat, the better it will be. (chapter 3)
4. Farmers are not very sophisticated (they live in the country, after all). (chapter 4)
5. Meat just happens. (chapter 4)
6. If it's in my supermarket, I can trust it. (chapter 5)
7. Cattle can't digest grain. (chapter 7)
8. Pigs are dirty. (chapter 8)
9. Sheep are stupid. (chapter 9)
10. Chickens are dumb. (chapter 10)

- Wondering how meat gets to your supermarket (and what happens once it gets there): See chapter 5.
- Curious about how animals become meat: See chapter 4.
- Eager to start butchering: See chapter 6.
- Ready to work on a beef carcass: See chapter 7 and the CD.
- Ready for a pork carcass: See chapter 8 and the CD.
- Ready to cut up an entire lamb: See chapter 9 and the CD.
- Interested in chickens (and who wouldn't be?): See chapter 10 and the CD.
- Wishing to make sausages: See chapter 11.

Our point is that we've designed this book as a tool (like Cole's knife, only not as sharp). Dip in and use what you need—no need to read the whole thing.

Introduction:

About Me, by Cole

I've been a butcher all my working life. Over those 40 years, I've watched my craft dwindle as once commonplace neighborhood butcher shops vanished from America's streets. My philosophy of the craft is based on personal observation (not formal education, which I didn't get a lot of), and took shape many years ago after I witnessed poor practices in raising livestock and processing and marketing meat. I learned long ago that honoring the animal, nose-to-tail eating, choosing locally raised meat, and eliminating harmful substances from the food chain aren't (as some may think) new-wave or counterculture ideas, but good commonsense agricultural practice. The more people understand the reality and impacts of large-scale conventional meat farming, the more they'll realize that this "counterculture" trend is a pretty good thing.

I'm proud of what I do and determined to keep butchery alive.

MY STORY

I was born in the small mill town of Sheldon Springs, Vermont—known in the late 19th and early 20th centuries for its invigorating waters. My father and uncles worked at the local pulp mill, which owned all the houses in town. It also owned the general store, which—when I was a kid—was leased to a guy who ran it as an IGA store. It was a much slower time. I still recall the mill whistle blowing at noon and everything closing down for lunch. It was 1956, and I was six years old.

My parents separated several times during my childhood, finally divorcing in 1958, at which point my mother bought her own house in her home town of Richford, Vermont. The price of the house was $600.

Mom was one of 17, and her siblings—our aunts and uncles—would gather at our house, drink my mother's homebrew, and play music. Everyone played an instrument. Mother was the only one

Here I am with my Olds Regency in Montana.

who never drank, but she played a mean piano and violin. This was our entertainment.

But by 1960, she could no longer afford to keep the old house and six remaining kids at home. So she placed the five youngest (including me) in an orphanage known as the Warner Home for Little Wanderers. We would be at the Warner home until 1964; for a very long time we five would be referred to as "those Warner kids."

It wasn't the worst place we could have been, plus it boasted a really nice piano in the parlor. Now, back at home we'd always had an old piano, and I'd fooled around on it as young as I can remember. So I wanted to play that orphanage piano. But the matron wouldn't allow any of the orphanage boys to play. Only the girls. She said boys who played the piano were sissies. Mind you, she loved Liberace.

I would sit on the stairs in the hallway and listen to my sister and all the other girls play. It frustrated me so much that I promised myself that one day I would learn to play better than any of them. I kept that promise.

My mother eventually became head cook at the Warner home and saved up enough money to move me and my siblings into an apartment over a mom-and-pop grocery store owned by Tony and Arlene St. Marie. Tony and Arlene became my godparents. It was Tony who gave me my first real job at the age of 14, washing meat trays and stuffing sausages at 20 cents an hour. I worked 54 hours a week in the summer, and during the school year on weeknights and weekends. I loved it.

One of the reasons I enjoyed meat cutting was that everyone treated butchers with respect. I had an uncle who was a butcher in a high-end market in Connecticut, and he was treated like a minor celebrity. This helped influence my decision to enter the trade. I supplemented my income from cutting meat by playing piano on the weekends at local bars.

At 15, I got a meat-cutter's apprenticeship at an IGA in St. Albans, Vermont. The owner, Jim Bray, was the nicest guy I had ever met. It was through Jim that I picked up meat-cutting skills as well as social and business skills and a serious work ethic. I moved on, after Jim sold Bray's IGA, to become meat department manager of Joe's Country Store, the original old company store in my hometown of Sheldon Springs. From there a few years later to a slaughterhouse in upstate New York and then to one of the supermarket chain stores.

After a few more years in the Northeast I wanted some adventure, so my kid brother and I decided to head to California and seek our fortune. A good friend of mine had moved there a year earlier and encouraged me to come out.

It was a great road trip on a very limited budget. My kid brother and I arrived in Los Angeles in early September 1978 at two in the morning. We sat around and drank beer and wine, shooting the breeze. I finally crashed on the sofa, to be awakened by the telephone ringing at 8 a.m.

I answered the phone and heard a woman say, "Did you apply for a meat-cutter job?" I couldn't possibly have, since I'd arrived six hours earlier. But, half asleep, I replied, "Yes I did." She said, "You have an interview at one this afternoon." I scrambled for a pen and paper to take down the directions. Then I woke up my friend and he helped me figure out exactly where I was going for this interview. I put on a suit and left early so I had plenty of time to find the place.

The meat-cutter job was at LaFrieda Prime Meats at the farmers' market in Los Angeles. This was a famous butcher shop, but I didn't know it. The owner was Tony LaFrieda, and his sister had offered to set up his interviews for him. Seems she had dialed my friend's phone number by mistake.

The cutting area at LaFrieda Meats was long and narrow with a 50-foot service meat case, kind of an aqua color. The walk-in refrigerator had glass on the side facing the customers so they could see everything. There was about an inch of sawdust covering the floor, typical of the 1950s and '60s—but this was 1978. So I was a bit surprised. You could look out under the green-and-white awning and see the sky.

A young man in the office asked if he could help me. I replied, "Yes, I'm here for the meat-cutter position." He looked at me strangely and said "Just a minute." He picked up the phone and I heard him say, "Uncle Tony, you're not gonna believe this but there's some guy here in a three-piece suit looking for the meat-cutter job." My first thought was, *What's so strange? I always dress up for an interview.*

Anyway, the young man said, "Uncle Tony will be down in about half an hour. Would you like a cup of coffee?" I accepted and started to read some magazines. After about 45 minutes the phone rang. The young man said, "Uncle Tony wants to talk to you."

I took the phone and a gruff voice on the other end said, "Where you from?"

I replied, "Vermont."

He said, "Where in the hell is that?"

"In the Northeast."

"How long you been cutting meat?"

"Eleven years."

He said, "I'm not coming down. I've got 800 bucks on a football game and my team's winning. Are you willing to bust your ass for 500 bucks a week? I mean no lunch breaks, no coffee breaks?"

And this is where I worked—it's the LA Farmers' Market in the 1970s. *Courtesy A. F. Gilmore Company.*

My top pay in Vermont as a meat-cutter had been $260 a week. So I replied, "Yes sir."

"Be at the market Monday at 8 a.m. and if you're not 15 minutes early, I will consider you late."

I was there at 7:30.

At first, Tony seemed like an ass to me, but he mellowed. Or maybe I did. On the third week, Tony threw the keys to the store and the combination of the safe at me and said, "I bought a pizza joint and deli 25 miles from here, so you're gonna run this place. Here's my phone number. If you have any questions call me. You're gonna get 100-dollar-a-week raise, plus a monthly bonus."

So at 600 bucks a week, plus bonuses, in 1978 I was styling. Working for this guy was the best education I ever got. He was good to me and came in periodically to teach me the finer aspects of the trade. Soon I began to see butchering as an art form. Because of the market's location near CBS Studios, many movie and TV stars shopped there. My customers included Raymond Burr, Gisele MacKenzie, Edith Head, Isabel Sanford, Perry Como, Bernadette Peters, Billy Crystal, and many more celebrities. I also did meat props—like roasts used in scenes on TV sitcoms like *Three's Company*. But that's another story.

By 1982, the bloom was off the LA rose. I was homesick for my family and Green Mountains, and decided I needed some Vermont. So in late '82 I returned to the East Coast. I worked in various markets around the state, then some larger chain supermarkets, then finally a local organic store as meat department manager.

At the organic store, working with meats free of growth hormones and antibiotics confirmed what I had known from my time in the big chains: The commercial meat products being sold to the average supermarket consumer were far from healthy. This is when I first began to investigate locally raised meats.

I began giving workshops and lectures and doing on-farm meat cutting for farmers. Then in 2009, I took a position at a wonderful old country store in Morristown, Vermont, owned by Joe Padulo and his wife, Lorraine. It was while I was at their store

Pig butchery class—that's Todd on the right.

that I began to focus on educating people about meat. Through Joe Padulo I met Michael Taylor and Karen Coshof, who produced a gourmet butchering DVD (*The Gourmet Butcher: From Farm to Table*) with me and my good friend Chef Courtney Contos. Which eventually led to this book.

I have three children: a son, a daughter, and an adopted son. My adopted son is assistant vice president of a major bank. My oldest son, Todd, works with me doing butchery workshops (he's also an awesome meat-cutter).

Todd has a terrific family: two beautiful grandsons whom I adore. Ben, the older of the two, and I share an interest in anything to do with cars. And his younger brother, Nick, and I share an interest in music. Nick plays guitar and keyboard, and I play the piano and accordion.

Now I butcher and teach. I lead workshops at several liberal arts colleges and grass-fed beef conferences, do presentations for the Northeast Organic Farming Association, and give private workshops. Onward.

A PASSION FOR THE CRAFT

From the time I first picked up a knife, I wanted to be the best I could be. I wanted to understand the anatomy of any animal I might encounter: beef, pork, lamb, venison, bear, and wild boar. Over the years, I've run into all of these and more, including moose, elk, and goat. Transforming them into gourmet cuts as pretty and appealing as anything offered by the most exclusive meat market became my goal.

As for customer service (to quote my old boss Tony LaFrieda), "Anyone can wait on an easy customer; it takes an artist to wait on an asshole." He was right. It is both a challenge and a reward to get a difficult customer to sing your praises.

When I consider the meat I process and sell, I think about its intimate relationship to the person who buys it. That meat will be going into that person's mouth and stomach—it will literally become

Teaching students how to handle a meat saw at a pig workshop at Sterling College, Vermont.

Buying Chicken Legs

I can't help being a practical joker, even when I'm waiting on customers. Fortunately, most of my customers have a good sense of humor. At the IGA where I worked many years ago, we had a customer named Fannie. The meat manager Moe always served her. Moe was at lunch one day when Fannie came in so I asked if I could help her. She replied "Moe always waits on me." I said "If you're not happy with my meat, I'll pay for it myself." She relented and said "I just want some chicken legs."

I asked her if she wanted the front legs or the back legs. She replied, "What's the difference?" I said, "Well, chickens run on their back legs, so the front legs are more tender." She said, "I'll take the front legs."

Then one day Fannie came in while I was at lunch and Moe was there. She told Moe that she wanted some front chicken legs. Moe replied, "You mean the ones down in the front of the meat case?" Fannie said "No, the front legs of the chicken. Cole said they were more tender because chickens run on their back legs." Moe began to chuckle. "How many damn legs does a chicken have?" he asked. Fannie thought about it a minute and started laughing. She said, "That son of a bitch has had me asking for front chicken legs for six months." We never let poor Fannie forget this one. And she remained a loyal customer until the store was sold.

part of them. I also know that small children consume meat, and that at very young ages their immune systems are not fully developed. I always assume that whatever I process may be eaten by children, and that drives me to be absolutely sure that it's handled under the strictest sanitary conditions and that it's as fresh as possible. It is also important to me to know how, where, and under what conditions the animal was raised. It gives me tremendous pleasure when a customer comes to me and tells me how wonderful that steak, roast, or a value-added product I prepared for them was.

MY MISSION

Over the past 15 years, I've become deeply concerned about the quality of the meat sold in the United States and how meat animals are raised, processed, and brought to market. What I learned and am still learning about the commercial meat sector was so startling, it nearly transformed me into a vegetarian butcher.

To use an analogy . . . when you buy a car, you don't blindly trust what the car salesman tells you (unless you're my cousin Delmer). You research the car—check its safety ratings, mileage, incidence of repair, and resale value history. If you're savvy, you also consult resources such as *Consumer Reports* and the Internet. You probably do the same when purchasing electronics or household appliances. Why? Because you want the best value for your dollar.

But when it comes to meat, most people just walk into the supermarket and throw items into the grocery cart, no questions asked. Consumers take for granted that the meat they buy is healthy and safe. But hold on a minute . . . this is going into your mouth and body. You are *literally* consuming the product you buy.

You wouldn't buy a new car if it only had three wheels, or a microwave that zapped you every time you touched it. Why shouldn't you have the same concerns about your meat? The reason is ignorance. The average consumer simply doesn't know what actually goes on in the industrial meat business. And the meat business likes this scenario just fine.

A strong artisanal culinary butchery sector would be a powerful bulwark against unsafe practices. A good butcher is an ethical professional who knows the provenance of his or her meats. I want to give everyone an understanding and appreciation of my craft and its culinary artists, and I want to celebrate and support our struggling small farmers and quality meat producers. So my mission is nothing less than to bring back culinary butchery—a craft that we must never lose. We must and will keep it alive for the next generation.

And Now . . . You

This book is about empowerment. It's about convincing you that you deserve—and should demand—more. In my occasional moments of sanity, I realize that all 300-plus million Americans won't be reading this. But you will. And I have a pretty good idea who you are . . .

You're a worried meat-lover seeking straight talk about how meat is raised and processed.

You're a savvy consumer who wants to save money on food and learn a new skill.

You're a small farmer who wants to extract more profit from your herd while at the same time producing the best-quality meat in a sustainable manner. You may also be thinking about adding some value-added products, or creating an on-farm shop.

You're a culinary student, cook, or chef who wants to understand the meats you'll be working with and the variety of cuts that a single carcass can provide.

You're an individual considering butchery as a career.

Or maybe you're just someone who's kinda curious.

This book is for all of you. Have fun!

Chapter One

What Is a Butcher?

Once, in America, there was a butcher shop on almost every corner. The tiny Vermont town where I spent some of my youth had three. To a 1950s homemaker, the neighborhood butcher was as important as her hairdresser. He didn't just sell meat; he taught her which cuts were best suited to different types of cooking: slow moisture cooking, dry cooking, or grilling. He would guide her through the quantity she needed for her family or guests. Eventually a special kind of relationship formed, built on trust. The butcher often knew the entire family. There was always time for a little conversation or even a joke or two.

I remember my mother saying that it was really her butcher who taught her how to cook meat. She'd order something, and he'd say, "No, that's not what you want, you want *this*." And he'd give her a cut of meat she'd never tried before and tell her how to cook it. It was always delicious.

Those days are pretty much gone—unless we want them back. Today, people are starting to question where their food comes from and how their meat is raised, and I think that's great. People are looking for real butchers again. Finding one may not be easy, depending on where you live, but don't give up. We're out there.

Old butcher shop. *Courtesy City of Toronto Archives (Fonds 1244, Item 3504).*

SEPARATING THE BUTCHERS FROM THE SLICERS

You may think that the guy or gal who cuts and packages meat behind the glass window of your supermarket meat department is a butcher. You're wrong. Those folks are meat-cutters. In the case of today's large supermarkets, they're really what I prefer to call meat slicers. Their training and knowledge are limited to a small set of skills that

Bertrand Makowka of Boucherie Viandal in Montréal is a master butcher who—with his brother Yves—is carrying on the family tradition of whole-animal butchery begun by his father 30 years ago.

they repeat over and over. They probably can't tell you the difference between one cut of meat and another, or how to cook it. And I'd bet that they probably don't know where the meat they cut comes from, or how to judge its quality.

There's a lot more a meat slicer doesn't know that a good butcher knows. A true butcher—and there are very few left—is someone who can take a live animal from slaughter to table. This means a person who knows how to butcher (kill) the animal, how to skin and gut it, how to divide the carcass into large basic cuts called **primals**, and how to subdivide these further into the myriad table-ready cuts you buy at the store.

A truly good butcher has extensive knowledge of all muscle groups in an animal . . . how hard each muscle group works and which muscles are better suited to different methods of cooking such as braising, grilling, oven roasting, broiling, pan searing, or dry uncovered cooking.

A good butcher is knowledgeable about specialty cuts and skilled at special preparations like frenching a pork loin or deboning a leg of lamb without opening up the leg. A good butcher has an eye for quality and presentation. (That old saying—Customers buy with their eyes—is true.)

It pains me to have to point this out, but *a good butcher is honest*. A good butcher never cheats or short-weights a customer, or sells meat of questionable quality or freshness. A good butcher gives customers what they've asked for and never switches out product (a trick I've seen way too often). A good butcher never substitutes one grade of meat for another.

A good butcher promises only what he or she can deliver (if a customer wants an order at a given time, the customer will get it at that time). And a good butcher never lets personal issues seep into work. Finally, a truly good butcher loves his or her job and has a genuine passion for meat and food in general.

WHERE DID ALL THE BUTCHERS GO?

Up to the early 20th century, a butcher was a butcher in the true sense of the word. Butchers would buy a live animal at auction; take it home or to their place of business; slaughter, skin, and gut it; break it down;

and try—in the short time available (no fridge, just melting ice!)—to sell it directly to customers.

With the advent of refrigeration and the ability to keep meat products longer, the trade divided into two separate crafts: the butcher working at the wholesale level in the slaughterhouse and the meat-cutter working at the retail meat market.

The butcher handled the initial steps of killing, skinning, and gutting the animal, then breaking the carcass down into the largest chunks: typically, beef quarters, pork sides, or—in the case of lamb—a cleaned whole carcass. The butcher would then sell these to a retail meat market. At the retail level, other people would further break down these large pieces into primal cuts (large sections of a carcass such as a whole loin of beef), subprimals, and table-ready cuts, displaying them in refrigerated cases. These people became known as meat-cutters . . . and a new trade was born.

The name *butcher* stuck around, but it's important to understand the difference.

I am a butcher because I can slaughter, skin, gut, break, and process meat in every stage: from large primals to ready-to-eat table cuts. The people in most supermarket meat departments and a surprising number of butcher shops are meat-cutters. Please understand that I do not mean to discredit meat-cutters. In fact, these days that's what I call myself, although I'm technically a butcher. My point is that you can't assume that the "butcher" in your supermarket knows much about the meat he or she is selling you.

Why is it so hard to find a good butcher nowadays? The reasons are many, but one important one is a lack of thorough, top-level professional training. Because there are fewer and fewer master butchers, the knowledge of their craft is disappearing. So naturally, we're seeing less-than-thorough training given by instructors who have had less-than-thorough training themselves.

I learned to be a butcher through an apprenticeship with an experienced butcher at the local IGA store in my hometown. We handled and sold beef and lamb from hanging carcasses only. Pork was purchased by the case—a case of whole loins, a case of fresh hams, and so on. Chickens came whole, packed in ice in thin wooden crates; from these we produced all our chicken parts: legs, thighs, drumsticks, as well as bone-in or boneless breasts.

In an average week, we would receive six to eight sides of beef in two deliveries. The store had three full-time butchers, plus two to three part-time meat-cutters. The store's slogan was *Bray's IGA . . . where you're always treated as the invited guest.* This was not just a meaningless phrase; we took it seriously.

It was this wonderful early training and supportive environment that helped me understand the value of customer service. We were happy to go to work; in fact, I often spent my days off hanging out at the store because of its great atmosphere.

In those days, an apprenticeship lasted two and a half years, during which the apprentice honed the skills of breaking down sides of beef and pork and whole lambs into primal cuts and then cutting those primal cuts into individual steaks, chops, roasts, and more for retail sale.

But apprentices had lots more to learn, too. They were expected to master basic aspects of the retail meat business, such as how to set up an appealing display, how to calculate profit margins, how to cook cuts of meats, and how to handle and cook organ meats like heart or tripe. They were also expected to understand and apply principles of good customer service.

A new apprentice began by learning how to clean and sanitize the equipment and meat-cutting prep area. Then he—and I'm using *he* because there were virtually no female apprentices then (things are changing now)—learned how to grind different meats, such as ground sirloin or ground round, to the customer's specifications.

The apprentice would move on to learning how to break primal cuts into retail cuts. Only then would he be taught the finer aspects of the trade, like butterflying a steak or chop, frenching a rack of lamb, or cutting a pocket into a cut of meat for stuffing. All of this training prepared a young person properly for a career as an independent butcher or for a management position in a meat or specialty market.

Meat Industry Metamorphosis

The decline in the number of butchers and increase in meat-cutters really came about during the 1970s. As large supermarket chain stores began to infiltrate my home state, small butcher shops and the mom-and-pop stores that sold meat found it harder and harder to survive. Some managed for a while by staying open on Sundays. But by the early to mid-1970s, supermarket chains were doing the same thing. The smaller retail stores tried to stop them by getting the state of Vermont to enforce existing statutes known as the Blue Laws forbidding the sale of most non-edible items on Sunday.

Soon supermarkets stopped receiving whole carcasses. Instead meat arrived already broken into primal cuts and packed in boxes labeled so that all the "butcher" needed to do was open the box and cut along the dotted lines. With this introduction of what became known as "boxed beef," the need for highly skilled employees began to drop off. Meat departments that had employed three full-time cutters could now do with one, and pay him or her much less because the work simply didn't require as much knowledge or expertise.

By the early 1990s, the average age of a true butcher was 59½ years old. In a trade that was once paid on par with electricians and plumbers, a meat-cutter now was lucky to get $12 an hour.

And even this is changing. I fully expect that soon all meat will be delivered to supermarkets vacuum-packed as retail cuts, thus completely eliminating the need even for meat-cutters or meat slicers. From the perspective of large meat processors, this is an ideal scenario because it reduces their costs and provides ideal inventory control. But in my opinion, it's not in the best interests of the customers. Like you!

The responsibility for the devaluation of culinary butchery lies mainly with the commercial meat industry as well as large supermarket chains. They have preyed on the ignorance of today's average consumer. The upshot is that the commercial meat industry in the United States has brought a once respected trade to near extinction.

Which begs the question: Which came first, the ignorant consumer or the venal meat industry? I'm exaggerating, of course, to make a point. Old-time butchers educated their customers. In my opinion, supermarket chains have spent the last 50 years de-educating customers. They couldn't resist it; the industrialization of the meat business combined with cost and volume pressures inherent in large-scale retail chain operations inevitably led to the devaluing of the butcher's craft, the "homogenization" of product variety, and a general dumbing down of expertise and knowledge.

Consumers aren't stupid, but gradually, in sync with the diminution of culinary butchery and the rise of heavily marketed commercial meat products, consumers have lost their best resource: the knowledge shared by an experienced culinary butcher. Who can blame folks for picking up their steaks at the supermarket when it's probably the only meat purveyor near to hand?

From your perspective as an individual customer, this shift in the retail meat world means the loss of service and choice, plus a loss of knowledge about the source and quality of the meat you buy. I'm guessing that it's this loss of choice and control that's one of the reasons you bought this book and are considering learning the art of home butchery.

WHAT EVERY TRUE BUTCHER SHOULD KNOW

Being a butcher is so much more than cutting meat. It requires great customer service skills, a good sense of humor, a great work ethic (including a clean, neat appearance), and the love of food. If you're doing butchery at home, customer service skills may not apply, but the rest still does, unless your family consider themselves your customers (just don't let them get too uppity).

As far as I'm concerned, anyone who wants to be a true butcher should have the following knowledge:

- Familiarity with the most common meat animal breeds and their attributes.
- Detailed anatomical knowledge of each type of meat animal (you'd be surprised how many "butchers" don't have this). This includes knowing where every muscle or muscle group is located in the animal and the best cooking method for it. And I mean every possible cut and option.
- Where the animal was raised, how it was fed, whether or not antibiotics or growth hormones were administered, and exactly where and how

Master butcher Ralph Citarella shows off an expertly prepared rack of American lamb. Citarella goes by the motto, *I am an artist and meat is my canvas. Courtesy Ralph Citarella.*

it was slaughtered. In my opinion, assurance that the animal was *humanely* raised and slaughtered is particularly important.

- How long a carcass was aged and what aging method was used (as well as ideal aging times and methods for each type of meat animal worked with).
- How the animal was finished (fed in the few months before slaughter). Was it grass-fed and grass-finished or grass-fed and grain-finished? Was it penned or pastured most of its life? I expect a butcher to understand the pros and cons of each method and not to be a slave to the latest culinary trend.
- How to deal with unpleasant things in an animal such as tumors, abscesses, or a dark cutter (see chapter 4 to learn what a dark cutter is), and how to tell the difference.
- The location of certain glands that can adversely affect the taste of the meat, and how to remove them.
- How to safely use meat-processing equipment.
- Proper sanitation methods: A butcher's work space and shop must be spotless at all times.
- Regional preferences and terminology for cuts of meat. For instance, tri-tips are common in California, Yankee pot roasts in New England.
- How to create an artful meat display, perform a cutting test, figure gross profit margins, and add value and quality to a piece of meat.
- Familiarity with different spices or side dishes to accompany a given species or preparation of meat.
- Cooking times for various cuts of meat as well as most suitable doneness—rare, medium, or well.
- A set of good recipes, as well as the ability to help a customer plan a meal around a specific cut. (I give my phone number to customers in case they get the least bit nervous when preparing a holiday or special meal for guests.)
- Knowledge of wines. A question often asked of a culinary butcher is, "Can you suggest a good wine for this cut of meat?"
- For value-added products, which ingredients—like gluten or peanuts—can cause allergic reactions.

This superb meat display is a daily showcase at MacDonald Bros. Butchers in Pitlochry, Scotland. *Courtesy Rory MacDonald.*

But wait, there's more! You need to top off all of this knowledge with a few other important qualities, including respect for other butchers' tools and work space, concern for your customers and their eating habits, a total commitment to please even the most difficult customer with a smile, and the very best service possible.

How to Recognize a Well-Run Meat Department

Customer service is part of good management, but there's more to consider. Let me introduce you to a well-run meat department . . .

Its physical space is designed for efficiency of movement and furnished to be ergonomically comfortable for its staff and its customers. All the equipment in this department is spotless, in excellent mechanical working condition, and regularly maintained, so accidents rarely—if ever—happen.

Every employee has been trained in the safe use of the equipment as well as proper cleaning and sanitizing procedures. They know to turn the power source off when they dismantle and clean the power saws. And they've been well trained in knife skills, knife safety, and the proper way to use hand tools and other shop instruments.

Every person in this meat department knows exactly what to do if someone gets injured and what precautions to take to eliminate any contamination caused by blood pathogens.

Cleanliness and sanitation are ongoing processes, and the department and its equipment are cleaned and sanitized throughout the workday, then fully cleaned and sanitized at the end of each day. The place is spotless. It's not just about appearance; proper sanitation is an important factor in the

The After-Life of Meat

Display life is the time period that meat holds a nice fresh color and remains visually appealing to the customer.

Shelf life is the period of time during which meat is safe to eat (in other words, no spoilage is present).

To demonstrate, this steak—in its original store packaging—was photographed over a period of nine days. Take a gander.

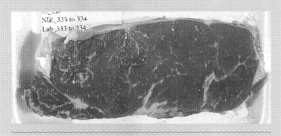

Day one. The steak is red and fresh looking; it's at the start of its shelf life, and will taste fresh and delicious. *Photo by Steven Shackelford. Courtesy USDA.*

Day six. The steak has slightly darkened, and brown coloration is creeping in at its edges. It is still edible, but has reached the end of its viable shelf life. I wouldn't buy it. *Photo by Steven Shackelford. Courtesy USDA.*

Day nine. Would you eat this? I wouldn't. This steak should not be on the store's shelf. *Photo by Steven Shackelford. Courtesy USDA.*

maximum shelf life of meat. The more bacteria are present on equipment, cutting surfaces, and meat cases, the shorter the shelf life of the product.

This meat department is very profitable because its managers understand how to calculate gross profit margins and verify them by performing accurate cutting tests to determine the true cost of their product after the removal of waste material like fat and gristle. They know the difference between display life and shelf life and understand that packaging and display are prime influencers of a decision to buy, because nothing stimulates appetite like a beautifully presented and decorated meat case.

That's why this meat department has two kinds of display case: a pre-packaged case and a service case. Only the most experienced employees work at the service case, where beautifully presented and absolutely fresh gourmet cuts are offered behind a glass window. Nothing is pre-packaged here. The butcher reaches into the case to get you the cut you've chosen or will cut it for you to your specifications. And if you ask, the butcher will offer recipes for the cut, as well as suggestions for presentation.

Another reason this meat department is top-notch has to do with ordering policies and ethical practices. Ordering is critical, because high inventory ties up money and creates older product that then has to be discounted or frozen (two practices I'm against).

Down the street from this meat department is a store whose meat department is—shall we say—a little more lax. They often end up with too much

Cole's Notes

Ever wonder why, when you go to a service meat counter (*service* means "staffed") and find two layers of meat in the case, there's pink paper between the layers? This is to keep the steaks from turning brown. Without the paper the meat would turn brown in less than five minutes.

This is an excellent example of a spotless value-added prep room. It's at Boucherie Viandal in Montréal.

inventory, so they freeze it for later rethawing. No problem, right? Wrong, because they then sell the defrosted meat as fresh. In Vermont (and most other states), this is illegal. Any previously frozen product must be labeled as such.

In the well-run department, the manager and staff know their customer base and track weekly sales. With this data in hand, they can plan their inventory so that they sell out of most products by Sunday and start fresh on Monday with a fresh delivery. Other deliveries are scheduled throughout the week—all to make sure that everything on offer is fresh.

They also allow for changes in demand. They keep their eyes on the weather, because cold, wet, or snowy weather keeps folks in the kitchen cooking roasts and stews. Nice sunny weather lures them out to the grill to cook steaks and hamburgers.

This place also creates its own value-added products such as stuffed chicken breast, roulades, and a variety of fresh sausage from scratch. Specialty oven-ready products help increase their gross profit margins.

Have you enjoyed your little tour? Bet you'll come back.

May I Serve You?
Defining Customer Service

In the retail meat business, providing friendly *quality* service is one of the most difficult and stressful tasks. Whenever I think about customer service, I remember what my former employer Tony LaFrieda used to say: "Remember, anyone can wait on an easy customer. It takes an artist to wait on an asshole!"

Tony was right that customer service is an art. I try never to think of any customer—no matter how challenging—as an asshole. I think of them as people just like me, but who have suffered multiple experiences with poor service and crummy products. Can't blame folks for being leery after they've been ripped off a few times.

MONTRÉAL, CANADA: BOUCHERIE VIANDAL

In the old working-class neighborhood of Verdun sits a gem: Boucherie Viandal, a true culinary butcher shop that is a mecca for meat-lovers. The shop was started by Polish master butcher Ted Makowka with his French (from France) wife, Raymonde Devroete, and now has absorbed the second generation: Sons Bertrand and Yves work side by side with their parents.

The establishment takes up two floors: butchery on the ground floor and an enormous kitchen upstairs—the province of Mme. Devroete—for the creation of the many value-added products sold here.

There are 17 full-time staff: 6 professional butchers, service clerks, bakers, and so on. The shop occasionally takes on apprentices, but they're hard to find. Not as hard to find as good butchers, though, according to son Bertrand, who's the general manager and a *very* busy man. He says:

We're a family business, so everyone has their own job. Except me—I seem to do everything. A typical day starts at 8 a.m. (we open at 9) when we prepare for the day, doing the meat prep, cutting steaks and roasts for presentation, and so on. It takes an hour to an hour and a half to get everything done; we have several self-serve and one full-service meat case, and they're all large.

Our customers are extremely diverse: French, English, Japanese, Korean (we've developed a profitable sideline preparing meats for Oriental cuisines). About 75 percent of our customers are individuals, but we also supply restaurants—from corner burger joints to high-end restos.

Twice a week, beef carcasses come in. We keep them in our own chill room and cut them down into primals and retail cuts. I'd say that most of our beef grades Canada AA. We also source fresh pork, lamb, and grain-fed chicken from local Québec farmers.

Our most popular item is our beef rib steak, but our value-added products also do well. Upstairs, we make meat and chicken pies, beef bourguignon, spare ribs, sausages, shepherd's pie, lasagna, soup, quiche, crepes, and beef, veal, lamb, and chicken stock.

Our entire focus is on quality, service, and professionalism. We have standards for personal appearance, generally changing two to three times a day so we're always spotless. Our customer service standards are very high. We expect all our staff to be polite and endeavor to understand what each customer wants, depending on what they're planning to cook; we have recipes and can explain exactly how to prepare the meat they're buying.

Our standards for sanitation and cleanliness are just as rigorous. Every day, we clean and sanitize the entire shop—this is absolutely basic. Twice a year, we do a major internal cleanup: the chill rooms, fridges, and so on.

You ask me what a butcher is. It's hard to answer. A true butcher doesn't happen overnight; it takes years to really learn the trade.

Ted Makowka.

Raymonde Devroete.

Appearance Is Reality

In my early days as a butcher, you had to be clean-shaven, or, if you had a mustache, it had to be neatly trimmed (I wouldn't purchase raw food of any kind from someone with unkempt facial hair). You had to be clean with well-groomed hair. A white shirt and black or navy blue tie was mandatory. Fingernails had to be clean. And all this was actually checked. It makes sense, doesn't it? After all, a butcher is handling your future dinner.

Service clerk Jennifer Sciou at Boucherie Viandal is ready for work—spotless and delightful.

My customers are the ones who really pay my salary. In my mind, they're all kings and queens—and that's how I treat them. My aim is to provide the kind of service I'd expect as a customer.

Never make customers wait! Greet them as soon as they walk in—even if you're with another customer—to let them know you'll be right with them. Offer extra service. C'mon, it doesn't hurt: "Would you like this freezer-wrapped or wrapped individually?"

Share your knowledge, share recipes, carry their packages, and watch your language with other employees in front of them. Imagine that with each new customer you are on a job interview—because in reality, you are. Your customer should always see you and your meat department in the best possible light. Be professional! Know what you're talking about and never deceive or cheat them. And finally, if you don't know the answer—say so. You can give them renewed faith in your industry.

Get to know them. Ask about their families. Listen to them. Treat them as friends and you will become friends. As you get to know them, share a joke.

By the way, for those of you who are not aspiring to become butchers yourselves . . . if you're not getting the kind of service described above, shop somewhere else.

THE RESURGENCE OF BUTCHERY

The industrialization of all aspects of our lives is deeply troubling. and nowhere more so than in food production. I think that people are starting to question the wholesomeness of the products they're being offered in the meat departments (and other departments) of supermarkets. Increasing media coverage of links between animal feed additives and antibiotic-resistant bacteria, the increasing frequency of product recalls, and other issues such as animal welfare are raising the awareness of consumers. Gradually, more people are making an effort to learn about how their meat is raised and processed. This is a good thing, because ignorance is not bliss; it's a possible case of *E. coli*.

I'm encouraged by the interest in traditional butchery, as well as in the recent enthusiasm for farm-to-table eating and local foods. Some small-scale butchers are forming business relationships with farmers to buy carcasses directly from them, without going through middlemen.

People often ask me how to find a good butcher. It's not easy, I reply. Try asking friends for a recommendation. If you can't get a personal recommendation, and you're making a "cold call" at a butcher shop or grocery meat department, start by asking lots of questions. Read this book, and you'll know what questions to ask, such as where the butcher learned the trade, and what he or she knows about sanitation laws and bacterial contamination. Request a tour of the facility. Remember—the butcher's meat goes into your mouth and the mouths of your children. Don't you deserve a sense of security about its safety?

My policy as meat department manager at the last few stores I worked at was never to sell meat from any producer whose farm I had not visited in person. When I can truly relay to the consumer what I know about the source of the meat I am selling firsthand, then I've done my job with integrity. These are all critical things that we should be thinking about as meat consumers.

Getting involved in the local food movement in your area can also help you find a good butcher. Just make sure that the one you find was trained by a professional butcher and went through the 2½- or 3-year apprenticeship with hanging carcasses. Don't accept second best. What you do *not* want is some self-taught individual who will more than likely stuff you with inaccurate information and incorrect meat terms that will only get you confused looks from that truly good butcher you are destined to find.

Good butchery is not rocket science. But . . . it's certainly not something learned in a few short lessons or by taking a few expensive workshops. It takes a few years of education in the trade to truly grasp the art of taking a whole animal from a hanging carcass to culinary delight on a plate.

Is Butchery a Man's World? Nope!

When I started in the butchering business, there weren't any female butchers—this was strictly a man's world. There were a few female meat wrappers. Things started to change in the 1970s in New England, but slowly, slowly.

Butcher-Farmer Partnerships

Butcher Ron Savenor of Boston's famous Savenor's Market is dedicated to developing direct relationships with farmers. So I asked Ron whether there's a way for butchers and meat farmers to help each other become more prosperous. Here's what he said.

It's a good question, and not easy to answer. I've made a conscious choice to make sure that the relationship between me and my farmers works for their benefit. This includes paying them COD when they deliver their meat. Technically, it's bad business for me to do this, but I choose to do it to support what they do. I work on a lower markup than I should, but I do it because I believe in the farmer and his or her product. I want to help them keep going. So it's more a personal than a business issue for me.

The process works best for my beef and pork suppliers. They're not interested in cutting up the animals, just in selling me the whole carcass. So they're on a "whole-animal" program.

It's challenging for me because Savenor's is a high-end place, so we sell high-end cuts very easily. We have much less demand for lower-end cuts like bottom round. We do have a terrific market for offal, like high-quality French beef cheeks, tongue, and so on.

The real trick is being able to use all the cuts.

Ron, at left, shaking hands with beef farmer Ray Buck of Archer Angus. *Courtesy Ron Savenor.*

LOS ANGELES:
LINDY & GRUNDY BUTCHERS

Founded by Erika Nakamura and Amelia Posada, Lindy & Grundy is a nose-to-tail butcher shop based on the principle of sustainability. The owners are proud that it's one of very few US butcher shops that works with whole carcasses; in fact, every animal is traceable back to the farm where it was raised. Erika and Amelia founded their business in part to create a stronger sense of community as well as raise awareness of the ethical treatment of animals. "We have a responsibility as omnivores and carnivores to acknowledge the fact that these animals used to be alive," Amelia explains.

Erika: "As a chef I started butchering smaller animals—chicken, rabbits, and whole fish. I began to develop a passion for whole-animal butchery. I soon realized that what I really wanted was to work in a more hands-on environment, interfacing with customers face-to-face. This led me to working in a local neighborhood butcher shop."

Amelia: "I met Erika at a drag show in Brooklyn actually! Once we got engaged we knew we wanted to open a family business together one day; we work so well together. I joined Erika at the butcher shop to learn the trade. The more we worked at this, the more we realized this was something we truly wanted to do. We opened our own place in 2011."

Erika: "Our animals are sourced locally, sourced sustainably, and are never fed antibiotics or hormones. They must be raised with full access to pasture and never fed any sort of grain. While we do prefer some heritage breeds over commodity breeds, we also believe that genetic diversity is imperative to a healthy animal. But what makes us unique is our focus on whole-animal utilization."

Amelia: "All our meat comes from very small family farms that raise all meat organically and sustainably. We only use whole carcasses at our shop; nothing comes in a box, which makes us one of just a handful of US butcher shops that start with whole carcasses. Every week we get two steers, seven pigs, four lambs, and 160 whole chickens (with head, neck, and feet). No boxed meat!"

Erika: "We have lots of specialties that we're proud of. Our Short Rib Royale, Tomahawk chop, porchetta, and a new item, lamb-chetta: the lamb loin and belly attached, deboned, and seasoned with traditional herbs and spices. Another specialty is our Smokey Sundays, where we offer hot-to-go BBQ featuring highlights from the week. We also offer catering, specializing in whole animal roasts."

The shop's poultry is slaughtered so that stress levels in the animals remain low, which leads to a much richer flavor. Same for the other meats; the duo works

Erika Nakamura (*left*), Amelia Posada (*right*). *Courtesy Jennifer May.*

directly with suppliers to promote humane methods of slaughter. "We choose slaughterhouses that are small," Erika explains. "We end up having a connection between the rancher and the slaughterhouse."

Erika and Amelia believe that creating a sustainable retail business is about more than products; fundamentally, Lindy & Grundy aims to create a strong sense of community. They work hard to foster a warm and open atmosphere for customers. Erika offers butchering and cooking classes throughout the year. The duo spend off-hours donating their services to their local community, sharing their craft by donating cooking classes to the neighborhood school system for school fund-raisers. "The business is built on our personalities

and politics, as well as getting out there in the world and being active," Erika says. "We try to be as involved as we can."

As for the future of culinary butchery, here's Erika's take:

"If the word *culinary* wasn't in there I would say no, there's probably no future, but because it *is* I say yes. Now restaurants have items vacuum-packed, sealed, and delivered to them ready to go. Along with other advances in the commercial environment, these have pushed artisanal craft-style practices like butchery out of the kitchen. But at the same time we're seeing a growing interest in whole-animal butchery. We hope people like us inspire others to take an interest in this."

Since then, I've worked with a few female meat-cutters—in fact, I worked for one who was a terrific cutter and a great boss. Like many other women, she had a rough time fitting into this male-dominated trade and suffered the discrimination that was once common. She was every bit as good as the male cutters I have worked with. I had tremendous respect for her.

Now times have finally changed for the better, and there's an increasing emergence of women in the trade. In fact, I've noticed that women are the majority in most of my butchery classes. It's about time!

Finding On-the-Job Training

If you want to make butchery your full-time profession, you'll need in-depth training. I suspect it will be hard to find. One big hurdle is not being able to get all the training you need in one place.

For example, if you apprentice at a slaughterhouse or custom-cutting shop, you *will* learn how to break hanging carcasses, produce the basic cuts offered by the facility, and package for the freezer. You *won't* learn the skills you need to operate a retail meat market, such as profit margin calculation,

how to perform cutting tests, value-added products, how to order professionally and efficiently . . . never mind the attributes of customer service.

And if you apprentice at a supermarket chain, you *won't* learn how to break down hanging carcasses into primals, subprimals, and retail cuts . . . because supermarkets no longer deal in hanging carcasses. You *won't* learn where all of the various cuts come from on an animal, because they don't know, either; it all arrives pre-cut in boxes. You *won't* get proper training calculating margins, since margins and pricing are determined at the corporate level.

How I Teach

I find that when I take on apprentices, the learning process takes nearly twice as long as it used to, because I can only expose them to a retail environment for a day or two a week. (I work part-time at a few retail meat markets for just this reason—to apprentice someone in the retail aspects of the meat business.)

To train apprentices in the hanging carcass aspects of butchery, I bring them to the on-farm cutting work I do throughout the year. This gives them a chance to learn how to break beef, lamb, or pig into primals and subprimals and retail or table-ready cuts—in the process learning where all the cuts come from on any given animal. I also teach them how to precision-cut any given piece of meat, how to properly trim, as well as the best methods of cooking various cuts.

It's possible that—with the dearth of expert butchers—better meat shops are willing to consider apprenticeships, and if this interests you, then get onto that computer and start finding them.

There are some self-taught butchers at high-end meat markets who charge as much as $2,000 a week to provide a few weeks of training. Frankly, from what I've seen, heard, and read, much of it is pretty inaccurate training. True culinary butchering isn't something you can learn in a few weeks of expensive meat-cutting classes. Anyone who believes it's possible is just throwing away their money.

I sometimes dream of opening a school for those interested in pursuing a career in the art of butchery or meat cutting. But in the meanwhile, I lead workshops on home butchery, and I've created this book and CD. My workshops vary according to venue and learner level. I teach at culinary schools, small colleges, food organizations (including the Northeast Organic Farming Association of Vermont), private farms, and "get-together" groups of meat-lovers. I get about two to three emails a week inquiring about workshops from folks . . . some want to become butchers, which—as you've probably figured out by now—really can't be learned in a workshop or two. But it gives them a taste of what's involved. Some just want to watch the process or purchase a workshop or class as a gift for a spouse, sibling, parent, or partner. (No more ties and scarves they'll never wear!)

Some people are seeking an internship—but here I'm limited. Although I help out at a few retail markets, it's tough to convince owners to allow an

Here's Sterling College student Nadine Nelson at one of my workshops, cutting her first pork carcass.

intern on the premises. There are several reasons: It slows production considerably, and there are issues with liability, worker's comp, insurance, and more. I have, however, interned a few people this way.

What I am observing is a growing interest in the art of butchery; beef, pork, and lamb. There seems to be more interest in pigs than anything else. This is probably due to the size and ease of raising swine. I see this interest in butchery classes and workshops sticking around for the foreseeable future, as more and more people want to know where their food comes from, how it was raised and treated, what it was fed, and whether or not it was administered antibiotics or hormones.

So given that, it is safe to say that as long as I am healthy and able I will also be teaching and giving workshops. I believe in knowing where my meat comes from, and I believe that educating people is key to the future of local farming and keeping the art of butchery alive in America.

Chapter Two

A Long Tradition

If you practice an ancient and honorable trade, it's good to know something about its history. It didn't seem like this book would be a complete guide to meat without touching on how butchery was practiced long ago. The story is fascinating. I'm no academic researcher, so this isn't a scholarly history (and you wouldn't read it if it were!). I've bounced through history and included the stories that most lit up my imagination. I've also tried to give a sense of how butchery progressed through the centuries from its primitive roots to the art form it is today.

I focused mainly on butchery in the West, since that's the tradition that's shaped my work. And I've included several sources in the resources section at the end of this book in case the stories here pique your interest to learn more.

Give me a mustache and I'd have fit right in with these 1890s butchers from the British Columbia Cattle Co. in Vancouver. *Courtesy City of Vancouver Archives (Bu P462).*

THE FIRST BUTCHERS

Butchery is as old as humanity. The earliest fossil records of our prehistoric ancestors, stretching back over three million years in East Africa, show clear evidence of early butchering. Cut marks on fossilized bones are the proof . . . the marks of these first butchers, who used crude stone tools (no more than rocks with sharp edges) to break down the carcasses of animals and crack their bones for marrow. And supporting this is the fact that the bones and the stone tools are found together, in what scientists call home base or kill sites—places where our early ancestors brought the carcasses of the animals they had scavenged or killed to be communally cut up.

More than three million years ago, some small, not-quite-monkey-like creatures in the Afar region of Ethiopia either found two dead antelopes, or somehow killed live ones, and cut them up with stones. These scavenger-hunters were one of our earliest ancestors, a hominid called *Australopithecus afarensis*. (*Hominid* is a term that refers to humans as well as any member of a species of extinct or existing animal that humans are most closely related to.)

These cut marks on fossilized bone have recently been interpreted as evidence of early hominids using stone tools to cut meat, circa 3.4 million years ago, in the Afar region of Ethiopia. *Photo © Dikika Research Project.*

The marks on the antelope bones suggest that these hominids were butchering meat 800,000 years earlier than any previous evidence of butchering.

There's other intriguing proof that early hominids ate meat—tapeworms! Apparently, hyenas, lions, African wild dogs—*and early hominids*—were all infected by the same species of tapeworm, which scientists believe could only be transmitted via shared carrion or animal kills. That means those early hominids were likely feasting on the remains of animals killed by hyenas and picking up tapeworms left on the carcasses by the hyenas—or the other way around.

The Significance of Eating Meat

The issue of meat eating in the early history of humans is a hot topic, scientifically, and some might say controversial. Scientific theories have ebbed and flowed. The debate among scientists began with a focus on hunting in early human history, based on an assumption that the traits required to identify food animals and create the social cohesion and communication needed to hunt them down were a key influence on the development of our brains.

But not all scientists agreed on the importance of hunting in brain development. Later evidence seems to suggest that the earliest humans weren't hunters, but scavengers, eating carcasses they came across or located via signs such as circling vultures. It wasn't necessarily rotting flesh they were feasting on, though.

Scientists use the term *power scavenging*, which essentially means driving away whatever creature killed an animal and stealing the carcass. There's evidence for this on the bones of an incredibly well-preserved young mammoth frozen in permafrost for over 10,000 years—marks of both lion bites *and* stone tools. The mammoth was young and fairly small . . . perfect prey for a lion. But humans then stole the carcass and removed the internal organs, ribs, and parts of its upper legs.

Some scientists argue that scavenging was early humans' primary way of acquiring meat; others argue that it was just one facet of a behavioral tool kit that also included hunting. Today the consensus is that early humans lived opportunistically—which makes sense, if you think about it; it's human nature. You come across a recently dead animal, you steal it from whatever's guarding it and take it home for dinner. Or you set about killing one yourself.

Comparison of marks on bones from about 10,000 years ago with marks on bones made by modern-day butcher's tools reveal that humans of that era were excellent butchers of mammoths, bison, **aurochs** (giant early cattle), deer, pigs, plus many smaller animals.

Meat and Human Development

I've often heard the argument that humans were never meant to eat meat. Turns out, meat might just be one of the things that made us human.

Scientists believe that becoming human was due to three seminal events:

- **Adaptation to living in the open:** Early hominids moved out of rich forests full of plant food into lightly forested savannas with less plant food but much more small game and carrion (new food sources).
- **Bipedalism:** Walking upright is not only an ideal way to support a larger head with its heavier brain, but also confers a powerful advantage in active hunting through the ability to walk slowly and steadily for great distances, tracking prey.
- **Encephalization:** The development of bigger brains began 600,000 years ago but really took off about 200,000 years ago. But brain tissue requires more energy than other body organs (in fact, 25 percent of all your energy is dedicated to your brain), so the nutrient density of food needed to increase. Many scientists believe that it was actually meat eating that made larger brains possible.

A Meaty Diet

The diet of most higher primates consists largely of leaves, nuts, and fruit. It takes a lot of leaves, nuts, and fruit to keep going. In fact, finding enough to eat takes up most of the day.

Some primates also eat meat; chimpanzees do, and so do baboons. But meat represents only 2 percent of a chimp's diet, and less than 1 percent of a baboon's. In contrast, most scientists believe that the diet of our very early ancestors was 40 to 50 percent meat. No matter the precise statistics, sometime in the last five million years our ancestors began eating more and more meat, until the proportion of meat in our diets began to make us a very different kind of primate.

What's That Thing on Top of Your Shoulders?

Encephalization refers to the quantity of brain mass relative to body size. Generally, the bigger the brain, the smarter the animal. Which begs the question: So what's "cephalization"? Simply put, it means having a head. Scientifically put, it refers to an evolutionary process in which nervous tissue becomes concentrated at one end of a creature, producing—*ta da!*—a head.

Meat is a much more concentrated source of nutrients and protein than plants, so it's a valuable prize. In prehistoric times, hunting large animals (the most rewarding source of meat) required social cohesion, communication, and group participation in hunting and killing animals and dividing and carrying away the carcasses. Our early ancestors needed to cooperate if they wanted to eat meat.

Imagine yourself back then. You're a hunter, so you're a man. As one individual, you know you're no threat to the mammoth you're hunting. You need allies. On the other hand, why join a group hunt if you're not guaranteed part of the kill?

At the heart of this dilemma are key concepts in human development: cooperation and reciprocity. You scratch my back and I'll scratch yours. Which suggests a social creature.

Another intriguing theory examines the way modern traditional hunters parcel out their kills. It turns out that the meat of the animal is uniformly shared among all the group, even if they play no part in getting it. Why? One of the reasons seems to be . . . sex. There is strong scientific evidence for the role of meat in securing sex, as male hunters shared the best pieces of their kills with sexually available females. Both sides benefited.

And there's another intriguing theory out there, too. A study by Swedish and German scientists has found a correlation between meat eating and early weaning in mammals. Let's put this another way: Young mammals stop suckling when their brains develop to a certain stage. And the more meat they eat, the earlier this happens. Comparing human babies with ape babies, weaning happens far earlier. The theory is that eating meat affected our evolution by facilitating shorter periods between births and thus helping kick-start our population boom.

BUTCHERY IN ANCIENT CIVILIZATIONS

Eventually the earliest hominids became extinct, and a new kind of humankind appeared. These are the Neanderthals you all read about in school—unless, like me, you were sketching cows.

Neanderthals were very accomplished hunters. They hunted all the animals—including the huge mammoths—and cooked them quickly over open fires, or slowly, in earth ovens similar to Hawaiian barbecue pits.

Gradually the Neanderthals died out (or, as many believe, interbred with early *Homo sapiens*) and modern humans arrived on the scene. Like the Neanderthals, these first true humans were hunters, although they included more small mammals in their diet (the Neanderthals were big-game specialists). They were also gatherers of wild grains, which they learned to grind into flour and bake. The result was the birth of agriculture. But that's a story for another book.

Tools Take Shape

Neve Yam, an ancient site just off the coast of Israel, offers clues to how early people cut up their meat. The site was occupied about 9,000 years ago. There, archaeologists found animal bones that reveal just how early butchers worked. The bones were located 3 to 10 feet underwater, lying together on the seabed in an area that also revealed the remains of ancient buildings.

This 18,000-year-old painting of a bull's head is part of a giant cave panel showing many ancient hunted species. Lascaux cave, Upper Paleolithic period. *Photo by HTO, Wikimedia Commons.*

Of the bones found, most were from domesticated animals, slaughtered and cut up on site with stone tools. Most of those tools were what scientists call "unmodified"—naturally found stones that could be used as tools. But some of the stones had been modified into more efficient flakes or blades.

There were two types of butchering evidence: slice and chop marks. Slice marks are narrow and linear. Chop marks are broad, short grooves. Try it on a couple of bones and you'll see for yourself (hopefully, with the reward of a nice steak after the experiment).

Almost all the cattle bones found were from the domesticated **Bos taurus** (modern cow). But one was from a *Bos primigenius*, or wild aurochs. The ancestor of today's tranquil cattle, the aurochs was a fearsome beast; this was the "sacred bull" of myth and ritual. There were also pig remains in the find, both wild and domestic. Looking at the percentage of animal species, it seems that the most common were cattle, then pigs, then sheep and goats.

So how were they butchered? The greatest number of butchering marks were found on jawbones, revealing that the butcher cut out and split the tongue and cheek meat, and extracted the marrow. The legs were then cut from the carcass and the meat removed. Here, too, evidence reveals that the marrow was extracted. And so on, all done very efficiently.

Most of the animals were cut up with blades rather than chopped . . . clearly showing different phases of butchering. Which makes sense: First the carcass was chopped up or disarticulated (head and limbs cut off) to create the large cuts that modern-day butchers call primal cuts. Then the carcass was cut into smaller pieces (a nice rib roast today, Mrs. Early Human?).

The First Butcher Shop?

Was Neve Yam an actual butchery operation? Or just a communal slaughtering ground where people processed meat? In the same region, not too much later in time, there is evidence of a butcher *shop*. Beidha is one of the very earliest villages identified, dated from 7200 to 6500 b.c.e. Its earliest stone houses were round and partially underground, and its inhabitants grew early kinds of barley and emmer wheat and herded goats. But they also gathered wild plants and hunted.

Around 6650 b.c.e., a fire destroyed the village houses and new ones were built. These were rectangular and included what were obviously specialized workshops, grouped in threes in alcoves of the ground floor of the houses. One contained rough knives of dressed stone and a pile of bones so expertly jointed that archaeologists believe the owner must have been cutting up meat for his fellow citizens.

Meanwhile, in North America

The very earliest Americans were superb butchers. Almost 10,000 years ago, across the Great Plains, they hunted mammoths and an early bison called *Bison antiquus* (now extinct). Later, they hunted the animal we all associate with early American history—the buffalo, or *B. occidentalis*. No secret there.

What's more interesting is the efficiency with which they butchered their *Bison occidentalis*, revealed by ancient kill sites across the Plains. A site found near the town of Kit Carson in Colorado revealed just how they did it.

This 8,500-year-old site reveals that the hunters stampeded a herd over a cliff into an arroyo. At the base of the cliff, archaeologists found the remains of 200 buffalo. The skeletons at the bottom of the pile of remains were intact, but the others tell an interesting story of butchering techniques.

First the people worked together to lift the huge carcasses out of the arroyo to a flat area, a few animals at a time. Then they rolled each animal onto its belly, cut the skin down the back, and pulled it downward to form a soft, clean mat to place the meat on (brilliant!). They then stripped away the tender back meat, after which they cut off the forelegs and shoulder blades to expose the prized hump meat, rib cage, and body cavity.

At this point, they stopped and ate the tongues of a few bison. We know this how? Because tongue bones were found scattered through the deposit, rather than in one place. Kind of like tossing a beer can out a car window. Next they went for the ribs, breaking them off near the spine with a hammer made of a bison leg bone with the hoof still attached. They then removed the hindquarters and stripped the meat from the legs for easy carrying.

What surprised the archaeologists who analyzed this site was the degree of organization displayed by these early butchers. And their tools? Well, 47 were found at the site. They included hammerstones and knives as well as stone scrapers for processing the bison hides. They killed their prey with stone arrowheads; 27 were found.

About 74 percent of the bison at the site were completely butchered, for a total weight—given the larger size of the animal compared with today's buffalo—of about 57,000 pounds. Plus another 9,000 pounds of edible organs and fat.

What's more fascinating is the fact that early Natives could completely butcher a bison in about *an hour and a half*. Given the number of skeletons found at the site, it seems that about 100 people got together to kill and butcher the animals—all in half a day.

King Tut's Butchers

Early Egyptians were accomplished hunters, with stone tools more advanced than anything Europe would have for thousands of years. They also raised many types of domesticated animals: cattle, sheep, goats, and pigs, as well as certain antelope.

The most important were cattle, which were considered sacred and worshipped. As early as 10,000 b.c.e. at a site called Tushka, wild cattle horns were found with two human burials, possibly as grave markers. This fits well with the much more ancient tradition of Neolithic cow goddesses as symbols of fertility (both cows and human females give milk and would be seen by early people as sources of life). The same underlying beliefs emerged in the later goddesses Hathor, Nut, and Mehet-Weret—all associated with cattle.

Later, in the time of the pharaohs, cattle were managed by specialists with systems of pens and open range. Top-quality bulls were selected for breeding, and an early papyrus called the Kahun includes a description of cattle diseases, providing evidence of veterinary knowledge. Cattle owners branded their animals, and a fragment of papyrus in the Museum Auguste Grasset in France describes legal proceedings against a man who burned his own brand over the legal owner's, making him one of the world's first cattle rustlers.

Cow Goddesses of Ancient Egypt

Hathor—depicted as a cow goddess with horns on her head—was one of ancient Egypt's most important deities. She predates the earliest dynasties and may have originated in early cults that worshipped nature and fertility, represented by cows. In ancient Egypt, she represented the sky, the sun, the queen, love, beauty, the arts, motherhood, and joy.

Mehet-Weret was the goddess of streaming water, linked to creation and rebirth. She was pictured as a cow lying on a reed mat, or as a woman with the head of a cow. Her name, which means "Great Flood," indicates she was goddess of the yearly Nile inundation, and linked to the concept of rebirth.

Nut represented the sky. She was depicted as a star-covered naked woman, body arched, facing downwards. As a great, solar cow, she was thought to have carried Ra up into the heavens on her back. In one myth Nut gives birth to the Sun-god daily and he passes over her body until he reaches her mouth at sunset. He then passes into her mouth and through her body and is reborn the next morning.

Egyptians ate male animals only; cows were too valuable as milk producers and breeders. Cattle were for food and for sacrifice, and herds were owned by temples as well as by private estates. Ultimately, cattle were status symbols, indicating how big a wheel you were.

It's clear from the Egyptians' language that cattle were central to life. There were separate words (for example) for cow, calf, yearling, two-year-old, adult, free-range, hornless, general stock, bull, wild bull, long-horned, stabled fattened cattle, and so on. I'm not sure we have as many today.

How did the Egyptians butcher cattle? Most ancient Egyptian tombs include a cattle-butchering scene, so we know exactly how they did it. They started by wrestling the animal to the ground and trussing it on its back. As assistants held the animal still, the butcher would cut its throat, and its blood would be drained by pumping its foreleg. The

All Hail the Great Butcher

Butchery was a recognized trade in ancient Egypt, with different levels. The most common word for butcher in Egyptian translates as "one who slaughters or makes sacrifice," followed by "director of those with whetstones." Above this was "butcher of the slaughterhouse."

Some butchers did well. Extremely well. From tombs and statues. their titles ring down through the ages: "Acquaintance of the King," "Priest of the King," "Great of Cattle," and "Overseer of the Rendering House" (oh my). The august holders of these titles probably didn't actually do any butchering themselves, but were in charge of certain sets of activities, and—as controllers of food—would have been persons to reckon with.

Wooden painted model, Egypt, circa 2400 b.c.e. Two men slaughter a cow at left. At the front right a woman brews beer. Others are baking. The pointed objects are jars. *Photo by Anna Ressman. Courtesy Oriental Institute, University of Chicago.*

animal was half skinned, then hung on a beam to finish skinning. Then the guts were removed and the heart sent as a offering to the gods.

Butchery tended to be a family affair, passed down the generations. A butcher might be attached to a large estate or travel between estates. Or he might also sell his services to a village or group of villages. Almost every temple had a butcher to take care of ritual offerings. There were also temple meat shops where middlemen bought cuts of meat to sell on to individual buyers. A tablet in Rameses III's reign reads, "I have made festive thy regular offerings with bread and beer, while cattle and desert game are butchered in thy slaughterhouse."

Whether there were butcher shops in ancient Egypt is more difficult to know. Models made by ancient Egyptians do show butchering, often adjacent to a bakery or brewery. Many seem to suggest that the actual butchering took place outdoors under coverings made of light reeds.

There were also temple slaughterhouses in ancient Egypt, to provide sacred animals for offerings to the gods. Excavations at the temple of Reneferef revealed a rectangular building over 4,000 square feet in size, full of tethering stones, flint knives, and cattle bones. Its walls were plastered and whitewashed, and its floors were clay. There were two stories, with butchery on the ground floor and processing above. In other rooms, excavators found a whitewashed butcher block made of mud brick, which would have had a wooden top. So there were clearly areas for killing, cutting, and storing.

The usual butcher's tool kit was basic: a knife, a knife sharpener, and some sort of bowl (perhaps for water). The earliest Egyptian butcher knives were made of flint, with the handle and the blade all one piece. Later, the handle was wrapped with leather thongs. Flint knives were used even after the discovery of metal. Flint was more efficient than metal; it didn't lose its cutting edge as quickly, so it didn't need sharpening as often. Later, in the New Kingdom period (between the 16th and 11th centuries b.c.e.), Egyptians did use metal knives made of bronze, an alloy of copper and tin. Not as sharp as iron or steel, they still cut quite efficiently. The consensus seems to be that throughout Egypt's history, both flint and metal tools were in use.

Ancient Rome: Greater Demand, New Tools

The Romans may have been the very first commercial meat processors . . . they were specialists, they used labor-efficient methods and custom-made tools, and they worked fast and moved a lot of product. Across the Roman Empire—particularly in its larger towns and cities—being a butcher was a good business, and butchers were beginning to specialize. There were specific Latin words for butchers (*boarii*) and pork butchers (*suarii*).

The Greek historian Plutarch, born in 45 c.e., wrote about the development of trade groups in Rome, placing their formation around the 7th century b.c.e. In the later days of the Roman emperors there were many trade colleges, including four public ones: the grocers, the bakers, the providers of lime for building, and the pork butchers. The members of these four public corporations received a fixed salary from the state.

Professional butchers owned shops where they slaughtered, processed, and displayed meat for their

These ancient pillars in Tuscany hint at the lost glory of the Roman Empire.

clients, suspending the carcasses as they worked on them. They had distinct tools for their trade, and developed fundamentally different butchering techniques from their predecessors.

Ancient art and architecture tell the story. A stone relief shows the inside of a butcher's shop, with pieces of meat hanging up for display as the butcher uses a cleaver on a large piece of meat. A woman is sitting and waiting for her order. Remains of what are certainly butcher's shops have been found across the Roman Empire. Shops were often built in rows, forming "retail streets" in the town center. Each shop was under separate ownership, though some were connected, suggesting expansion.

As populations began to increase and people became more affluent, there was pressure for butchers to find ways to process animals more efficiently to meet growing demand. Butchers began using sharp metal meat cleavers to break up carcasses. This wasn't quite the same as a modern cleaver; the Roman cleaver's edge was slightly curved, suggesting that it was a "double-use" tool. It could chop, but could also slice. Fast and efficient.

To save space and dismember carcasses more efficiently, butchers used a pulley system to hoist the slaughtered animals and suspend them in midair to be worked on. There may even have been larger multi-employee shops where two or three master butchers worked side by side, each cutting up one animal.

Their techniques show increased efficiency. For example, hind legs were cut off neatly at the femoral head (the ball-shaped top of the thighbone) with five cuts and two chops. And rather than paring away the shoulder meat from the spine on each side of the animal, butchers used a cleaver to cut the spine (with the meat) away from the shoulder blade, which is a much faster method.

Some have suggested that this evolution of butchery had a great deal to do with the need to keep Rome's huge and far-flung army fed. Provisioning remote towns would necessitate a more efficient approach to acquiring and processing meat.

Latin Roots

Macellum is a Latin word meaning a butcher's stall, meat-market, or provision-market. The plural is *macelli*. The word is still with us: *macelleria* means butcher shop in Italian.

The Rise and Fall of the Cleaver

After stone tools, the cleaver is considered the oldest butchering tool. Cleavers were most popular in the 17th and 18th centuries. The cleaver's use fell in the 19th century as the bone saw was adopted; saws are much more precise, and faster.

Another interesting feature of Roman butchering was the importance of marrow. Often, archaeological evidence shows that limb bones were collected from butchers' shops and sent on to other sites for marrow extraction. Romans may have used marrow to make lamp oil, soaps, cooking fat, and possibly glue.

MIDDLE AGES TO MODERN TIMES

We leap ahead now to the Middle Ages and the emergence of government-regulated butchery operations in Europe. Before 1200, there were well-known and -patronized butcher shops across Paris whose owners were recognized pillars of the community. Some shops were privately owned; others were owned by powerful religious communities such as the Templars and the abbey of Saint-Germain-Saint-Denis. The butcher's trade included purchasing and keeping livestock, slaughtering and processing animals, and, of course, selling meat. Master butchers might keep and graze as many as 200 cattle and even more sheep to satisfy their clients' demands.

Le Ménagier de Paris, written in 1393 by a 60-year-old citizen as a domestic guide for his 15-year-old wife (!), contains a list of butchers from whom he bought all his meat. Here—in his own words—was what he expected his young wife to absorb (and you can tell exactly what sort of man he was by what he thought she needed to know):

> *Porte de Paris: 19 butchers, who sell in an average week 1900 sheep, 400 cattle, 400 pork, 200 veal; Sainte-Genevieve: 500 sheep, 16 cattle, 16 pork, 6 veal; Le Paris: 48 sheep, 10 cattle, 8 pork, 10 veal; Sainte-Germain: 13 butchers, who sell in an average week 200 sheep, 30 cattle, 30 veal, 50 pork . . .* [and so on, market by market, for all of Paris].

Just so she knows exactly how to purchase her meat, her husband describes it in mind-numbing detail:

> *The Paris butchers hold that a cow, according to their style of talk, only has four principal members: that is to say the two shoulders, the two thighs, and the body at the front along its length, and the body at the rear along its length.*
>
> *For, with the shoulders and thighs removed, they cut the cow in two and make of the front one piece, and of the rear another; and thus the body of the cow is carried to the stall, if the cow is small or medium: but if it is large, the front part is cut in two lengthwise, and the rear part also, to carry it more easily.*
>
> *Thus we now have six parts of a cow, from which the two briskets are removed first, and*

then the two supports that hold it which are a good three feet long and half a foot broad, coming from below and not from above. And then they

Why Is a Butcher a *Butcher*?

The word *butcher* seems to have originated around 1300, from the Anglo-French word *boucher* (still in use in French today) and from the Old French word *bochier*, meaning "butcher" or "executioner." It is thought to have derived from the Old French word for "male goat," *bouc*, and the Frankish and Celtic words *bukk* and *bukkos* (also referring to a he-goat).

Following the Rules

In medieval Hungary, as in most of Europe, guilds controlled the trade. There, slaughterhouses had to be situated near water, and only fresh meat could be sold. "Meat sighters" inspected all live animals and meat products. After slaughtering an animal, the butcher had to wash before he was allowed to enter the street. From the late 1500s, meat had to be placed on ice in summertime, and a would-be butcher's knowledge of his chosen trade had to be demonstrated through a master's examination.

There were penalties for breaking rules, too. From the medieval statues of the town of Ipswich, England: "All butchers, both residents and outsiders, should also take care not to display for sale the meat of diseased animals, or that is rotten or smells bad. Any such meat shall be confiscated on the first occasion; on the second occasion the meat shall be confiscated and its seller sent to the pillory. On the third occasion for the offender, the meat shall be confiscated and the seller shall give up his occupation in the town for a year and a day."

cut the flanks: and then to the sirloin which is not much over three fingers thick or two.

Then, to the loin which is closest to the spine, which is as wide as a big fist; then to the fillet which is called the numble, *which is about a foot long and no more; and one end is at the neck and the other is at the kidney, and it is the right of him who has the feet of the cow to flay it, and sell it in a little stall below the large Butcher's; and it is of small value.*

And there you have it, young wife. (How long do you figure it took for her to have an affair with one of the butchers?)

In France, the right to become a master butcher was limited to a small number of families (men only!). As Paris grew, it took more and more meat to feed its citizens; in 1637 that totaled over 3,000 sheep, 511 oxen, 600 pigs, and 300 calves . . . for a population of just 100,000.

In the 14th and 15th centuries, many master butchers had employees who actually cut up the carcasses. The butchers concentrated on wheeling and dealing with livestock brokers. The biggest butchers were wealthy and very influential. This was partly because of their guilds or corporations—organizations formed to protect the craft and its practitioners and to ensure that they continued to get top rates for their products.

Trade guilds could be established only by the highest authorities and were under strict legal control. Membership was often limited to the sons of guild masters, or to assistants marrying the daughter or widow of a guild master. Masters appointed by the guild inspected the quality of meat sold in shops; the guild imposed heavy fines on those who didn't live up to quality standards. To control unfair competition, guilds sometimes set an upper limit on the number of livestock that one butcher could purchase.

How rich were these medieval butchers? Early French legal records give some insight. A legal suit brought in 1383 against the widow of Guillaume de Saint-Yon, a wealthy Parisian butcher, describes the extent of his wealth.

These paintings from the *Tacuinum sanitatus*, a medieval handbook on health (circa 1501), show how an 11th-century butcher would kill a cow (*left*) and a pig (*right*). *Courtesy Wikimedia Commons.*

Saint-Yon owned three profitable market stalls. He owned a house in Paris, plus four country homes "richly furnished" with gilded furniture and silver cups. His wife possessed a treasure trove of jewels, belts, purses, long and short dresses, plus three fur coats. He was able to grant each of his two nieces substantial dowries and in his lifetime rebuilt his Paris house to the tune of what would today be about half a million dollars.

How could butchers become so wealthy? Well, if you lived in town and had some money, you ate meat. A lot of meat. In fact, presenting your dinner guests with a nice roast was a sign of status.

Initially, meat markets and butchers were located outside city walls. This made sense because the trade of butcher encompassed slaughter, and animals were driven live to meat markets. Noise, manure, flies—you get the picture. Plus, meat scraps were thrown straight out the front door of butcher shops, to the delight of wandering pigs and rats.

But cities grow, and "outside city walls" soon became "inside." You can often trace the origins of meat markets by city names. In France, there are many *rues de la bûcherie* street names (meaning "street of the butchers").

English Victuals

The story of Smithfield Market—Britain's famous meat center—is an interesting tale that begins in the Roman Empire. At that time, it was customary

A typical 16th-century meat stall, with all its lovely offerings, including sausages, pigs' trotters, a cow's head, and even some fish. *A Meat Stall with the Holy Family Giving Alms*, by Peter Aertsen, 1551. *Courtesy North Carolina Museum of Art, Raleigh, NC.*

to bury the dead of Londinium (London) in a flat grassy field outside the city walls. But the Romans left, and memories grew short.

When the Normans arrived in 1066, the field was being used for crops; a small livestock market had sprung up beside it. In 1174, administrator William Fitzstephen described the site as "a smooth field where every Friday there is a celebrated rendezvous of fine horses to be sold, and in another quarter are placed vendibles of the peasant, swine with their deep flanks, and cows and oxen of immense bulk." Smooth Field's growth was further stimulated in 1381 when the Corporation of London banned the slaughtering of animals within the city walls.

Gradually Smooth Field morphed into Smithfield, which expanded as London's population grew. In 1710, it was fenced to control wandering livestock; by 1726, writer Daniel Defoe wrote that the market was "without question, the greatest in the world." By 1750, yearly animal sales at Smithfield were 74,000 cattle and 570,000 sheep.

As London grew, of course, the market was no longer "outside city walls." Soon people began demanding that it be moved farther away to protect public health (and keep sensitive eyes and ears from witnessing the inhumane conditions inside its fence). They had a point: By the mid-1800s over 220,000 cattle and 1.5 million sheep *per year* were being driven—mooing and baaing—into a tiny 5-acre area in what had become the very heart of London.

Protests and petitions carried the day, and in 1860 the Metropolitan Meat and Poultry Market Act created the Central Meat Market at Smithfield, in the eastern part of London (it's still there and you can visit it; we've included its website in the resources section of this book). Work began immediately and was completed in 1868.

Designed by Horace Jones, who also designed London's famous Tower Bridge, the market was an impressive monument to Victorian England. A huge construction of stone, slate, cast iron, and glass, the market covered 8 acres. Two main wings surround an interior Grand Avenue embellished

A Place to Sell Beasts

August 1644: An Ordinance for the better Regulating and Leavying of the Excise of Flesh, within the Cities of London and Westminster, Suburbs, and Lines of Communication: "And be it Ordained, That no Grasier, Butcher, or other person whatsoever . . . shall bring any live Beeves, Muttons, Veales, Lambes, Porkes . . . either by Land or by Water, into the Cities of London, Westminster, their Suburbs, and Lines of Communication, do or shall presume to put any such Beast to sale, in any place . . . but in the open and usuall place and market in Smithfield, upon pain of forfeiture of such Beast, or the value thereof."

More Than a Meat Market

Smithfield's open space attracted many high-profile events such as Richard II's famous 1390 tournament, attended by sixty knights and sixty noble ladies, and organized by Geoffrey Chaucer. Later, Smithfield was the place where dissidents were hanged, and later still (in the 16th century) swindlers and forgers were boiled to death there. Smithfield, Rhode Island, was settled in 1730 by British colonists from Smithfield, London.

with statues, towers, and bronze dragons carrying London's coat of arms. A railway track brought animals directly into the market (the railway is no longer there, since most of the meat is now trucked in). Today, the old tunnels are mysterious regions of storage rooms, parking lots, and dim basements. And, perhaps, ghosts?

Smithfield remains a powerful force—the only great London market remaining in the city—although not in its original Victorian condition. A

THE OLDEST BUTCHER SHOP IN THE WORLD:
R. J. BALSON & SON

Courtesy Oliver Balson and Richard Balson.

Richard Balson, runs the UK shop with Rudy Boulay, a former French pastry chef. At busy times such as holidays, other family members help out. Oliver Balson describes the business:

I work closely with my uncle in terms of recipes, but our materials and methods are very different. Our UK meats are sourced from local sustainable farms throughout the county of Dorset. Whole carcasses are brought in and that is where the craft of butchery comes in, in terms of proper cutting and dissecting of the animals to ensure maximum quality with minimal waste. There was a time when hunters would bring in their day's spoils, such as wild boar or pheasants, for us to purchase. But much of that has been put to an end due to increased health regulations.

The shop typically offers over 20 varieties of sausages at any given time—some of our more unusual and popular sausages are an Ostrich and Cranberry and a Lamb and Mint. Our Wild Boar sausage is a hit as well, as is our dry-cured, hand-rubbed bacon. We also make all sorts of meat pies, pasties, et cetera. Our customers are mostly locals and regulars, but we do get out-of-towners, on holiday, et cetera, some just coming by to see the shop because it's also part of a special postcard series, Heritage Collection of Historic Places.

In the England market town of Bridport is a family butcher that has been in operation since 1535. Let me repeat that—1535.

To give you some perspective, when John Balson first opened his meat stall in the Bridport town market, Henry VIII was the king of England. Balson & Son is the longest-running butcher shop in the world, and the United Kingdom's oldest family business.

The shop hasn't moved much; it's in the same town where it began 500 years ago. And for all those 500 years, Balson family members have been the butchers, passing the skill down through the generations.

Oliver Balson runs Balson's fairly new (since 2007) American operation. His uncle, master butcher

The Horseback Delivery Boy

Oliver Balson remembers: "My grandfather used to tell us about how when he was a boy his dad would wake him up and tell him to fetch a horse from the field behind the shop. So he'd go and put a bridle on the horse and ride it for meat deliveries."

The shop circa 1880. *Courtesy Oliver Balson and Richard Balson.*

In the United States, we source our meat from a couple of Midwest meat distributors—all as primals, not whole carcasses. Our most popular product is our Traditional Pork Banger—we get great reviews from those looking for an authentic English/Irish sausage. Our customers are mostly ex-pats—English/Irish origin—looking for a taste of home. We get regular emails from people saying they've finally found proper Irish/English sausages in America.

Our UK shop has been sourcing grass-finished beef for ages, having established relationships with small local farms, so it is a viable option. We believe the quality of the beef does indeed increase on an all-grass diet.

All across the UK we hear about butchers closing down—with supermarket competition and convenience, it is a challenge—but we try to offer a specialty that they cannot get in the supermarkets— we deliver for elderly and shut-ins and make coming into our shop an experience—many of the customers are regulars, who like to have a little gossip and leave with a laugh.

1957 fire destroyed most of it, and it was rebuilt in 1963 under a huge domed roof.

Smithfield is not a retail market; its clients are the city's butchers, shops, and restaurants. To see it in action, visit early; trading hours are from 4 a.m. to noon every weekday. More than 30 wholesale traders sell beef, chicken, pork, duck, game, goat, goose, ham, lamb, mince, halal, offal, sausages, turkey, veal, venison . . . and cheese.

Butchery in North America

There's not a lot of information about butchery in early North America. We do have one interesting account from 1653 that describes the arrival of 100 men sent to New France to help the struggling colony of Ville-Marie (now the city of Montréal). On board was a butcher, Guillaume Gendron dit

La Rolandière, who settled on nine arpents (an arpent is slightly larger than an acre) in the new settlement, where he fathered two daughters, Marie and Catherine.

There were never many butchers in New France. In 1666, nine butchers served a population of 550 people. Fifty years later there were only four, yet the population had risen to 1,574. How come?

In the earliest days of the French colonies, every domestic animal had to be imported from Europe; there simply weren't enough in the new land to breed into viable herds. Reaching a "tip-over" point where raising meat animals would become self-sustainable would demand not only time to build up the necessary breeding stock, but also considerable investment in feed that didn't come easily. Keeping chickens was cheaper and made more sense, because they could be quickly raised for eggs and then meat. Feeding and managing cattle was much more problematic for a struggling family. Cows were valued more for their ability to produce milk, butter, and cheese than meat. In addition, along the wild shores of the St. Lawrence River, constant threat of Iroquois attack kept people from spending much time growing feed for beasts. So raising meat animals was not as common as it would later become.

Most relied on wild meat. Families would send their menfolk—and sometimes womenfolk—into the forests to bring back deer, moose, bear, beaver, and squirrel (small game was often hunted by boys and girls). Almost all families did their own butchering—they had to.

The Butcher's Cart

In the 19th century, Québec butchers would travel country roads and city streets in horse-drawn carts looking for clients. These carts were still plying the streets of Montréal in the 1950s.

But there were certainly some trained butchers in the colonies. Butchers killed the animal by bleeding, then skinning it. Cowhide would be made into leather. If it was a pig, the butcher would singe off the hair and scrape the skin clean. Then he would attach the animal by its hind legs to a rack and cut it into quarters.

The meat would be turned into raw cuts, blood pudding—still very popular in Québec (and Louisiana, part of its Cajun heritage)—or prepared dishes to sell at the market, the only place where a butcher was allowed to offer his products. In 16th- and 17th-century Montréal, that market was Place Royale, the economic hub of the city. Now the site of the Pointe-à-Callière Museum, every year during the final weekend of August, Place Royale returns to its past with a re-creation of the early market that once thrived there.

Like butchers in Europe, butchers in New France had to abide by government regulations that limited the number of butchers, regulated the price of meat, and established the allowable interval between slaughter and final sale. There was a system of fines; any butcher who sold bad meat or meat from an animal that had died from disease could lose his livelihood. And every slaughtered animal was inspected by the king's solicitor.

How Swine Came to America

The first New World pigs arrived with Christopher Columbus—not in 1492, but on his second trip in 1493. This original group of eight pigs landed on the Caribbean island of Hispaniola, where they quickly settled in and multiplied freely. They became a kind of living livestock resource for early explorers, rounded up as necessary and loaded onto ships bound for other places. Going to Mexico? Need some very fresh ham?

But Hispaniola is not America. So how did pigs get to the United States? They arrived with the Spanish conquistador Hernando de Soto, who in 1539 set out to explore the continent now called America. De Soto landed at Tampa Bay with 13 pigs, which quickly did what pigs do so well—have more

Tiny Ossabaw Island, off the Georgia coast, is home to a pig called the Ossabaw hog that many believe to be a direct descendant of the original pigs brought to America by de Soto. *Photo by Jeannette Beranger, The Livestock Conservancy.*

pigs. De Soto and his men ate the pigs as fresh meat, then salted it for later consumption. Of course, news—and the taste—of this wonderful critter soon percolated out to the local Native people, who became so obsessed with this new meat (bacon!) that the worst attacks on de Soto's expedition had nothing to do with killing strange white men, but everything to do with acquiring pigs. When de Soto died just three years after he landed in Tampa Bay, hundreds of pigs were rooting around the South.

In the early 1600s, more pigs came over with the British to places like Jamestown, and with the Spanish to New Mexico. And—again—the pigs multiplied. They prospered because forests provided rich pickings in chestnuts and acorns and pine nuts. Pigs were perfect. You didn't need to feed or tend to them because they managed quite well by themselves wandering through the woods. Then in early autumn, the fattened pigs could be rounded up, slaughtered, and salted to be consumed throughout the cold winter.

But—as I pointed out earlier—pigs kind of expand, numbers-wise. Soon feral pigs were eating so much of the grain crop that the colonists of early New York were required to insert a ring in their pigs' snouts to control rooting. And—you'll love this—on Manhattan Island, rampaging feral pigs got so out of control that authorities built a long wall on the northern edge of the colony to keep them out—which eventually became Wall Street.

19th-Century Rural America

Like most of the New France homesteaders, farmers in 19th-century North America did their own butchery. They had to; farms were often remote, and a traveling butcher would have been much too expensive (if one could even be found).

They began, of course, by killing the animal, usually by stunning it, then slitting its neck to kill and bleed it. Then they removed its feet with a cleaver—assuming it was a cow (feet were usually left on sheep and pigs). The animal was hung by its hind legs and its head removed. Then the carcass was split in two along its spine, probably with a handsaw. In the third stage of the butchering process, the carcass was subdivided into the large primal (wholesale) and small ready-for-cooking cuts of meat.

The challenge was preserving the meat. Many methods were used to keep meat palatable, including salting (salt-cured) and smoking (smoke-cured).

The Native Americans had used smoke to cure their meats; they'd hang the meat in their tepee above a fire. The settlers followed suit, hanging their meat above a smoking fire pit for about a month. The smoke not only dried out the meat, but also added a delicious flavor, and people quickly discovered that different kinds of wood produced different tastes.

In the salt-cure method, meat was first rubbed, then completely covered with salt and left in a cool area for a few weeks. When the meat was completely dry, it was washed and stored to age.

Pigs have always been symbols . . . of all kinds of things. Take a look at this 1884 United States map with the states personified as hogs. I have no idea why. The text with the map reads: "Nicknames of the states. H. W. Hill & Co. Decatur Illinois, sole manufacturer of Hill's hog ringers, Hill's triangular rings, &c. Printed in our own advertising department; copy of this map mailed for 5 one cent stamps."

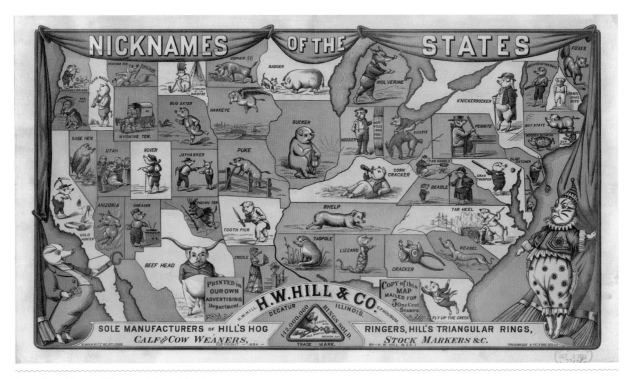

Image from Library of Congress, ID pga.03942.

Pigs in Manhattan

As late as 1842, there were still 10,000 pigs in New York City. That year, Charles Dickens wrote in his *American Notes for General Circulation* that "free roaming" pigs were numerous just north of Wall Street.

Once more in Broadway! . . . Take care of the pigs. Two portly sows are trotting up behind this carriage, and a select party of half-a-dozen gentlemen hogs have just now turned the corner.

Here is a solitary swine lounging homeward by himself. He has only one ear; having parted with the other to vagrant-dogs in the course of his city rambles. But he gets on very well without it; and leads a roving, gentlemanly, vagabond kind of life, somewhat answering to that of our club-men at home. He leaves his lodgings every morning at a certain hour, throws himself upon the town, gets through his day in some manner quite satisfactory to himself, and regularly appears at the door of his own house again at night . . .

He is a free-and-easy, careless, indifferent kind of pig, having a very large acquaintance among other pigs of the same character, whom he rather knows by sight than conversation, as he seldom troubles himself to stop and exchange civilities, but goes grunting down the kennel, turning up the news and small-talk of the city in the shape of cabbage-stalks and offal, and bearing no tails but his own: which is a very short one, for his old enemies, the dogs, have been at that too, and have left him hardly enough to swear by.

He is in every respect a republican pig, going wherever he pleases, and mingling with the best society, on an equal if not superior footing, for every one makes way when he appears, and the haughtiest give him the wall, if he prefer it . . . They are the city scavengers, these pigs. Ugly brutes they are; having, for the most part, scanty brown backs, like the lids of old horsehair trunks: spotted with unwholesome black blotches. They have long, gaunt legs, too, and such peaked snouts, that if one of them could be persuaded to sit for his profile, nobody would recognise it for a pig's likeness. They are never attended upon, or fed, or driven, or caught, but are thrown upon their own resources in early life, and become preternaturally knowing in consequence. Every pig knows where he lives, much better than anybody could tell him. At this hour, just as evening is closing in, you will see them roaming towards bed by scores, eating their way to the last.

Your Personality in Swine Terms

Perhaps you're a *gas hog* or a *road hog*. Or you occasionally go *whole hog* or *hog wild*. Perhaps you *hog your food*, or drive a *chopped hog*. Or try to *hogwash* your audience . . . or like to be *hog-tied*.

As pioneers began heading west, pigs joined them in their prairie schooners. Soon the number of American pigs had expanded to such a degree that a new industry was born: commercial pork processing.

The industry quickly centered in Cincinnati, which by the mid-1800s was also known as Porkopolis. Cincinnati's access to river transport and hog farmers made it the ideal location. Pigs were shipped on the river, debarked, and herded through the streets to the slaughterhouses. Or huge herds might be driven as much as 1,000 miles by hog drovers, traveling up to 8 miles a day. As many as 70,000 pigs reached markets that way each year.

Not every Cincinnati citizen was proud to be from Porkopolis. But they didn't need to lose sleep over it, because by 1861 Chicago had replaced Cincinnati as America's biggest meat processor, becoming "Hog Butcher to the World."

Eventually, the pork-packing industry moved farther west, and today, Iowa is the top pork producer in the States.

The Pace Quickens

Change came quickly as the wilderness of the early colonies was transformed. People spread westward, towns and cities mushrooming in their wake. Folks had more money, which supported their increasing taste for meat. Satisfying this growing consumer demand for meat would require raising far more animals, processing them far more efficiently, and getting perishable meat to distant markets far faster.

In the United States of the mid-19th century, new approaches to raising meat animals emerged—in particular the western system of range-raised cattle driven in huge herds to centralized railhead towns for transportation to the big city markets of the East.

These large food markets were part of city life and, for most urban dwellers, were the only way food was bought. It wasn't much different from the early Smithfield days in London: Every city had at least one—and usually many more—food markets, satisfying the growing and increasingly diverse population of the young United States.

Meat, being highly perishable, was a focus of efforts to improve the supply and demand system represented by central and western meat animal producers and eastern consumers. We know quite a bit about mid-1800s city markets because of a remarkable man who devoted his life to making them work for the city of New York. As for the supply side, it was revolutionized by another remarkable man—certainly the individual most responsible for our current industrialized meat sector.

The Market Man: Thomas De Voe

Colonel Thomas F. De Voe was appointed to the influential posts of superintendant of markets and collector of revenue for the city of New York in 1881. He was probably the perfect man for the job. Born in 1811 in Yonkers, he was apprenticed as a boy to a butcher. After a stint in the militia, De Voe started his own butchery business, setting up a stand in the newly opened Jefferson Market. In 1840, the butchers asked him to represent them on a committee formed to work with New York's board of aldermen on setting up regulations for the city's food markets. Later he was instrumental in reforming the way the city's food markets were managed, including getting certain city officials fired who'd been supplementing their personal income by blackmailing stall owners.

But Colonel De Voe is best remembered for his writing, which he always signed "Thomas F. Devoe, butcher." Here's an excerpt from the preface of his book, *The Market Assistant*:

> . . . *But butchers here, like other men,*
> *Have common sense and sense of pain;*
> *These weigh the meat, and you must know,*
> *The meat side of the scale is low*
> *And wants your care to balance it,*
> *If you would have your proper weight,*
> *Or else two pound of beef, you'll see,*
> *Will just two pound off ounces be*
> *The rich, who buy a stately piece,*
> *Will scarcely know their meats decrease;*
> *But 'tis the poor, who little buy,*
> *That miss their meat, and wonder why.*
> *'Tis thus with some—but not with all—*
> *For many, from the loaded stall,*
> *With balance even, weigh the meat,*
> *Too honest to defraud or cheat.*

In the 1860s, meat was available all year round at the markets of American cities. De Voe refers to beef, mutton, lamb, veal, and pork being "always in season." Butchers hung their meat for days or weeks in a cool place like an icehouse or deep well. Meat that had to be shipped any distance was kept as cold as possible, then wrapped in linen "around which should be placed cabbage leaves" and further wrapped in cloth, thus allowing it to stay cool for 6 to 10 hours. For poultry, De Voe recommended removing the guts and inserting a wrapped piece of charcoal into the cavity. Charcoal is well known as a purifier; in Europe, it's prescribed for human digestive problems. I've tried it and it works, even though it was a bit like eating crunchy charcoal briquets and turned my teeth black.

The Meat Supplier: Gustavus Franklin Swift

Gustavus Franklin Swift is the man to thank (or revile) for today's meat industry. Swift was born in 1839 on a small farm in Cape Cod, where his family raised cattle, sheep, and pigs. At the age of 14, he went to work for his older brother, who ran a butcher shop. At 16, Swift was convinced by his father not to move to Boston through a gift of $20 to begin his own meat business. Gustavus used $19 of it to buy a heifer, which he butchered and sold—netting $10, which he used to buy another heifer. Then another. And so on, until his uncle lent him $400 to open his own butcher shop in Eastham, Massachusetts. Hard work and business acumen soon turned the business into a roaring success. But young Gustavus was just getting started.

In 1869, he opened a meat market in Clinton, Massachusetts. His market was radically different from others of the day, displaying a variety of cuts that people could look at and select from. And it was spotlessly clean—a rare phenomenon at that time. Soon Swift's little market was reaping $40,000 a year (in today's dollars that would be $800,000 to $1.4 million per year).

Not enough for our boy Gustavus . . . he aimed higher. He became a cattle agent—buying animals for resale—and moved to Chicago, the ideal point for supplying western livestock to eastern markets. At that time, meat was shipped on the hoof. Cattle drives brought cattle to the Midwest, where they were loaded onto cattle cars and shipped live to eastern slaughterhouses. Not efficient; many animals weakened by long cattle drives simply died in the train cars.

Swift's breakthrough was the realization that the most efficient way to operate was to slaughter cattle at one central location, then ship only

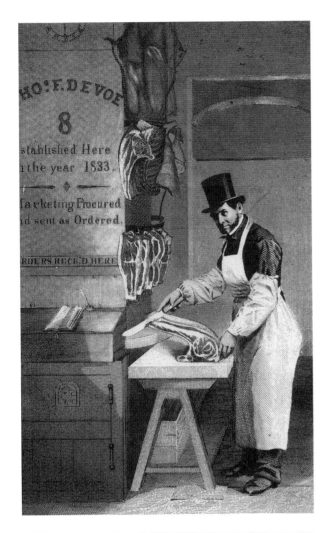

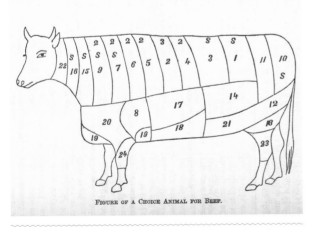

FIGURE OF A CHOICE ANIMAL FOR BEEF.

This is the original frontispiece (*top*) and a cutting diagram (*bottom*) from *The Market Assistant*, by Thomas De Voe, Cambridge, 1867.

processed meat out to stores and customers. But this method required a way to keep the meat cool during shipping. How?

Easy—invent a refrigerated train car that circulated cold air over ice. Doable—but it wasn't a slam-dunk. People were used to buying fresh beef shipped in alive by train, then butchered at their final destinations. There were actually boycotts of Swift's meat, with protestors claiming that eating meat that wasn't slaughtered locally would kill you!

Consumer resistance didn't last long. Swift's ad campaigns and deals with local butchers soon overcame it, and an exclusive deal with the Grand Trunk Railway sidestepped the exorbitant prices a

We might find this steer a little stringy; it had probably been driven many hundreds of miles to get to the slaughterhouse. *Courtesy Jack White Photograph Collection, Special Collections, The University of Texas at Arlington Library, Arlington, Texas.*

cartel of the other railroads had decided to charge him for shipping pre-cut meat.

In 1885, Swift and Company incorporated. It soon became a force to contend with—not only because of the centralization of the operation, but also because of Swift's focus on reinventing every aspect of his business. His obsession with cleanliness resulted in much less spoilage. He found ways to use slaughtering by-products to make other products like soap, glue, fertilizer, and so on. And he vertically integrated his operation—one of the very first to do so. Swift and Company departments handled every aspect from moo to chops: sourcing and buying cattle, slaughtering and packing, shipping, sales, and marketing. He can really be credited with establishing the basics of 20th-century assembly-line production, and profoundly influenced other industrial pioneers like Henry Ford.

Soon Swift owned plants in St. Louis, Kansas City, Omaha, and Fort Worth as well as other cattle cities. And then . . . on to the world. The company exported beef in refrigerated ships to its own distribution network in Japan, China, Singapore, and other countries.

Swift turned his prodigious business drive to mutton and pork as well. By the beginning of the 20th century, Swift and Company was worth $25 million. And when he died, his company was worth over $105 million; three years later, over $250 million. Swift's imprint is everywhere on the modern North American business system—from assembly-line production to the creation of distribution networks to the adoption of new technologies.

Today the meat and poultry industry is the largest segment of US agriculture, contributing over $800 billion to the economy. The commercial meat sector is a huge machine built on science and technology and powered by production, slaughtering, processing, and distribution systems that Gustavus Swift would be proud of.

How meat gets to consumers' tables today is a complicated and, in some ways, disconcerting story, which I tell in chapter 5: "The Way the Meat Industry Works."

Chapter Three

We Love Meat

We eat meat because (1) we like the way it tastes; (2) it's a great delivery system for protein, which our bodies need; and (3) it's kinda everywhere. We eat meat raw, whole, ground, cut into little pieces. We eat it broiled, boiled, fried, stewed, steamed, grilled, braised, dried, marinated, smoked, spiced, fricasseed, on spits, in pits of hot stones, and nailed to planks in front of fires. We are very inventive, because each different cooking method produces a subtly different taste, a different eating experience, a new path to culinary glory.

WHAT IS MEAT?

What constitutes "meat" depends on cultural as well as geographic factors. Common meats come from cattle, pigs, sheep, goats, game animals such as elk and deer, and birds. But some North Americans also like to eat possum, moose, bear, squirrel . . . and other critters. In Africa or parts of the tropics, some people eat bush meat—wild animals like monkeys, apes, or other mammals. In France, some people eat horse; in the Andes, guinea pigs. In this book, I focus on the basics: beef, pork, and lamb/sheep (with a dash of poultry).

Ah, meat! Star of steak houses, fast-food joints, and home dinner tables. *Photo by (top) Archie MacDonald52, Wikimedia Commons (bottom) Renee Comet, National Institutes of Health.*

A different kind of meat. Beijing snack food. Flash-fried scorpions. Really.

Biologically speaking, meat is the muscle tissue of a food animal. Of course, more of an animal can be eaten than just its muscles. We also eat **offal** (also called variety meat or organ meat) like liver, kidney, heart, sweetbreads, and so on. These, too, are meat.

Most animal muscle is made up of about 75 percent water, 19 percent protein, 2.5 percent intramuscular fat, 1.2 percent carbohydrates, and about 2.3 percent other nonprotein substances like amino acids and minerals. Muscles are composed of bundles of cells called fibers. Each fiber contains filaments made of two proteins: actin and myosin, which give the muscle its main structure. Other muscle components include connective tissue like collagen and elastin.

Meat also contains fat—adipose tissue—which the animal uses to store energy, and intramuscular fat, which contains things like cholesterol and other substances.

MEAT QUALITY

I'd like to begin by exploding some assumptions you probably have about meat. For example, what does the sign WE SELL PRIME BEEF mean?

Nothing.

Prime is an adjective. It's an opinion, or—in the case of a meat market—probably a deliberately misleading statement.

But *prime* is also one of the classifications of beef defined by the US Department of Agriculture (USDA). In the United States, only the USDA can certify beef quality grades, such as prime, choice, select, and on so. So calling beef "prime" is meaningful only if the label or sign says USDA PRIME BEEF. In Canada, beef is graded by the Canadian Beef Grading Agency, a not-for-profit

Am I choice or prime? *Courtesy Dickinson Cattle Co.*

organization accredited by the Canadian Food Inspection Agency of the federal government. (See chapter 7 for a comparison of US and Canadian beef quality grades.)

You might also assume that the leaner, redder, and fresher a piece of meat is, the better. Again, not always true; it might be pretty damn tough. The quality of beef is mainly determined by two primary factors: fat marbling and aging. You already know that fat tastes great. Most of the wonderful experience of meat eating comes from our craving for fat and satisfaction when we get it. And in most meat—particularly beef—that satisfaction is delivered by internal flecks of fat within lean meat called **marbling**. In fact, when beef is graded into its various categories (prime, choice, and so on), much of the rating depends on the degree of marbling.

Aging is *very* important. Meat is not ready to be eaten right after slaughter. It needs time to become tender, which happens as connective tissues within the muscle break down. Aging is that breakdown process. The ideal aging period is 21 to 24 days. I explain the proper way to age beef in detail in chapter 7.

Let's consider the basic grades of beef. The lowest grades, up to select, are all pretty lean. In my

The Meaning of Meat

The word *meat* itself has uncertain roots. Some claim its origin is the Old English word *mete* or *maet*, which refers to any food. Others say it goes back farther to the Indo-European word *mad*, meaning "moist" or "wet." Up till about the 13th century, *meat* simply meant food—any food. Then gradually, the word began to be associated just with animal flesh.

opinion, select at its best lacks flavor and tenderness. For example, two steaks sold in most supermarkets are the New York strip and the rib eye (also known as Delmonico). The price per pound today is usually $7.99 to $8.99 for select grade. This is a pretty hefty price for a steak that is not going to be very satisfying to eat.

The same steaks in a choice grade will cost about $9.99 to $14.99 per pound. The choice-grade steaks will have much more flavor and tenderness, but will often lack consistency. You may buy great ones 9 times out of 10—but then one will disappoint you.

Stepping up one subgrade level to premium choice, we fall into the **branded beef** programs.

There are more than 100 such programs across North America. The first brand identity was Certified Angus Beef. Another branded program is Chairman's Reserve Certified Premium Beef. Two Canadian examples are Canada Gold Beef and Spring Creek Premium Beef. Each program selects for a specific set of quality characteristics.

Several companies brand their own selections of premium choice, claiming that it is hand chosen—and usually it is. I have always found it to be extremely consistent in both flavor and tenderness. Also, from a butcher's perspective, it has a nice and consistent yield. The same two steaks will, in the premium choice grade, range from $10.99 to $15.99

See the difference between the USDA prime steaks and the Australian Wagyu steaks? The Wagyu is so marbled, it's almost white. *Courtesy Ron Savenor.*

per pound. I think the quality of premium choice is worth the price.

Now up to the top—prime grade. This is not easy to find in retail supermarkets. If you can find it, it will be considerably higher in price than all the other grades. Personally, I find prime a little high in fat content for my digestive system. So for me prime is a waste of money. But that's me; you might love this special treat.

Is there anything higher than prime? Since the grade prime contains many additional subgrades, the answer's yes. A lot of meat-lovers salivate over the extra-high marbling found in Japanese Kobe beef. And yes, you can find it. This glistening wonder is known (outside Japan) as Wagyu beef—and it can cost $40 to $65 per pound. Some adore it (if they can afford it!). Others, like me, avoid it.

To sum things up, here is my opinion of the various grades of beef.

- Standard to select: Chewy and flavorless.
- Choice: Inconsistent.
- Premium choice: Very consistent with good yield, flavor, and tenderness.
- Prime: Too much fat (for me), too rich and expensive.
- Kobe/Wagyu: Like owning a Rolls-Royce: handles like a '59 Cadillac, but gets lousy mileage. Remember, this is only my opinion!

What Grade Should You Aim For?

Frankly, I'm not sure that grade matters much. This somewhat surprises me. I used to be a firm believer in premium or choice beef. However, long involvement with the local movement has given me the opportunity to sample a lot of local meats (often as the farmer was cooking it to serve for lunch while I cut his carcass up). I've done some rethinking about the importance of grading, particularly in beef.

I'm now more convinced that—where beef is concerned—quality has more to do with the breed,

the type and quality of its feed, the soil it was pastured on, how humanely it was raised, how it was slaughtered, how it was aged, and how it was cut and processed. I have had very lean meat that was surprisingly tender, and I have had extremely marbled meat that was chewy.

Keep in mind that different breeds of animals have different attributes. Some are more efficient at converting feeds like grass to meat and fat than other breeds. Some breeds do better in harsher climates than others. There are just too many variables, so it is important that you do your research.

This is my opinion, and I'm a plain kind of guy. Please don't take my word for all of this—go ahead and sample prime and Wagyu. You may love them.

Don't Assume It's Aged

Another assumption many people make is that meat from animals raised on a small-scale, local farm without additional hormones or antibiotics will always taste great. I'm sorry to say this, but that's not a given, because the eating quality of meat—how it tastes, how tender it will be—has to do with very specific factors that are only partially related to how the animal was raised. As I mentioned above, one of those factors is whether the meat comes from a carcass that was aged properly.

No matter the quality of the animal or degree of marbling, *without proper aging it will not be tender.* So if you're buying a beef carcass from a local farmer to cut up yourself, it's up to you to ask the farmer how long that carcass has aged—don't make the assumption that every farmer fully understands the importance of aging a carcass. Much of the time it's not ignorance on the farmer's part; it's simply an issue of practicality. Most small farmers don't have the means or facilities to chill-age full carcasses.

Bottom line . . . if you just pick up a freshly slaughtered carcass from a farm, take it home, cut it up, and put it into the freezer, it will be shoe leather. (And a tip here: If the first steak is tough, the entire animal will be tough.)

Hanging beef carcasses to age for up to three weeks or so before cutting up the meat for cooking and eating may seem strange, but it's an essential part of the process of ensuring tender meat.

Ideally, you want a carcass that has aged for 14 to 24 days in a chilled facility. If this hasn't happened, then I recommend that you arrange for it to be done. You may have to find a locker plant where you can rent a cooler. (For more details on the logistics of handling animal carcasses before butchering, refer to chapter 6.)

My point is that somebody along the way has to age that carcass. And while we're on the subject, it's also up to you to verify how the aging was done. Proper aging should be done at 32 to 38°F (0–3°C) without any temperature fluctuation.

Tenderness, Taste, and Value

Do you want an incredibly flavorful piece of meat that's so tough you'll be picking it out of your teeth for a week? Probably not. Do you want a butter-tender piece of meat that has absolutely no flavor at all? Definitely not.

You want an affordable piece of meat that's nice and tender and has lots of flavor: a piece of meat with excellent mouth-feel. This is a common term used in the food industry, and it means exactly what it says. That meat should feel good in your mouth. It should be juicy, flavorful, and tender with enough texture to make chewing a pleasure, not a chore.

Most folks believe that only the high-priced cuts like New York strip, rib eye, or filet mignon will be tender, flavorful, and juicy. This couldn't be farther from the truth. There are a number of other cuts, particularly in beef, that are not only more flavorful, but more tender as well.

And guess what? They don't come from the loin or the rib. They come from the chuck. For example, take the teres major (never heard of it, right?). Also known as the petite tender, this is the second tenderest piece of meat you can get from a beef carcass. It sits just under the shoulder blade. It's extremely tender, juicy, and flavorful with an incredible mouth-feel.

Another cut from the shoulder blade is the flat iron (I'll bet you have heard of this one). Cut slightly differently, it's sometimes sold as a blade steak. As a flat iron, it is extremely juicy, flavorful and tender, also with a nice mouth-feel. The flat iron can also become an incredible and very affordable oven roast if you layer one on top of another. I think this is better than prime rib, with more flavor and tenderness.

So yes, you *can* have value, tenderness, and taste without mortgaging your house. Stick with me through the chapters to come (and the CD) as I tell you and *show* you how to find these cuts.

More Tips About Tenderness

I will sometimes see how easily I can poke my finger through a piece of meat to tell whether it will be tender or tough. If you can't handle the meat before you buy it, then look for consistent white lines of fat marbling in the muscle of the meat and about a good ⅛ inch of exterior fat around the side. Also verify the grade of meat before you buy. If the meat is graded choice and there's a good degree of marbling, you can expect it to be reasonably tender. If it's graded select (a lower grade) and is very lean with little marbling, chances are your piece of meat will be on the chewy side.

Since grading is voluntary, it's doubtful that carcasses offered for sale by an individual farmer will be USDA-graded. In this case, you might ask for a sample of meat to take home and cook before committing to buying a whole carcass, or ask the farmer how he or she considers the animal would grade, or ask for the chance to inspect the carcass yourself to check the degree of marbling (keep in mind that generally, the more marbling, the higher the meat will grade).

Location, Location, Location

In my butchering classes, I teach my students a very simple rule of thumb. The closer to the head or the feet of the animal a muscle is located, the

Cole's Notes

A **locker plant** is a refrigeration establishment with rentable lockers for food storage.

To make a flat iron roast, take four or five flat iron steaks, lay them one on top of another, and tie them together with beef twine. Trim around the tied parcel to make it uniform, then roast uncovered as you would any roast. You're welcome.

What's a really lean meat? Rabbit. It's one of the leanest meats as well as one of the healthiest. Rabbit is very low in fat and cholesterol and very high in protein.

harder that muscle needs to work to carry the animal around and get it through an animal-type day (grazing, trotting through the pasture, butting heads, et cetera) Harder-working muscles will naturally be tougher. It's sorta like those guys you see at the gym—the ones crunching 40-pound arm weights and grunting theatrically. Next time you

To many people, this is the essence of ecstasy—a honking big ol' steak.

SCOTLAND: MACDONALD BROS. BUTCHERS

Up a side street in the pretty town of Pitlochry sits MacDonald Bros. Butchers. From the street, it seems a modest enterprise, but appearances are deceiving: This is one of the UK's finest butchers. Rory MacDonald is the proprietor—the third generation in a business that goes back to its founding by his grandfather in 1928.

Yes, there's a counter where local customers are served, but this business is much larger. As Rory explains, "We have a large base of local customers but we also rely heavily on the tourist trade. We supply wholesale to the hotels, restaurants, and guest houses in the region and also have a very active mail-order business, which covers the whole of the UK."

From the street, you see two doors. One takes you into the butcher shop; the other, into the meat pie and specialty product shop. Behind these public areas are the private inner workings. Here, whole carcasses

are brought into the store's cooler room: "All our beef comes to us dry-aged for 10 days; then we dry-age it for an additional 21 to 28 days."

Rory says that while he has seen a decline in culinary butchers in the past 10 years (which he attributes to supermarkets), there's increasing interest from the public; people are beginning to recognize that no supermarket can deliver the quality and variety of a culinary butcher. He also notes that good butcher shops are especially popular in country areas, because country people are knowledgeable. In his case, they also know where his meat comes from: "Our beef is all Aberdeen-Angus from three farms within 40 minutes' drive from the shop. We have been buying from those farms for nearly 50 years. Most of our beef is grass-fed but as you can understand, that's not practical at all times of the year so other feeds are used during the

The Craft of Butchery

"When a customer buys a piece of beef for a special occasion, you advise on the cut, size, cooking, and you know it will be fantastic," says Rory MacDonald. "They then return and rave about how well it was received and that it was the best beef they ever tasted! That is what I like best about what I do!"

Rory MacDonald.

❖⋯▶ ◉ ◀⋯❖

Haggis

Haggis is Scotland's national dish, immortalized by Scottish poet Robert Burns in the 18th century by his poem "Address to a Haggis" ("O what a glorious sight / Warm-reekin', rich!"). Haggis is made by mixing sheep's offal (lungs, hearts, and livers) with suet and oatmeal, then stuffing the mixture into the sheep's stomach and boiling it all for three hours. And it's delicious. Honest. In Scotland, you often see it on restaurant menus as "haggis, neeps, and tatties," which means haggis with turnips (neeps) and potatoes (tatties). This is even more delicious.

A final note: When asked what it takes to make a great haggis, Rory replies:

"A master butcher!"

winter months. Our lamb is from Aberfeldy, which is the village I live in—15 miles from the shop. Our venison, which is red deer, is sourced from the hills around us in Perthshire and Angus and is entirely wild. Our pork is from Aberdeenshire, the main pig-breeding area of Scotland."

When meat enters the shop, it's in carcass form. Rory and his staff first break the carcasses down into primal cuts, then gourmet products. Every week, Macdonald Bros. processes four cattle, 18 to 20 lambs, and about four pigs. He mentions that many shops buy boxed beef, but he won't.

The shop has won many awards, including being named UK Champion by the Scotch Pie Club for its sausage rolls (pastry stuffed with pork).

Rory says that people love the haggis and black pudding (blood pudding), as well as the steak pies. Literally everything the shop sells—from gourmet meats to meat pies, sausages, lasagnas, and other specialties—is produced on-site. The place is bigger than it seems.

As for the establishment's products, here's an extremely condensed list: meat pies; sausage rolls; steak pies; venison pies; beef lasagna; stroganoff; lamb curry; wild red deer steaks and chops; cooked meats (beef, ham, pork, tongue, turkey, smoked duck); all beef, pork, and lamb cuts; haggis; white and black puddings; lamb tikka masala; pork Cajun schnitzel; back bacon; potted meat; every kind of sausage; and smoked venison.

When did you last find *all this* in your local supermarket?

see one, ask to feel his bicep—it'll be hard as a rock. Now transfer my tortured analogy to a meat animal. Lots of exercise = toned, hard muscles. Tough meat.

But this doesn't mean these tougher cuts are no good: You simply need to prepare them properly to counteract the toughness. Cuts of meat from these muscles are best suited to slow moisture cooking like braising.

The farther away from the head or feet a muscle is located, the less involvement it has in carrying the animal around and being part of its daily routine. So meat from these muscles will be more tender and thus better suited to dry cooking methods like uncovered oven roasting, grilling, broiling, or pan frying. Note that this rule applies to *every* type of meat animal.

So typically, cuts of meat from the round or rump in beef, the ham in pork, or the leg in lamb will be leaner cuts. Neck meat will also be lean.

Cuts of meat from the chuck or shoulder in beef, lamb, or pork tend to be fatter. The fattest cuts will be from the belly of pork, the breast on lamb, and the plate on beef (where some of the short ribs come from). Brisket of beef also tends to contain a large amount of fat.

But as in life, there are always exceptions to the rule. Cuts like flat iron and teres major come from the head area, so according to my location-location-location formula, they ought to be tough. Only they're not. So sue me.

This rule of thumb is important to everyone who likes to eat meat—not just to butchers. Knowing where each retail cut you buy comes from on the animal's body offers you insight on how to cook that meat. For a blow-by-blow description of cuts of meat and where they come from on a cow, pig, and lamb, refer to chapters 7, 8, and 9.

Cooking Methods for Tender Results

Don't be fooled by the less expensive cuts of meat. They can be as tender and flavorful as the more expensive cuts *if cooked properly*, and these cuts are a great way to stretch your food dollar, as well as impress your family and guests!

Here's a brief list of cuts and their optimal cooking methods, using beef as our exemplar:

BEST FOR SLOW MOISTURE COOKING (BRAISING, STEWING, AND THE LIKE)

- Chuck roast.
- Shoulder or arm pot roast.
- Neck roast.
- Beef shanks.
- Short ribs.
- Bottom round.
- Brisket.
- Stewing beef (which can come from any of these cuts).

BEST FOR DRY COOKING (UNCOVERED OVEN ROASTING, GRILLING, BROILING, PAN FRYING)

- Rib roast.
- Rib eye.
- Delmonico steak.
- New York strip.
- Tenderloin (also known as filet mignon).
- Porterhouse steak (basically a bone-in New York strip steak with the tenderloin intact).
- Sirloin steak.
- Round tip, also known as sirloin tip.

Avoiding Bad Meat

How can you tell when meat is going "off"? It'll start losing its color. Red meat will start turning brown. It'll have a sour smell. If it's really bad, it'll get a glossy golden-colored film on it. Pork tends to turn more toward the green side (ick). It'll get grayish first, then have a really rancid smell. Chicken will get a sheen. Starting to feel queasy?

Chicken—in fact, all poultry, like turkeys and their ilk—very often carries salmonella and must be carefully handled. In addition, poultry generally gets handled a lot. So whenever I work with raw poultry, I sanitize everything it touches (surfaces, faucets, and so on).

Old-Time Tips

Before the days of refrigeration, spoiled meat was a pretty common problem. In Thomas De Voe's *The Market Assistant*, an 1867 guide for householders and shoppers, he recommended two methods of dealing with dubious meat:

Tainted meat or game may be restored as follows: wrap it up in a fine linen cloth; have ready a vessel of cold water; take a shovelful of live wood-coals and throw in; then put the meat or game in and let it lie under the water for five or ten minutes. After taking it out, all the offensive smell will be removed.

The fly which blows the meat is known as the green or meat fly. They are always, in warm weather, found wherever there is fish or flesh—slaughter-houses, markets, larders, pantries, etc.—which they frequent for the purpose of leaving their eggs in some moist crevice in the meat. These eggs will hatch in a few hours, so that live maggots are seen to creep. Many housekeepers imagine that meats in this state are spoiled, and unfit to be used, but such is not the fact, as a little vinegar or salt and water will wash all signs away.

Professional butchers also know that whenever you change species (poultry to beef, beef to pork, et cetera), you must either change cutting boards or sanitize the cutting environment thoroughly.

I always cook poultry the day that I buy it—I don't store it in the refrigerator (even overnight). I always wash chicken in cold water, because using warm water will tend to activate any bacteria on the bird, and may even start the cooking process.

Poultry's tendency to carry bacteria is one reason that responsible butchers don't cut chicken on the same bench where other meats are cut (unless the bench has been thoroughly sanitized between meats). If you cut beef on a surface where raw chicken has been, that beef could end up contaminated with bacteria. And if that beef were then cooked only to rare, some bacteria (such as salmonella) might survive the cooking process, and you could end up with stomach upset, diarrhea . . . you get the picture.

So sanitization is extremely important, and today any good butcher understands this. Wasn't always so. I've heard older butchers say, "I wonder how many people we made sick in the old days . . ." when it was common to cut chicken on the same bench as meat.

Storing Meat Safely

In Europe, most consumers buy their meat on a daily basis, and many folks in the United States are doing the same. But if you don't have the time to shop daily, or prefer to buy all your meat for the week at once, you can freeze part of it and put the rest into your fridge's meat drawer.

A good fresh piece of meat—such as a steak or roast that was cut the day you purchased it—is good for about four days. Keep in mind that at supermarkets, large cuts of meat arrive at the store vacuum-packed and will hold for up to six weeks with no problem—as long as the vacuum seal isn't broken. But as soon as that seal is broken in order to cut the meat into smaller pieces for sale to consumers, the meat begins to deteriorate. So if they're cutting that piece for you today, I wouldn't keep it more than four days. To learn more about freshness issues with supermarket meat, read "Is Your Butcher Being True to You?" in chapter 5.

If you know that you're not going to use fresh meat within three days, then you can certainly wrap it and freeze it in a vacuum bag with as much of the air sucked out as possible. Properly packaged, you can safely store the meat for up to six months in your freezer. (See "Wrap Like a Butcher!" in chapter 6 for step-by-step instructions on how to wrap meat well for freezing.)

WHAT'S THE BEST TEMPERATURE FOR BEEF? PORK? CROCODILE?

Storing meat is fine, but at some point you'll probably want to eat it (*just a thought*). I get lots of questions about cooking temperatures for meat. Kinda matters, 'cause we've all suffered through one of those disastrous dinners involving steak cooked to a crisp, or a roast bleeding onto the table. The USDA has developed guidelines for cooking temperatures of the various meats, and I urge you to consult these.

Having said that, let me tell you that I don't follow USDA guidelines for cooking temperatures except for poultry, eggs, and ground meats *whose source I don't know*. I feel comfortable with this because I know the provenance of every piece of meat I consume: where it was raised, how it was raised, when and how it was slaughtered, and so on. I'm comfortable cooking it as I like it. This is probably an example of "don't do as I do."

For beef, lamb, and veal, the USDA recommends an internal temperature of 145°F (63°C). I prefer rare at 125 to 130°F (52–55°C). If you prefer medium rare, cook to 130 to 140°F (55–60°C). For medium well, 150 to 160°F (66–71°C). And if you prefer your meat well done, I can't help you, because I would never order or cook meat well done. My preference is rare, and it can be difficult to convince a restaurant—hampered as they are by health inspection regulations—to serve you a truly rare ("blue") steak.

If you are cooking burger from ground muscle meat that you are *certain* comes from a healthy local source, I recommend 140 to 145°F (60–63°C). For any other (unknown) source, 160°F (71°C) is safest and is the temperature recommended by the USDA.

All poultry should be cooked to 165°F (74°C), and fish to at least 145°F (63°C).

I get a lot of questions about pork. Specifically, the correct internal temperature to cook it to before serving. I'm vigilant about buying only the best meat from a properly raised animal (which is why I like to know about the farmer behind the product), so an internal temperature of 145°F (63°C) is what I recommend. This gives a tender, delicious result. However, most people prefer to cook pork to a higher internal temperature of 155°F (68°C) . . . it provides peace of mind. And I agree. If you're uncertain about the quality of the meat, err on the cautious side.

I can't tell you about crocodile. The Vermont crocodile is endangered, and hunters can no longer purchase croc permits from the state.

Reheating Red Meats

Here's how to reheat red meats like standing rib, steak, eye round, or flank steak so you don't lose taste or quality. The assumption for these instructions is that the meat was cooked only to the rare stage at first serving.

For *oven* reheating, remember these mantras:

- The *larger* the piece of meat, the lower the heat and the longer the time.
- The *smaller* the piece of meat, the higher the heat and the shorter the time.

For a 2-pound piece of red meat, preheat the oven to 350°F (177°C), put the meat into an ovenproof dish, cover with aluminum foil, and place in the oven. After 10 minutes, flip the meat and give it another 10 minutes, removing the foil for the last 5. This will give you a medium-rare result.

Reheating on the *stovetop* is actually best for smaller pieces of meat. Place a pan on the burner, bring to high heat, and add a tablespoon of butter or oil. Then "fry" the meat for about 2 to 5 minutes a side (depending on thickness). For a flavorful jus, deglaze with ½ cup red or white wine and reduce by half. Then turn the heat off, add 2 tablespoons of *very cold* butter, and swirl in. You'll love me for this.

Microwave reheating: *Don't do it* (unless you love rubber).

Chapter Four

Getting Meat to the Table

Sometimes it seems that there are as many ways to raise meat animals as there are farmers. It's likely that close to 90 percent of the meat eaten in North America comes from animals raised in large-scale operations—what's generally known as conventional meat production. This is the meat that's usually sold in supermarkets, and served in most restaurants. I want to stress that there's nothing wrong with conventionally raised meat; furthermore, the production methods used to raise it are, *on the whole*, safe and humane.

Worldwide, most people like to eat meat. Conventional meat production evolved and exists to feed these billions of humans. The system is tuned for volume, throughput, and the delivery of affordable meat to as many consumers as possible. It will not be going away soon, if at all.

This impressive steer is being raised on natural grassland. *Courtesy Dickinson Cattle Co.*

Since you're reading this book, you are probably interested in getting your meat from a different source. Perhaps you've read or heard reports about the evils of conventional meat production. Perhaps you're interested in trends like locavore eating, or grass-fed beef, or pastured pork, or animal-welfare-approved raising methods, or heritage animal breeds.

I applaud your curiosity and willingness to explore this very complex world. However, one extremely important phase of meat production is rarely discussed, and that's slaughter. Meat doesn't "just happen"; animals die. If you want to become an educated carnivore or bought this book to explore home butchery, it's important that you understand the process of turning live flesh into food.

HUMANELY RAISED AND HUMANELY SLAUGHTERED

These are the goals to strive for, whether you're considering raising your own meat animals, buying a carcass from a farmer to butcher yourself, or buying a beef quarter that will be prepared for you as retail cuts. If the farmer acts in the best interest of the animal, it will live a healthy life and its death will be as stress-free as the farmer can make it. Read the stories of the farmers later in this chapter; their concern for their animals is obvious, and I know from personal experience (no, actually, tasting) how good their product is.

Raising the Best

What does it take to raise the best? The answer you get may depend on who you ask, but the basic factors that matter most are genetics, feed, and environment. If you were to eat mostly potato

Just Normal Animal Life

Farmer Walter Jeffries defines *humanely raised* this way: "To me, humanely raised means that the animals are raised outdoors in a natural setting where they can graze and roam uncrowded in the fresh air and natural weather, and where they can act out their natural behaviors and interact with other animals. Gentle handling and no unnecessary interventions like tail clipping or teeth cutting. This does *not* mean life will be all cozy by urban standards or that there will be no stress. Some stress in the course of life is normal and natural . . . animals must put in some effort to climb up the mountain to the high pastures, they have to sort out where they rank in the pecking order, et cetera. That's normal animal life. What they don't need is being kept in a climate-controlled, manure-fume-filled, crowded pen where they never see the light of day, eat in an open pasture, or feel the wind. To do this properly means managed rotational grazing, not a pen. It doesn't take a lot of land or time for managed rotational grazing and it isn't hard to do—even on ¼ acre. The idea is to move the animals. This improves both the soil and the animals."

Walter Jeffries and Spot the pig. *Courtesy Walter Jeffries.*

chips, fries, and hot dogs, you'd probably not be in top shape. Like people, animals are what they eat. That's why livestock farmers need to devote so much attention to the minutiae of pasturage, types of feeds, grazing methods, "ideal" (which depends on region and climate) mixes of forage, and so on. You'd be very surprised at how much science is involved in raising a healthy meat animal, and how complex meat production actually is. It's a sophisticated business.

There's a direct link between good meat and healthy animals; this is why I stress the importance of butchers developing personal relationships with the farmers who supply their meat. Good meat comes from good farmers. I wanted to introduce you to some, so Karen Coshof and I interviewed a few to find out how they do it. But first, a quick explanation of some basic farming terminology, so that when you meet farmers, you'll understand what they're talking about.

Genetics

Ever been to a track meet? Notice the difference between the shot-putters and the high jumpers? Shot-putters are big, wide, thick-necked, muscled; high jumpers are impossibly tall, thin, and slight . . . with legs that go on forever. It's almost as though each athlete evolved into the ideal physical type for the sport he or she competes in. Which is not precisely true, but I'm trying to make a point here.

Genetics determine much of our makeup: Will we be tall and thin? Short and stocky? Tend to put on fat easily? Able to eat all day and never gain an ounce?

The same is true for meat animals: Different genetic and environmental conditions shape the way an animal is configured and what it's best suited for. From a farmer's perspective, it becomes obvious that the breed of animal perfectly suited to milking, such as a Holstein or Jersey cow, is not the ideal breed for meat production. And vice versa.

Although they do produce wool, Texel sheep have been selectively bred to produce excellent and lean meat. *Photo © 2012 TC Pet Photo. Courtesy Patrea L. Pabst.*

Genetics—the traits that "form" the final beast—are the first key to raising the best meat animal. And this includes not only physical qualities, but also important adaptive qualities; for example, how well will this sheep or cow do in a cold climate? A warm one? How will it do on the type of fodder available on a particular farm? If a farm is characterized by mountainous terrain and rough high pasture, better not select a breed that can't take care of itself without human intervention. And so on.

So the first task for a livestock farmer is to do some homework and select the breed that is genetically best suited to the farm's particular environment—its "farming condition."

The second step in raising the ideal animal is to breed selectively—using only the best animals for breeding in order to gradually "evolve" the ideal animal for your conditions and objectives. This is easy to say, hard to do.

As farmers consider breed choices, they assess a number of factors:

- Is it a *meat* breed? (Some cattle breeds are good as dairy animals, but not for meat; some sheep breeds produce great wool, but not great meat; some chicken breeds are broiler birds, others are for egg production.)

- What are the breed traits?

  1. Appearance (size, color, conformation, et cetera)
  2. Growth rate
  3. Ease of giving birth and (for sheep and pigs) average number of lambs/piglets per litter
  4. Temperament and ease of handling
  5. Ability to thrive in specific environments (hardiness, resistance to heat or cold, resilience, and so on)
  6. Meat qualities, taste, degree of fat (marbling qualities), carcass yield, et cetera

All are necessary inputs for selecting the best breed for a specific farm, a specific farmer, and a specific goal.

Breeds Matter

Vermont beef farmer Paul List has this to say about the importance of breeds: "When was the last time you saw a dogsled pulled by basset hounds, pugs, and poodles? *Breeds matter.* Holstein, Jersey, and the other dairy breeds are excellent milk producers—for hundreds of years, this is what they have been bred for. But they can't compete with breeds bred for beef production. And buyer beware . . . I've come across a lot of small beef producers who are really dairy farmers getting rid of bull calves and nonbreeding heifers by selling them as local grass-fed beef."

Forbs and Legumes

A **forb** is a nonwoody, broad-leafed plant that isn't a grass. Examples of forbs are dandelions, goldenrod, and milkweed.

A **legume** is a flowering plant that produces seeds in pods. Beans, lentils, peas, and peanuts are legumes, as are the animal fodders alfalfa and clover. Legumes are valued because they're protein-rich, thrive in drier soils, and yield well without the need for fertilizing with nitrogen.

To learn more about the wide range of breed choices, check out the sections on breeds of meat cattle, pigs, sheep, and chickens in chapters 7, 8, 9, and 10.

Pasture Terminology for City Slickers

To understand how meat animals are raised, you need to know some vocabulary. **Pasture** is land for grazing animals. When we talk about a farmer's pasture, we're generally referring to an enclosed field planted with **forage**: plants intended to be eaten as **fodder** by animals like horses, cattle, sheep, pigs, and so on.

Pasture is different from **rangeland**, although both may be used for grazing animals. A pasture is managed by the farmer, who may seed and fertilize it, while rangeland is left in its natural state.

Pasture isn't just grass; it's a mix of grasses, legumes, and forbs. Farmers select grasses and other plants according to their geographic region, micro-climate, annual rainfall, type of animals grazed, and so on. So all that green out there didn't just happen—it's been designed. Green couture.

Hay is the pasture's fodder harvest, cut and dried for use as animal feed. The quality of hay depends on many factors: moisture (too little, and the nutri-tional value suffers; too much, and the freshly cut hay may spoil), ripeness of the plants, and so on. Haying is serious business, with dawn-to-dusk work to get the crop in while the weather's good.

Hay is not **straw**. Straw is a by-product from cereal plants like wheat, barley, oats, and so on. After harvesting and threshing (removing the grain from the stems), the dry stems and dead leaves become straw, often baled for further use as animal bedding, baskets, or fodder (in restricted amounts due to straw's poor digestibility and low nutritional value).

Silage is fresh, high-moisture fodder that's fer-mented and stored in a silo or other system that keeps air out. Silage can be made of many plants, like oats, alfalfa, grass, and more. If you've seen large round bales covered with white plastic by barns or farm fields, those are silage bales.

Green chop is fresh-harvested forage fed wet directly to livestock. Green chop feeding is common in developing countries, where labor is cheap and abundant. It's like room service for cows.

Rotational Grazing

The term *rotational grazing* is one you'll hear much of as you learn about the world of animal management.

Sometimes termed cell grazing or management intensive grazing, it's actually a very simple con-cept: Move animals around on the land so that grass (or forage, as we now know to call it) is harvested—that is, eaten—quickly, then allowed to regrow.

This grass mixture includes timothy and perennial ryegrass. Timothy is often used in pastures that are both hayed and grazed. Ryegrass is extremely nutritious.

Fescue mixed with the perennial forage legume alfalfa (the three-leafed plants).

This forb is milkweed. While it's often viewed as a weed, increasingly farmers are finding that animals enjoy eating it.

Above photos by Jenn Colby, pasture program coordinator, UVM Center for Sustainable Agriculture.

This is done by dividing a large pasture into several smaller paddocks (cells), and moving the animals from cell to cell. The animals graze one paddock, giving the land in the others time to recover—this is also called resting the land. Generally, the length of time a paddock is grazed depends on the size of the paddock, the growth stage of its forage, and the size of the herd.

This may seem self-evident, but research has found significant benefits—both for the animals and for the land—in rotational grazing. Resting the land gives vegetation time to renew energy and grow faster. It also deepens root systems, creating a pasture that is stronger overall. Animals do better eating tender young plants, which are more nutritious, thus reducing the need for supplemental feed. And concentrating the herd within a smaller area forces animals to be less choosy about the plants they're eating . . . so, fewer weedy species. Keeping noncrop land strong and healthy reduces erosion and soil loss. It's win–win.

STRAIGHT FROM THE FARMER'S MOUTH

So now you've had a speed-learning introduction to farm lingo and you're ready to appreciate the wisdom of some great farmers from Vermont to Texas . . .

Vermont: Paul List, Grass Roots Farm

I got into beef farming out of a desire to invest in my own future as well as the future of Vermont, whose agrarian tradition is under threat. I saw fallow lands that could be productive grazing pastures and a huge market for high-quality local

Paul List with a few of his grass-fed and -finished Lowline Angus beef cattle.

Lowline Angus is a specialty beef breed that originated in Australia. They are unique and quite rare in America.

beef. I also believe that changing weather patterns will make drought in the Southwest persistent.

Vermont has everything going for it: abundant water, good grass, a healthy environment, and the infrastructure [barns, fences] of a slowly dying dairy industry in reasonably good shape, with people who still know how to farm cattle.

My farm is a patchwork of other people's property, thus the name Grass Roots Farm. To succeed, I need the cooperation of many individual landowners. I lease some land where my main herd is, but most of my hay crop comes from property I maintain for the grass. It's a great arrangement that saves me from having to buy land. The property owners get what they want: beautifully groomed land and lower taxes due to "current use" tax status [a Vermont program that provides tax benefits to landowners who keep their land in agricultural or forestry production]. I get what I want: quality feed. It's win–win.

I work exclusively with Lowline Angus cattle. I am currently at about 180 head—which makes this probably the largest Lowline herd in the United States.

I manage the land naturally: no chemicals and no tilling. I enrich the soil by spreading it with my own composted manure plus wood ash from a local wood-burning power plant. I use no artificial nitrogen whatsoever.

I sell my beef by word of mouth and only direct to the consumer. I've shipped to Beverly Hills, California, and sold to my neighbors. I won't sell more than I can sustainably grow. My aim

is to develop a customer base of about 250 families, although I am interested in expanding the Lowline breed. I'm not the least bit interested in feeding the world, just a couple hundred families. So far I've sold all the beef I can raise.

Vermont: Walter Jeffries, Sugar Mountain Farm

I can sum up our philosophy by saying that it's humane, outdoors, natural, minimal intervention, and that we concentrate on breeding for success. There are a whole lot of reasons I like pasturing better than confinement. First, I don't like seeing animals in confinement. But there are more reasons. I live here. I raise my children here. I want them to have quality food. I want them to be able to participate in the good life.

I always wanted to farm. I had cousins who raised livestock and I knew I wanted to do the same thing. Land is expensive so I knew I'd have to do something else first to be able to afford it. I invented some high-tech and consumer products, created a manufacturing company, published a magazine, and spent all the money I made on a farm.

We started with rabbits, then chickens, ducks, and sheep. In the spring of 2003 we got our first pigs. Turns out we were very good at raising pigs on pasture without buying commercial grain or hog feed. People told me pigs can't eat grass but fortunately the pigs didn't know that. People also told me that pigs prefer to root. In fact, they'd much rather graze.

Our homestead is a system of many parts. Big Ag is all about specialization and monocropping. That is the opposite of how we do things at Sugar Mountain. We have built a system where animals co-graze, because it's more effective to keep multiple species together. That's because they each eat different things: Poultry naturally hunt parasites that are a problem for grazers. By rotating them

Left to right: Will, Ben, Holly, and Walter (*back*); Hope with stuffed tigers (*front*). *Courtesy Walter Jeffries.*

Courtesy Walter Jeffries.

together we take advantage of these natural and sustainable behaviors. That makes farming easier, and the end product the customer sees is better while doing less damage to the environment and actually improving the soil, air, and water.

Pigs are a good balance. They go through generations quickly, and feeding solely on pasture, they're ready for market in just seven to eight months.

I researched diet and nutrition to figure out what we could combine with pasture to produce a balanced diet for them. Growing pumpkins and other things helps. So does the whey we get from a local creamery after they make their cheese—whey is a wonderful traditional food for pigs because dairy is high in lysine and adds a boost in calories. Combining pasture and dairy results in an excellent diet and fast growth, plus the pork tastes delicious—slightly sweet in flavor with wonderful fat. The local chefs rave about it. We had become pig farmers.

Our herd developed randomly, based on the local availability of animals. With each generation we bred the best and ate the rest. Our pigs include Yorkshires, Large Blacks, Berkshires, and Tamworths. We also have a touch of Gloucester Old Spot and Landrace in our pigs. [See chapter 8 for more information on pig breeds.]

I like crosses selected for our needs rather than specific breeds; combining, then breeding

the best of the best over years, improves our herd genetics so that they perform well in our climate and taste delicious.

There is no perfect pig. It depends on variables like climate, temperament, taste, growth rate, litter strength, and so on. I use 27 criteria to rank and evaluate pigs. About 5 percent of the gilts (females) get a chance to breed and about 0.5 percent of the males are kept as breeders.

None of this happened overnight. Rather it was a slow, gradual, systematic progression. We now have about 300 to 400 pigs on the farm. There are three big boar plus about 60 sows and their offspring: piglets through finishers. They're out grazing on pasture all year round. In the warm months, they graze about 70 acres. In the cold months, they have about 4 acres of winter paddocks they rotate through, and we replace the standing pasture with stored hay.

Getting to where we are has been a journey, and we have many years to go to get things where we want. We deliver fresh (not frozen) meat, sausage, bacon, ham, and our unique all-natural smoked maple hot dogs to our customers on our weekly route. About 88 percent of our business is to local stores and restaurants with the remaining being direct to individuals for half pigs, whole pigs, roasters, and some weaners and the occasional breeder animal.

Courtesy Walter Jeffries.

Courtesy Walter Jeffries.

Virginia: Craig Rogers, Border Springs Farm

I call myself "the accidental shepherd." My business started because of a hobby run amok. My wife and I saw a sheepdog trial while I was a professor of mechanical engineering at Virginia Tech. We were amazed. The beauty of the ballet of sheep and dog captivated us and inspired us to buy a Border collie and a farm.

Well, you can't train sheepdogs without sheep. So we bought six and I started competing in sheepdog trials. I soon noticed that the winningest owners had the largest flocks (more practice), so . . . more sheep. Soon we were surrounded by them. What to do? Aha. Meat.

I tried a few things, but wasn't happy with the route of traditional lamb production, or with selling to local restaurants (who, frankly, didn't know what to do with it). I needed to find people who would honor my animals, so I began contacting top chefs directly.

Now I sell to the United States's finest chefs: people like Bryan Voltaggio, Richard Blais, Sean Brock, Mike Lata, and others. Many have won James Beard Foundation Awards and have appeared on Iron Chef America, Top Chef, *and other shows. My lamb is often featured on the James Beard Celebrity Chef Tour. So yes, you could say that my life is governed by insatiable passion.*

But I know what I do. I'm a shepherd—even says so on my card. We now have 1,500 sheep grazing many hundreds of acres. The farm's run by me, my wife, two farmhands, and our dogs. We have four sheepdogs (including Jake, who was the top all-around sheep and cattle dog in North America in 2006), seven livestock dogs, plus a few agility dogs my wife trains. We also have about 20 pet sheep—runts which in any sane operation would have been culled, but which my wife (Chief TLC Officer) won't let me slaughter. We treat our animals well—our farm is Animal Welfare Approved.

Our sheep are pastured on ordinary grass and finished on high-sugar grasses to put fat

Courtesy Craig Rogers, photo by Peter Taylor.

Craig Rogers says: "Every good farmer I know has a true passion for one animal. Mine's sheep." *Courtesy Craig Rogers, photo by Peter Taylor.*

Wendy and Cody, on a very cold day. *Courtesy Wendy and Cody Fulwider.*

onto the carcasses. Most grass-fed animals are too lean; it's one problem of the grass-fed movement that's led to (overstating this a little) an attitude of "You mean if I just feed them grass I can sell them for a premium?" The idea is to create a grass-fed animal with enough fat for flavor and tenderness—you have to farm for flavor, not for pounds.

The future? I have a retail shop, The Shepherd's Larder in Union Market in DC, and soon will be opening a restaurant called the Lamb Bar, also in DC. As for my philosophy? Just farm right—be a steward of the land, a compassionate custodian of your animals, and make your sliver of the planet better and happier with what you produce.

Wisconsin: Wendy and Cody Fulwider, Pinnacle Pastures Farm

Wendy Fulwider was born on her parents' dairy farm but left with her son, Cody, when it became clear that traditional dairy farming was dying. Still, she loved farming life. She went to school, eventually getting a PhD in animal behavior and management. When her father could no longer continue farming, she and Cody decided to return and turn the dying dairy farm into a direct-to-customer meat farm that

offers grass-finished cattle, pastured pork and lamb, turkey, geese, guinea hens, and chickens.

If you like farming and livestock, it's in your blood. Cody and I live and work on the farm full-time. We decided early on that to be profitable on a small operation, you've got to cut out the middleman and produce a unique product.

Our clients are mainly city folks concerned about healthy meat and animal welfare.

We sell only to the ultimate consumers . . . people who're actually going to eat the meat. Every animal is "pre-sold"—assigned to a customer—and when it's slaughtered, it's divided according to their wishes and cooking preferences. We're fortunate to be working with a small slaughterhouse that will work directly with our customers to make sure they get exactly what they want.

Our animals are selected for docility, maternal qualities, and the ability to flourish on pasture. We began with Hereford Angus crosses; now we're on our second British White bull, a breed that's gaining popularity with grass-finishing farmers. Our pigs have to be good moms and mustn't sunburn easily since they're outdoors all the time. We mainly use Durocs, Berkshires, and Hampshires. Our sheep are Katahdins.

When I think about what I do, I flash back to a conversation I had with a fellow farmer, who

The Secret to Food Security

Farmer Wendy Fulwider muses, "People worry about food security. In my opinion, if you want to ensure the security of the food chain, you should have lots of little farms."

said, "We don't have blood flowing through our veins—we have dirt." He was right. When you're farming, you work really hard, but you can see what you've done. There's a visible result and you know that you have accomplished something.

My ultimate aim is to enlarge the meat operation to sustainability, and to return my parents' farm to dairy production—only this time, organic.

Ohio: Dickinson Cattle Company

Dickinson Cattle Company, the largest of the farms featured in this book, is a family ranch in Barnesville, Ohio, run by three generations of—you guessed it—Dickinsons. If you're a Texas Longhorn fancier, you'll know of this ranch: It's one of the largest Texas Longhorn producers in the world.

The ranch was started by Darol Dickinson and his wife, Linda, in 1967 with six cows and a bull; it's now 5,000 acres. Darol is general manager and Linda runs the office. Their son Kirk handles the high-tech side of things, and their other son Joel is herd manager. In the summer two more staff run the operation's ranch tour business and its on-ranch store. Two cowboys handle day-to-day cattle work. Eight grandkids ages 9 to 18 do ranch work, each according to his or her skill level. "When a whole family chooses to work in the family business, it means there is a joy and satisfaction that spans

Shadow Jubilee held widest horn-spread record for six years (89.74 inches). *Courtesy Dickinson Cattle Co.*

the generations," Darol Dickinson says. "Yes, our family enjoys raising high-quality cattle."

"The only two profitable parts of the cattle business are retail beef and registered breeding stock. Each person likes a certain breed. We've found that minor breeds are more profitable than well-known, overproduced breeds. We respond to those desires and provide cattle for all needs."

The operation raises its cattle in two regimens. Cattle that provide beef for the ranch's beef operation (which includes highest-quality steaks) are grain-finished. Cattle destined for the ranch's "heat-and-serve" canned beef and ground beef products are grass-fed and -finished. No matter which final destination they're bound for, none of the cattle are fed hormones, ionophores, or antibiotics.

Unfortunately, sooner or later meat animals who don't graze fiber—grass—will become more and more expensive due to the cost of grain. In the future people worldwide will eat less meat—and not by choice. It is important that meat producers not live in denial, but change and adapt to available fibers for growing programs. We believe Texas Longhorn cattle are the best browsers of all breeds and will produce the most efficiently on lower-cost fiber. In the future, the heroes of the meat industry will produce human consumable food without human consumable input. Cattle that require great labor and grain input will become less valuable.

Vermont: John and Rocio Clark, Applecheek Farm

Organic grass-fed beef, veal, pastured pork, heritage chickens, ducks, guinea hens, organic dairy products . . . and emu oil!

Turn up a hidden side road off Route 15 in northern Vermont and you drive into what surely

Applecheek herd.

John Clark of Applecheek Farm.

the ecological capital in our soils. Grass-fed beef is the "new red meat," and we are taking it very seriously. We use heritage breed genetics such as Devons and Shorthorns to raise meats that acquire excellent marbling, texture, and flavor.

We take pride in raising and finishing our beef on grass. Our cattle consume large amounts of beautiful, nutrient-dense grass without any pesticides or harmful chemicals. Our grazing system is done in a managed rotational pattern that builds ecological resilience and fertility within our soils. In return, this method sequesters carbon at a very high rate, unlike nonsustainable farming operations.

Texas: Pam Malcuit, Morning Star Ranch

cannot be farming country. The road passes a few small houses tucked into increasingly dense forest. Nearing the top of a winding hill, a side road carries a hand-lettered sign with an arrow. As you turn, the ground suddenly levels into beautiful flat pastureland that stretches away around you. In the distance, you can see the summit of Mount Elmore with its fire tower. Applecheek Farm is a surprise and a perfect example of what used to be common in this state—the Vermont hilltop farm. This is the home of the Clarks, who've been farming in this place since 1965.

Applecheek Farm is a diversified organic farm run by John Clark for his father, John Clark (second generation, same name). Everything thrives on the rich, sandy loam, which—strangely for Vermont— never tosses up stones like most northeastern fields. The farm is involved in agritourism and regularly holds community events that help teach people about organic agriculture.

Our philosophy is simple. We strive to produce food that delivers optimal nutrition and restores

It seems we were always raising something—five kids, Yorkshire terriers, mastiffs. We bought this ranch after my spouse retired from the air force and started with Beefmaster cattle. Those guys have just enough Brahma in them to make them have no respect for fences. When one of our fence wires came down, all my Beefmasters disappeared onto my neighbor's 2,000 acres. Now we raise Dexters. They like it here. If there's a wire down, they just stand around and look at me as if to say, Are you gonna fix this, or what?

Why Dexters? 'Cause they're smaller and easy to manage. They need half the equivalent pastureland, half the feed, and are great for older farmers (ahem). We've run as many as 100 Dexters on our small pastures. We can fully manage them ourselves with a utility vehicle and a bucket of feed—no need for horses or herding dogs.

Dexters produce high-yield carcasses of lean, tender, fine-grained beef with excellent flavor. We raise them on pasture with additional supplementation in the form of mixed-grain pasture cubes. The steady diet seems to keep them from making too much fat the last few months.

I like the taste of grain-fed beef; grass-fed seems a bit gamey to me. But that's personal preference because I have many Dexter friends who do strictly grass-fed with excellent results.

I know there are people out there who say cattle shouldn't eat grain. Cattle have eaten grains and vegetables and whatever they could forage for eons. The only time they have digestive problems is if someone suddenly changes their diet. We do not, however, believe in using any steroids or growth enhancements with our animals. We do inoculate, because if we didn't they could catch diseases from neighboring cattle. We have 140 acres and are surrounded by other people's cows. It is just good sense to provide your animals with protection from diseases. Our animals are healthy, hardy, and seem pretty happy.

We sell our steers on the hoof for others to raise and feed as they wish. A Dexter steer will bring you from 400 to 500 pounds of great beef; slightly smaller cuts, but great taste with nice marbling, and minimal extra fat. The nice thing about their size is that it is easy to fit in your freezer—you don't have to find someone to buy the other half of the steer.

We have greatly enjoyed our time here on the ranch raising our little cows. Cattlemen at the local veterinary hospital used to tease me about my "midget" cows until we discussed feed and acreage needs and the prices we get for our animals. Some of those big ol' Texas cattlemen have actually switched over to these smaller animals now, and they are not the least embarrassed about it . . . though a few may claim they are "the wife's" little cattle!

Vermont: Mike and Julie Bowen, North Hollow Farm

North Hollow Farm in Rochester, Vermont, is a family affair, like so many farming businesses across America. Mike and Julie Bowen and their

MornStar Bowie, MornStar Inky Dink, MornStar Brody. *Courtesy Pam Malcuit.*

A Dexter family group on Pam's ranch. *Courtesy Pam Malcuit.*

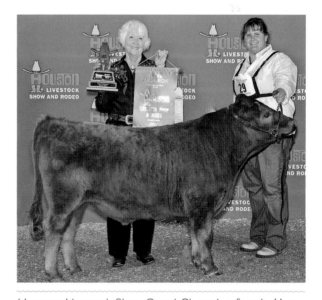

Houston Livestock Show Grand Champion female MornStar Ruby. Breeder Pam Malcuit; owner/exhibitor Elissa Emmons. *Courtesy Pam Malcuit.*

son-in-law Bryant work the farm all year, adding three or four hands in the summer months to help with haying. The farm was started by Mike's father Carroll Bowen, who bought its original 200 acres for $2,000 in 1948. Today, the Bowens farm 1,000 acres. The farm runs a cattle herd of about 500 head of cattle plus pigs and goats. Mike Bowen describes the operation.

North Hollow herd coming home on a misty day. *Photo by Angelique Lee.*

In 2003 we stopped growing corn and moved to an all-grass program. No chemicals, GMO feeds, growth hormones, or antibiotics. We felt that for the good of the soil and for the animal and human health benefits it was the better direction to take.

We breed Angus, Hereford, and Charolais cross cows with British White and Red Devon bulls, since research has shown that the older English breeds do well on grass.

All of our beef is processed at a small USDA-inspected plant in Vermont. We transport our animals to the plant ourselves in a trailer they are used to, so they arrive unstressed.

The plant we use has a maximum capacity of eight cattle a day. What this means is that the inspector carefully watches each individual carcass. The slaughterhouse routinely tests for E. coli. We wait for the test results before sending out the meat to our customers.

Our pigs are a mix of Tamworth and Old Spot. In the summer, they're outdoors; in winter they have a nice warm barn with plenty of room to trot around in. Pigs are very social animals and prefer being together in a large area. They're also naturally clean—they reserve one area for sleeping and another for eating. Their bathroom is always as far away from where they sleep as they can make it.

North Hollow's customers are diverse. About 70 percent of our meat is purchased by stores—mostly co-ops and natural food stores (and one chain store—Mac's Convenience Stores). The other 30 percent is sold directly by mail order to people who have gotten discouraged with commercial meats and want something different.

Our most popular item is our grass-fed ground beef, but lots of people love our nitrate-free bacon, sausage, kielbasa, ham, and franks. If you visit our online store, you'll see that the range of products is quite large. We're offering organic whole chickens, various cuts of lamb and goat, beef in every shape (ground, steaks, roasts, et cetera), pork products, beef organ meats, maple syrup (this is Vermont!), and even our own alpaca yarns. We also will sell sides of beef and pork.

We love farming for lots of reasons. We like to keep land used so it doesn't revert to brush. We also love animals and all the work that goes along with farming; you're never at a loss for something to do. In farming you never get to a point where you've achieved all your objectives; there are always new things to try like improving genetics, or growing more grass per acre, or finding ways to further reduce stress on our animals. We can always work toward doing things better.

DOING YOUR HOMEWORK

After reading these stories, you may be wondering, *How can I find a great farmer like that to buy from?* Well, they're all listed in the resources section at the back of this book, but you may be in a region they don't serve, so what to do?

One way to identify farms that raise and slaughter humanely is to find one that has been awarded a seal of compliance from an animal welfare association. Animal Welfare Approved (AWA) is a program in which farms are annually audited for compliance with extremely stringent animal welfare standards. Those that pass are allowed to display the AWA seal. The certification is offered to farmers free of charge, so that the process can't be tainted by financial influence. There are other humane certification programs out there—just go online and search.

It's also critical to ask questions—talk to the farmer face-to-face, or by phone or email. Reading this book will give you plenty of ammunition for developing a list of questions. And I'll help you out here, too. Here's a checklist of things to ask a farmer before you commit to your first carcass:

- How long have you been raising animals?
- What is the breed of animal you use? (For info on meat breeds, see chapters 7, 8, and 9.) Do you breed your animals yourself, or purchase young animals for raising?
- Do you use any growth hormones, feed additives, or nontherapeutic antibiotics? If so, why?
- How do you raise your animals? What do they eat?
- At what age do you slaughter them?
- Are they humanely slaughtered? (See the next section of this chapter to understand the true meaning of this.)
- How is your meat inspected? Federal or state? (Either is fine, by the way.)
- Is it USDA-graded? If not, how well is it usually marbled? How do you believe it grades?
- Is it aged? If so, for how long and where?
- Can you give me the names and phone numbers/ email addresses of previous purchasers? (If a farmer won't give you references, find a different farmer.) Do you ship? How is the carcass/primal packaged?
- Do you stand behind your product? In what ways?

Finally, how can you guarantee that the carcass you're purchasing is high quality and high yield? This is a tougher question. First, get to know your farmer and—once you've found someone you trust—stick with him or her. But frankly, to really determine that the animal you're buying is a high-quality animal, visit the farm and inspect it before you commit. By the time you've finished reading this book, you will have a good grasp of the differences that genetics, breeds, raising methods, and treatment make in determining meat quality. The way an animal looks, its conformation and body shape, fat covering, and so on can be eyeballed "on the hoof" (so to speak), and I strongly recommend that you do this. If the farmer is advertising Black Angus, the animal should look like the Black Angus breed photo in chapter 7, and so on for all the meat breeds. And if you can't visit the farm, ask for a photo of the animal you're buying, with a description of its size and approximate weight.

IN AT THE KILL

Nobody really likes to think about killing animals, but it's at the root of meat eating. I recognize that most people are uncomfortable with the realities of slaughter, but the truth is that what you don't know about the process of killing a meat animal may not hurt you but could hurt the animals you're eating. So how about we all man-and-woman-up and learn about it?

Until about 50 years ago, no laws governed meat animal slaughter. Eventually, however, public concern about cruelty to animals, coupled with meat producers' realization that unstressed animals produce better carcasses, led to the passing of national "humane slaughter" acts.

In the United States, any animal killed in a USDA-inspected plant must be slaughtered according to the Humane Methods of Livestock

Slaughter Act. In Canada, slaughter is governed by the Meat Inspection Act. In the UK, by the Humane Slaughter Act. And so on, country by country.

Legislation is clear and consistent, basing regulation upon the core principle that any animal that is slaughtered must be protected from avoidable excitement, pain, or suffering. Most animals are slaughtered in some sort of inspected facility. A small percentage are "custom exempt" in rural areas where farmers raise their own animals and pay to have them processed. A much smaller percentage are killed on the farm where they were raised for consumption by the farmer's family; this is called on-farm slaughter. You may also hear the term custom-cutting shop. This is not a slaughterhouse; it's a meat facility that specializes in processing animals that have been either farm-slaughtered or killed at a slaughterhouse and then delivered to the custom-cutting shop.

There are two basic approaches to slaughter. The first—**stun/kill**—involves stunning the animal before any further action is taken. The animal must be unconscious and unable to feel pain before it is actually killed.

The second is called **ritual slaughter**. It involves methods attached to particular religions, such as Jewish kosher dietary law and Muslim halal dietary law. Both involve cutting the animal's throat deeply from side to side with an extremely sharp knife, so that all major arteries and veins are severed and the animal becomes unconscious and quickly bleeds out.

Stun/Kill Methods

Let's begin with the stun/kill system. Before killing, the animal must be totally unconscious (in a state of surgical anesthesia). How is this achieved? Four ways are accepted: chemical (CO_2), mechanical—captive bolt, mechanical—gunshot, and electrical. Not all methods are used for all animals.

Chemical

Animals enter a closed chamber into which carbon dioxide has been introduced. Since CO_2 is heavier than air, it settles, pooling in the lower level of the chamber. The animals enter and quickly become unconscious. Plant operators are required to monitor the level of CO_2 concentration in the chamber to make sure it's kept constant. This method is often used for pigs.

Mechanical—Captive Bolt

There are two types of mechanical captive bolt stunners—**penetrating** and **nonpenetrating**. Both use gun-like mechanisms that fire a bolt or shaft out of a muzzle. The effectiveness—and humaneness—of this method depends on the operator, since not only must the bolt hit the animal's skull in exactly the right place, but it must also be fired with the right amount of force (via adjusting the air pressure or detonation charge). Some slaughterhouses use "double knocking"—two rapid blows—to avoid the catastrophic effect of an animal regaining consciousness as it's being hoisted and bled.

In the penetrating captive bolt method, the bolt penetrates the animal's skull. Unconsciousness is caused by physical brain damage.

Many meatpacking plants use the nonpenetrating captive bolt method. A nonpenetrating bolt differs from a penetrating bolt in that its head is flattened. When fired, the bolt doesn't penetrate the skull and the animal loses consciousness from the

This is one type of a penetrating captive bolt gun.

Mad Cow Disease

Bovine spongiform encephalopathy (BSE), commonly called mad cow disease, is a fatal animal disease in which the animal's brain degenerates. In 1996, a new type emerged that could be transmitted to humans. As a result, the World Health Organization issued recommendations requiring that:

- The inclusion of ruminant material such as beef by-products is prohibited in ruminant feed (no more feeding cows to cows).
- Visible nervous and lymphatic tissue is removed at slaughter (the disease is not present in muscle tissue).
- Any substance for human use that might contain bovine substances (such as cosmetics and vaccines) is prepared in countries with a BSE surveillance system in place.

To protect against BSE transmission, air-injection cattle stunning is no longer allowed. This was a method where compressed air was injected into the cow's head as part of the stunning process. The pressure of the compressed air often resulted in portions of brain tissue being forced into the carcass. And if that cow was BSE-positive, well, you get the point. It's repulsive to think about, but you've got to know the facts. In my opinion, you should not eat the brain of any cow stunned with a captive bolt of any type.

A Farmer's Stunning Story

Pork farmer Walter Jeffries relates this true story about what his wife experienced after being stunned in an accident (in which Jeffries played a somewhat embarrassing role). Here's how Jeffries tells it:

I once killed my wife (an accident, I brought her back), by hitting her on the head with a large wooden fence post (she walked into it, honest!). She was instantly stunned and dropped limply to the ground. She has no memory of approximately 30 seconds before and after the moment of stunning. That time simply does not exist for her. She felt no pain. She says that experience convinced her that stunning is totally humane. As she puts it: "You could have bled me out right then and I would never have known it." Stunning with captive bolt and being bled out is completely humane. I've watched hundreds of animals humanely slaughtered this way and it works.

impact of the accelerating bolt. Again, it's obvious that the operator needs to know what he or she is doing—especially here, where if the animal is not immediately bled, it may regain consciousness.

Mechanical—Gunshot

Guns are also used to stun animals. The caliber needs to be selected carefully to make sure that one single gunshot will produce unconsciousness. Since gun stunning leaves bullet fragments behind, US regulations forbid eating the brains, cheeks, or head trimmings of animals stunned this way (the tongue is okay).

Electrical

Electrical stunning is used for hogs, calves, sheep, and goats, but not generally cattle. There are two methods: head only and cardiac arrest. Head only induces an epileptic seizure, resulting in insensibility to pain. Cardiac arrest stunning also induces a seizure, as well as a heart attack. This causes pain, so it is critical to make sure that the animal's head is stunned before the chest. Obviously, where the electrodes are placed matters. This varies from one slaughter plant to another—electrodes may be positioned on either side of the head (head stunning), on both head and chest (cardiac stunning), or first the head, then the chest (two-phase stunning). The electrical current has to be powerful enough to guarantee unconsciousness from stunning through bleeding. Too high an electrical current can damage capillaries, resulting in multiple pinpoint hemorrhages in the muscle tissue. This is commonly referred to as "splashing" or "speckling."

Ritual Slaughter

Kosher and *glatt kosher* are specific terms applied to foods allowable under the dietary laws of the Jewish faith. *Halal* is the specific term applied to foods allowable under the dietary laws of the Muslim faith. These dietary laws cover all foods, not just meat. Let's take a look at how the terms apply to butchering animals.

Kosher

Kosher meat harks back to the Jewish Bible (the Torah)—specifically, Deuteronomy 14:3–10, which states:

> *These are the animals you may eat: the ox, the sheep, the goat, the deer, the gazelle, the roe deer, the wild goat, the ibex, the antelope and the mountain sheep. You may eat any animal that has a split hoof divided in two and that chews the cud. However, of those that chew the cud or that have a split hoof completely divided you may not eat the camel, the rabbit or the coney. Although they chew the cud, they do not have a split hoof; they are ceremonially unclean for you. The pig is also unclean; although it has a split hoof, it does not chew the cud. You are not to eat their meat or touch their carcasses. Of all the creatures living in the water, you may eat any that has fins and scales.*

But simply eating one of the allowable animals doesn't make meat kosher. The animal must first be slaughtered in a specific way known as ritual slaughter. Jewish ritual slaughter is known as **shechita**, and the person who performs the slaughter is called a **shochet**.

A shochet is a butcher who must also be a pious man trained in Jewish law, particularly relating to **kashrut** (dietary laws). The shochet kills the animal with one quick, deep stroke across the throat. He uses a razor-sharp blade, which cannot have nicks or unevenness. The method is painless and causes rapid unconsciousness.

Back to Deuteronomy: "But be sure you do not eat the blood, because the blood is the life, and you must not eat the life with the meat. You must not eat the blood; pour it out on the ground like water."

Since throat cutting is the surest and fastest way to drain the animal's blood, shechitah is the best method of adhering to the biblical commandment.

Glatt Kosher

Lots of people think *glatt* means a higher standard of kosher, but this is inaccurate (although generally accepted). The Yiddish word *glatt* means "smooth," and refers to the condition of the animal's lungs when it is inspected after slaughter. If the lungs contain adhesions or other defects, the meat will not receive the certification of glatt kosher. As simple as that. Glatt kosher cannot be applied to chicken, dairy products, or fish. If you see cheese labeled glatt, it's not.

Halal

Halal means "permissible" in Arabic. In butchery terms, it is applied to meat slaughtered according to Islamic dietary law. Like kosher, halal slaughter is done by cutting the animal's throat. However, halal requires praying to Allah at the time of slaughter.

Muslims are taught by the Qur'an (Koran) and the teachings of the prophet Mohammed that animals should be treated with respect and well cared for. Muslim law regarding how animals are killed is known as **Dhabihah**, and is quite specific.

The animal must be treated gently and should be offered water at the time of slaughter. Out of mercy toward the animal, the knife must be extremely sharp, not serrated, and should be kept hidden until the last moment. Slaughter must be done by an adult Muslim, Jew, or Christian (termed "People of the Book" in Arab culture).

Preferably, the head of the animal should be positioned to face Mecca; the animal is then killed in a way summarized by the word *Ihsaan* (in a beautiful, caring way)—limiting suffering or pain. This is done by cutting the jugular vein swiftly to cut off oxygen to the brain and pain receptors, then

What Is Ritual Slaughter?

Don't be influenced by the term—there are no people dancing around the animal, nor is the animal considered some sort of sacrifice. *Ritual slaughter* simply means killing an animal in a manner required by a specific religion. The meat of ritually slaughtered animals is deemed acceptable for consumption by people who practice that religion.

If you see any of these on a meat package, it means that the meat has been certified kosher.

This is one example of a halal label.

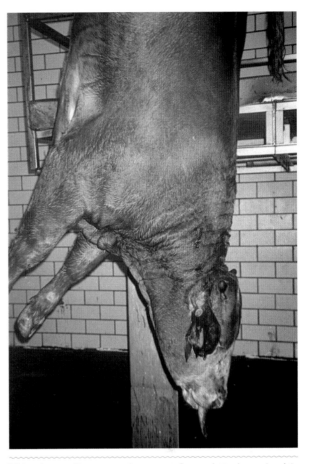

This photo of a stunned cow confirms that the animal is truly unconscious. *Courtesy Dr. Temple Grandin.*

waiting as blood completely drains out (like Jews, Muslims are forbidden to consume blood).

The Kill Floor

It is quite bracing to consult the USDA Food Safety and Inspection training manual for meat facility inspectors. Here's what it says regarding the signs used to verify that animals are insensible to pain (unconscious) following stunning (I've shortened this, but am quoting accurately).

- The head dangles from a flaccid neck.
- If the animals are suspended, the head should hang straight down.

- The eyelids should be wide open and pupils fully dilated.
- There is no vocalization—mooing, bellowing, baaing, or squealing.
- You may observe movement of the head and neck. This can be because of reflexes caused by random firing of damaged muscle neurons. *It may also be voluntary movement because the animal is regaining consciousness* (my italics).
- A stunned animal that has regained consciousness may vocalize.
- It may also show a "righting reflex" (the physical actions taken by an animal to move itself into a normal lying, sitting, or standing posture). For example, a conscious cow hanging from a bleed rail will show a contracted back, stiff extended neck, and rigid extended forelegs as it tries to pull itself into a normal upright position.
- When assessing unconsciousness, you need to observe the animals at different places along the bleed rail. For example, you could observe animals after they have been hanging on the bleed rail for several minutes.

Verifying a Slaughterhouse

I recognize that reading and thinking about the stunning and killing of animals is unpleasant. This is why I insist on checking every slaughterhouse or on-farm slaughter specialist I work with. I want to know how they treat animals. And you should, too.

When you're researching slaughterhouses, above all, ask about the facility's philosophy regarding humane kill. (What the law requires is clearly described above.) Here are some questions to ask. I urge you to challenge any evasive answer you receive.

- What slaughtering method is used? (Captive bolt, gunshot, et cetera.)
- How regularly is slaughtering equipment (stun guns and the like) verified for accuracy, consistency, and mechanical function?

The intake area at a large slaughter facility. The curved chutes are less stressful to the animals. *Courtesy Dr. Temple Grandin.*

The curved chute calms cattle; hard sides make them feel secure. *Courtesy Dr. Temple Grandin.*

The right way. Handlers stand calmly away from the moving cattle; cattle move quietly through the gate. *Courtesy Dr. Temple Grandin.*

- How experienced is the slaughtering staff? How are those involved in the actual killing of the animals trained? Are they regularly monitored?
- When do most of the animals arrive? (If a large percentage are unloaded after normal working hours, it may be a ploy to evade federal inspectors.)
- Is there an automatic water source in *every* holding pen? (This is required by law.)
- Is feed available for animals held for more than 24 hours? What kind of feed? (USDA inspection policy stipulates that feed must be appropriate for the age and species of animal.)
- If animals are going to be held overnight, are their pens large enough for them to lie down without having to lie on top of one another? (The fact that this must be stipulated sends chills down my spine.)
- How are pens and driveways—especially the single-file chute and stun box—constructed? There should be no sharp corners, protruding objects that can cause injury, or things like loose boards or broken ramps. (Allowing this is also a federal violation.)
- Is there good footing for the animals?
- Are disabled or nonambulatory animals segregated in pens that protect them from poor weather? (This, too, is a federal requirement.) It is also against the law to drag disabled animals, or move them with equipment such as bucket lifts. Instead, they must be stunned before moving.
- How are animals handled—in a calm, quiet manner, or are they being forced to run? (The latter is against federal policy.) Are noncompliant implements such as baseball bats, shovels, sharp prods, or whips used? (These are *definitely* against the law.)
- If AC-current electric cattle prods are used, what charge are they carrying? (This mustn't be higher than 50 volts.) Canadian legislation states that "no goad or electrical prod shall be applied to the anal, genital or facial region of a food animal."
- What kind of restraint system is in use? (The stunning area should be designed and constructed to limit the free movement of animals.)

- How are animals tagged? This is necessary to ensure effective tracking should there be a later health issue, particularly to benefit the farmer who raised that animal. It's no secret that some meat-processing plants are less than conscientious (to be polite); one farmer told us about a smokehouse that took his pork for smoking and gave him back what was clearly the bacon from another animal—it was a different size than the piece of meat he had delivered to them.
- Cooler capacity and aging are important. Beware of a facility that hot-cuts the meat (does not chill it after slaughter). Good meat needs time to chill and ideally time to age. Not only cattle, but also swine.

Above all, do you get the impression that the facility is driven by throughput to the detriment of proper procedures? What's the attitude of the managers? More to the point, *how have they reacted to your questions?*

Yes, It's Sad

Pam Malcuit of Morning Star Ranch reflects on the poignancy of taking animals to the slaughterhouse: "Many of my cattle are pets that come up for a neck scratch. We have some that we show, and I currently have a bottle baby who is certain that I am her mother. The show critters are spoiled rotten—lots of handling and grooming. The problem is when your show steer reaches maturity, it is time to take that lovely animal, who has brought home grand championships, been spoiled with TLC, elevated to prima donna status, and who gives you big cow kisses in the show ring, to the meat processor. We use a local slaughterhouse, which "gets" that it breaks my heart to lead my lovely tame steer into their holding pen. They do it cleanly, quickly, and efficiently . . . minimizing the sadness as best they can."

You're Not Alone . . .

It's no one's idea of fun to pose questions about animal treatment standards to a slaughterhouse owner. But keep in mind that you're not the only one who is concerned about humane treatment and asking questions. There are many organizations out there concerned with the same issues. Among them is the US-based Animal Welfare Institute (AWI), a charitable group whose fundamental aim is to improve the welfare of farm animals. Its approved stunning methods are essentially identical to federal regulations, with the exception of captive bolt stunning for hogs (which it does not endorse). The AWI also urges the adoption of gas stunning for poultry (a method used by many other countries).

Also weighing in is renowned animal behavior specialist Temple Grandin; you've probably heard of her. Dr. Grandin is a doctor of animal science and a professor at Colorado State University. She's also an author and consultant on animal behavior to the livestock industry.

Here's some of what she has to say about slaughtering practices (adapted with permission from her website):

Penetrating captive bolt: *A penetrating captive bolt stunning gun kills the animal and renders it instantly unconscious without causing pain. Practical experience in slaughter plants indicates that cattle shot correctly with a penetrating captive bolt sustain irreversible damage to their brain and will not revive.*

Heavy mature bulls are more difficult to stun with this method. Practical experience in plants indicates that heavy bulls are most effectively stunned with either a perfectly maintained, cartridge fired, penetrating captive bolt stunner, a fire arm with a free bullet, or one of the new powerful pneumatic penetrating captive bolt stunners.

For large bulls and heavy livestock such as bison, some plants routinely shoot them twice with a captive bolt. To verify that 95% or more are rendered insensible with one shot, the auditor

or inspector should check for signs of return to sensibility BEFORE the second shot is fired.

Nonpenetrating captive bolt: *If a nonpenetrating captive bolt is used the animal may revive unless it is bled promptly. Some European regulations require that animals be bled within 45 to 60 seconds after captive bolt stunning. This is especially important after use of a nonpenetrating captive bolt method.*

This is the reality of meat production, and we need to face it squarely. Perhaps it will help to know that those in charge of verifying the ethics, practices, and safety of America's slaughterhouses take the welfare of animals very seriously. I found it interesting to read the USDA's meat inspector training manual. Here's an excerpt:

If you observe a humane handling noncompliance, you must take immediate action. . . . The first thing to think about when you observe a humane handling violation is whether there is

Vouching for the USDA

Farmer Walter Jeffries of Sugar Mountain Farm has plenty of face-to-face experience with USDA inspectors. "The USDA's number one goal is to make food safe, and they do an amazing job of that," Walter says. "Virtually all of the inspectors and midlevel people I've dealt with for almost a decade are honest and working toward that goal. I think the public expects too much; people don't take enough responsibility for themselves. Food needs to be correctly handled and cooked. Everyone—farmers, truckers, processors, sellers, government, and consumers—has to do their part to make food safe at their level. Reducing the number of links in the chain would help. Keep in mind that according to the CDC [Centers for Disease Control and Prevention], most foodborne illness is caused by raw fruits and vegetables."

immediate harm done to the animal. If it is being harmed, your first duty should be to ensure that the animal doesn't continue to be harmed. For example, if you observe an employee driving livestock with an instrument (the edge of a shovel, a pointed metal prod) that can cause injury, you must stop that action from continuing. Your action or inaction should not result in further or continued inhumane treatment to the animal. So, take care of the animal first.

A further note on this: Under certain conditions, an inspector or veterinarian can suspend the operations of the entire processing facility until the plant has made assurances the action will not occur again.

Let's conclude here by asking the fundamental question: Are these methods humane? And the consensus is that, when practiced correctly, along with appropriate handling and transportation methods, stun/kill methods are humane. Put a big emphasis on *when practiced correctly.*

The Debate About Ritual Slaughter

Now I'm going to step into a real minefield and ask: Is the practice of slaughter by throat cutting truly humane? How can we know?

The answer is equivocal: I'm not sure we can. All I can do is to provide you with both sides of the debate and allow you to make your own call.

The slaughter of conscious animals has been largely abandoned in the 20th century. Consumers expect animals to be stunned before death. However, observant Muslims and Jews will only eat the meat of animals that are not pre-stunned. Positions on religious slaughter vary around the world—in the United States, ritual slaughter is defined as humane under the Humane Slaughter Act—but some other countries have restricted or banned slaughtering unstunned animals.

The US Animal Welfare Institute has made its position clear: "AWI does not approve of cutting an animal without prior stunning, as is the practice with Halal and Kosher slaughter." Most animal advocacy groups in the United States and European Union consider slaughter without prior stunning to be inhumane.

However, here's what Dr. Temple Grandin says about ritual slaughter, an issue she has researched at length:

When I've seen shechita on a cow done really right by a really good shochet, the animal seemed to act like it didn't even feel it—if I walked up to that animal and put my hand in its face I would have got a much bigger reaction than I observed from the cut, and that was something which really surprised me.

From an animal welfare standpoint, the major concerns during ritual slaughter are the stressful and cruel methods of restraint (holding) that are used in some plants. Progressive slaughter plants use devices to hold the animal in a comfortable, upright position. Unfortunately, there are some plants which use cruel methods of restraint such as hanging live animals upside down.

Don't Stress Me!

There's actually visual proof of poor slaughtering practices. In beef, it's revealed by meat that is noticeably darker in color than normal—known in the trade as **dark cutter** beef. You probably have not seen it, because producers don't want you to.

Normal beef brightens in color when it's freshly cut and exposed to air. This is called **bloom**, and it turns the meat a bright cherry red—exactly what we're seeking in our fresh beef.

But dark cutter beef doesn't bloom; instead, it turns a dark purplish red. It is often sticky in texture, has a reduced shelf life, and will cook poorly, becoming very dry. Basically, you wouldn't want to eat it

This color change is a physiological reaction. When an animal is killed, a chemical reaction immediately begins converting glycogen (a substance similar to starch) in its muscle tissues into lactic acid. The acid causes the pH of the muscle tissue to decrease. It's the decrease in pH that produces bloom—that nice cherry-red color. What we

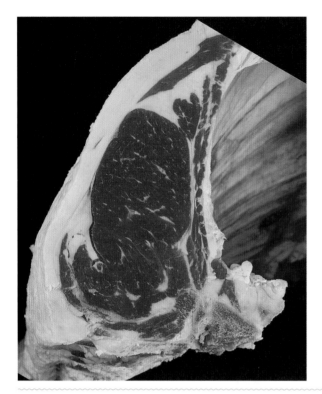 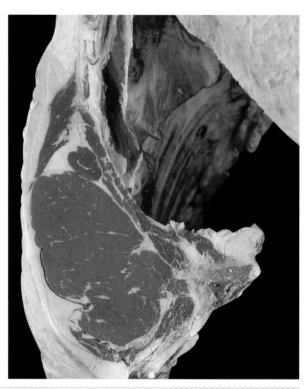

Compare the color of a rib-eye cross section (*left*) from a carcass classified as a dark cutter with a rib-eye cross section that has normal lean color (*right*). *Photos from the Meat Evaluation Handbook (2012) courtesy American Meat Science Association.*

want in our beef is a pH in the range of 5.3 to 5.7. If pH is higher—over 6—a dark cutter is likely.

How does a dark cutter happen? It's caused by a decrease in the amount of glycogen in the muscle *prior to slaughter*, and is directly linked to stress that affects the mobilization of muscle glycogen—the energy store—in the live animal. So—you ask—what would account for low muscle glycogen?

Poor (stressful or painful) handling practices. Stress during transport. Mixing different groups of animals together. Even sudden weather changes. Ultimately, anything that causes a stressed animal to draw on its glycogen reserves.

An interesting note: In the United States (unlike Canada), dark cutter beef isn't generally thrown out. Instead, it's used in the food service industry, where meat is cooked before you see it. But you still have to eat it!

What about pigs and poultry? Well, a similar process happens here, only the term is different—it's called PSE meat, which is short for "pale, soft, exudative meat." Doesn't that sound delicious? PSE occurs in pork and poultry and is characterized by a pale off-color and poor consistency. Simply stated, it's mushy. Once again, glycogen levels are the culprit (as well—in pigs—as a genetic predisposition). And once again, the condition is related to stress.

Expect Inspection

Meat inspection isn't optional. Every piece of meat butchered in the United States for resale must be inspected by the USDA, or a state program determined to be "at least equal to" the USDA, to verify that it's fit for consumption. Same in Canada, inspected by the Canadian Food Inspection Agency (CFIA).

So any animal raiser who boasts that his or her meat has been "inspected" is just trying to take advantage of your (assumed) lack of knowledge.

How to Vet a Slaughterhouse

Farmer Paul List describes how he satisfied himself that a slaughterhouse met his standards:

Choosing a slaughterhouse is absolutely critical, because all your efforts can be ruined if it's poorly run. I vetted my slaughterhouse over the course of an entire year before I committed a single animal. It's run by a retired meat inspector, and his staff have a long and honorable legacy in the trade. The owner invited me to visit and watch as animals were slaughtered so that I could see the process myself.

Before I did, I did two things. First, I turned up unexpectedly to see the holding pen. It was clean as a whistle. Then, on several different days, I secretly followed the driver who picks up my animals (in my alter ego—James Bond). He was careful and considerate of his cargo and drove smoothly and gently, not bouncing or knocking the animals around. This may not sound like a big deal, but it is. If the animal is knocked around it's stressed, and if it's held in a dirty pen overnight it's not happy. Nor would you be.

When everything goes right the animals are properly treated during the transport and slaughtering process. The kill is instant and totally without pain or stress. And that's what I need to be sure of to continue in this business.

You may find it strange, but my cattle are like pets. People ask, "How can you become attached to your cows when you have to kill them?" It's true, I know all my animals by name and can tell which is which in the dark (and these are black animals with no markings to speak of!). I tell people that it takes personal commitment and an emotional investment. It isn't easy and I don't think I'll ever become numb to the end process. But the alternative is to turn a blind eye, and that's when animals suffer. If I couldn't provide a happy healthy life and a painless humane end for my animals I wouldn't do this. And I wouldn't eat meat.

Deciphering Meat Inspection Seals

Meat products that have passed inspection are marked with an official seal. The seal will display an establishment number that identifies the exact slaughtering facility in which the animal was processed. In the examples shown here, *EST* stands for "establishment." The *P* in the bottom seal stands for "poultry." The number is the establishment number, which can be used to identify the company and address of the plant where the animal was processed. If it is a USDA-inspected animal, the seal will read USDA. If it's state-inspected, the seal will carry a state abbreviation, such as TX for Texas. You can look up establishment numbers numerically on the FSIS website through the web address http://www .fsis.usda.gov/wps/portal/fsis/topics/inspection /mpi-directory.

Unless an animal is slaughtered on a farm for the farmer's own consumption, it will probably be killed in a processing plant. In the United States, all such plants must be inspected every day by certified meat inspectors.

There is one exception. An animal can be slaughtered in a custom-cutting facility without daily inspection as long as the animal is "pre-sold" as a live animal, the custom-cutting facility provides the slaughter service, and all products resulting from that slaughter are marked NOT FOR SALE.

All inspectors are employed by the federal government or a state government and not by the individual slaughtering facilities. Meat inspectors who work in beef, pork, sheep, and goat facilities will have completed livestock slaughter training, while those who work in turkey- or chicken-processing plants will have completed a similar course in poultry slaughter. An in-plant supervisor certifies that the inspector is trained and permits him or her to conduct independent inspection duties.

There are three types of meat inspectors:

- Veterinary inspectors.
- Food inspectors.
- Consumer safety inspectors.

Veterinary inspectors must be college-trained veterinarians. Food and consumer safety inspectors are generally laypeople with experience handling meat. The food inspectors are under the direction of a veterinary inspector. Consumer safety inspectors receive additional training and work in more specialized processing establishments where valued-added products are made (specialty ham and salami, ready-to-eat items like jerky, and so on).

What Does a Meat Inspector Do?

I asked a federal inspector and a state inspector what they actually do. Here's what I learned.

Each inspector starts by performing pre-operational sanitary and cleanliness inspections to verify that the establishment is following its written

What's a Prion?

A prion is an infectious agent made of misfolded protein that can make normal proteins misfold themselves; thus acting like a virus. BSE (mad cow disease) is a prion disease. All known prion diseases affect the structure of the brain or other neural tissue, and all are untreatable and fatal.

Sanitary Standard Operating Procedures guide. Then the inspector carries out an antemortem (before death) inspection to check that arriving animals don't have diseases that are difficult to detect postmortem (after death)—for example, central nervous system diseases such as rabies or BSE.

Inspectors check that correct humane handling activities are being practiced. These include protection from bad weather, proper off-loading from vehicles, access to water (and food if held longer than 24 hours), effective stunning, observations for slips and falls, handling of disabled animals, and facility deficiencies that could cause injury to the live animals (broken pipes or boards, protruding nails, and the like).

The inspector monitors the sanitary dressing (cutting) procedure of all livestock. As the slaughterers go through the process of skinning and eviscerating the carcass, the inspector verifies that they use procedures that minimize the chances for contamination, such as sterilizing knives after certain cuts, trimming visible contamination before washing, and so forth.

The inspector also verifies that controls for bovine spongiform encephalopathy are being followed. Facilities that slaughter cattle over 30 months of age must have procedures in place to remove specified risk materials (SRMs)—nerve tissue where prions that cause the disease are most likely to be found—such as the spinal cord and other nerve clusters.

He or she also verifies that the facility is following the procedures laid out in its Hazard Analysis Critical Control Point (HACCP) plan. The plan is a series of specific procedures to identify and control all physical, allergenic, chemical, and biological hazards that may compromise the safety of the final product. HACCP comes into play before and during the production process, and thus is quite different from final inspection. Think of it as a science- and measurement-based food safety system.

In operations where large tractor-trailers are bringing in the animals, inspectors usually stay at the same location all day, all week. Smaller operations receiving just 1 to 100 animals a day often slaughter only one day a week. An inspector will travel to several small facilities during the week. A food (meat) inspector may spend his entire day on the slaughter floor or line, while a consumer safety inspector will perform other related duties (salmonella or *E. coli* testing, et cetera). The slaughter facility will also have a veterinarian assigned as its supervisor or scientific expert to manage the regulatory operations.

Eyeballing Viscera

Being an inspector is not for the weak of stomach. Inspection is done before and after slaughter, and every inspector is trained to recognize signs of disease in both live animals and carcasses. Live animals that are ill and unlikely to recover from a disease are not permitted for human consumption.

After slaughter, carcasses are carefully inspected. Small nodules in the carcass may indicate a cancerous condition. Or there may be an abscess, where the animal's immune system has tried to fight off infection by walling it off to prevent it from spreading throughout the body. Sometimes the infection can be removed by trimming or removing a large portion of the carcass. Inspectors have the authority to stop the line if carcasses are not properly presented for inspection. The facility must "present" a carcass with its viscera (abdominal contents including the major organs such as heart, liver, and lungs) and lymph nodes displayed in a consistent manner so that the inspector doesn't waste time searching for lymph nodes in order to perform a dissection for evidence of disease. The kidneys must also be removed from their attachments so they can be examined.

In a very small operation, the inspector will perform most of the cutting him- or herself to obtain the required "inspection view." In larger plants, an employee does the cut-down so that the inspector can quickly move from carcass to carcass. In even larger operations, there may be multiple inspectors at work: one looking at the viscera, one at the head, and a third examining the sides of the carcass.

If signs of disease are found, even if the inspector cannot identify the exact disease, he or she can certainly judge whether or not the carcass is fit for human consumption. However, inspectors do not have authority to condemn an entire livestock carcass; only a USDA veterinarian can make that call. Note that due to the economic value of a carcass, plant owners often request a second opinion from the on-site USDA veterinarian or another USDA vet for a final decision; this may happen several times per day depending on the type and age of the animals.

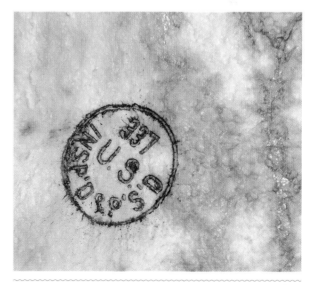

Approved! A side of beef stamped with the USDA Food Safety and Inspection Service inspection stamp. *USDA photo by Alice Welch.*

Farmer-to-Consumer Sales: Legal?

The USDA states that "all meat offered for sale must originate from a federal or state inspected slaughter facility." This seems to suggest that you cannot purchase meat directly from a farmer. But you can. There are two exemptions that apply to direct sales from farmers: the custom exemption and the retail exemption.

The Custom Exemption

Let's say you decide to buy a cow from a local farmer and have it slaughtered so you can fill your freezer with beef. You or the farmer arrange for a slaughterhouse to slaughter the animal. This is often called custom slaughter, and the animal does not have to be inspected, thanks to the custom exemption from USDA regulations.

To be granted a custom exemption, a processor must satisfy several criteria:

- Slaughter must only be for the personal use of the *owner* of the animal.
- All resulting products (whether it be a whole carcass, primal cuts, or what have you) must be marked NOT FOR SALE *immediately after slaughter*.
- The slaughtering facility must maintain accurate production and business records.
- The animal and/or product must be prepared or processed in a sanitary manner.

But what if you and some buddies are going to split up the meat from one animal? Or what if you're sharing an animal with people you don't know? Does the exemption still apply? Fear not. So many have asked the same questions that states have developed clear guidelines that explain what you can and can't do. Here's how most of them interpret the exemption:

Can more than one person own an animal? Yes.

Do they have to know each other? No.

Can a processor offer quarter or half carcasses? Yes, providing that the entire animal is owned by its ultimate owners *before slaughter*.

What if an animal is slaughtered on the farm and brought to a processor for further processing? Okay, as long as each piece of the final product is marked NOT FOR SALE and the farm has its own approved custom slaughter facility. (In the state of Vermont this requires a license issued by the Agency of Agriculture.)

What are "accurate records"? The slaughtering facility's records must document all safety measures (water, sewage, and chemical). If the facility slaughters beef, it must provide records describing how high-risk materials such as the brain and spinal cord are disposed of, and it must confirm that it only slaughters ambulatory animals. Business records should include the numbers and kinds of livestock slaughtered, quantities and types of products prepared, and *names and addresses of the final owners* of the livestock and products.

What does "the animal and/or product must be prepared or processed in a sanitary manner" mean? The USDA Food Safety Inspection Service (FSIS) will conduct periodic visits to custom facilities to ensure they meet sanitation regulations, such as washable and sanitizable walls, floors, ceilings, and equipment; hot and cold potable running water; pest control; proper employee hygiene; and more.

The Retail Exemption

Retail businesses such as stores, on-farm shops, and market stands are exempt from federal inspection, so long as they meat they sell comes from animals slaughtered under either state or federal inspection. A retail shop *cannot* legally sell meat to other retail establishments or to distributors or wholesalers.

Since its regulations stipulate that shops must sell only to the final consumer, the USDA has set limits on what they call the "normal retail quantity" that a customer can purchase at any one time. At any one time, you, the consumer, are allowed to buy—and you, the on-farm seller, are allowed to sell:

- 300 pounds of beef;
- 37.5 pounds of veal (why the half pound, I wonder?);
- 27.5 pounds (there's that half pound again!) of sheep or lamb;
- 100 pounds of pork; and
- 25 pounds of goat.

The details of the retail exemption could fill their own book. And all laws and regulations are regularly rewritten; otherwise, what would politicians do? That said, here are some of the most salient questions and answers about the retail exemption.

How often can somebody buy a "normal retail quantity" of a given meat? There's no restriction. Theoretically, you could buy 100 pounds of pork this week, and 100 pounds next week. You just can't buy over 100 pounds at a time.

What if I want to buy an entire 800-pound beef carcass? Then the order has to be divided into several transactions. Apparently (but this may vary from state to state), this is rarely enforced. Check in your own state please. (Note that if a retail store is processing entire animals for one customer, it's generally moving into custom processing and therefore not operating under the retail exemption for these transactions.)

Can I, the shop owner, ship retail-exempt products to out-of-state customers by US mail? Yes, but not if the product originates from a state-inspected facility.

Can I run more than one store—say, an on-farm shop and a roadside stand or booth in a farmers' market? Yes, but you can't have more than two markets open at the same time. Check this regulation, too, in your own state.

Got It All Straight?

If you're buying a carcass or primal cut through one of the custom exemptions, remember that proof of the animal's ownership is critical, generally attested to by a certificate of ownership (looks nice over your desk). Remember that one animal can be co-owned by several people; in such cases, all the co-owners have to sign the certificate of ownership. And getting very specific here: These rules might vary from state to state. Sorry.

From the perspective of somebody who wants to buy a cow, pig, or lamb to butcher at home (or share ownership with a group of friends), this means that *yes, it can be done.* Hooray. You will need to find a farmer and purchase the animal. The farmer will raise it and have it slaughtered for you.

I'm giving Vermont Meat Programs Section Chief Randy Quenneville the last word: "Remember, if the animal is slaughtered on the farm there must be an *approved facility.* Hanging it from the bucket of a tractor with no control over flies, dirt, and dust on a windy day, or stuff stirred up by the milk truck traveling 40 miles an hour down the dirt road, is not considered sanitary conditions!"

For Farmers Who Process Their Own Meat

I should point out that farmers who slaughter their own meat may use it for their own consumption *but may not sell it.* Only USDA- or state-inspected meat may be sold to arm's-length consumers.

Farm slaughter practices vary from farmer to farmer. (I can only speak for the area I live in but am sure it's similar elsewhere.) Some farmers do the actual slaughter themselves; others opt to have an experienced butcher or slaughterer come to the farm. Here in Vermont, I know of guys who travel from farm to farm to slaughter animals, and most do an excellent and humane job. The method they use is either a stun gun or small-caliber rifle.

After slaughter, the animal is bled, gutted, and skinned. Even pigs are usually skinned, although some guys still scald the carcass to scrape the pig bristles off—leaving the skin and fat as usable commodities. Pigskin has a variety of uses. It can be used to wrap a roast for cooking to add flavor and retain moisture, or it can be used to make pork rinds. Pig feet can be cooked in a variety of recipes or pickled. Note that for pigs or lamb, long hanging times are not necessary. Both animals need to be hung for at least 24 hours to lose body heat and firm up; ideally, the internal body temperature should be 40°F (4°C) or below but not freezing.

Beef *must* be hung and aged for (at the very least) 7 to 24-plus days, depending on the covering of fat as well as other factors discussed further in chapter 7. With little or no external fat, the outer meat will become dry and discolored and will need to be trimmed off. You may see mold on the outer layer of fat, but this isn't an issue since much of that outer layer will be trimmed off in the cutting process. Beef should be hung in a controlled temperature of 32 to 34°F (0–1°C). A chill room would be ideal accommodation but is rarely in a farmer's budget. If you can't afford to install a chill room at your farm, your best option is to find a locker plant that rents chilling lockers.

After cutting, you can wrap meat in plastic-coated freezer paper, as shown in chapter 6. Wrapped properly, meat can be kept frozen without freezer burn for up to a year. If you have the capability to vacuum-pack it, its freezer life might go to a year and a half.

CLAIMING YOUR MEAT

Another aspect of working with a slaughterhouse or custom-cutting shop is communicating with the staff about how you want your animal cut up. That's done via a cut sheet.

Cut Sheets

When a slaughterhouse or custom cutting shop is going to cut up your animal, they present you with a cut sheet that lists the various options you can have. In my opinion, most cut sheets are designed for the benefit of the slaughterhouse or custom-cutting shop.

I think that they're primarily designed to speed up the process of cutting so that the facility can process more animals per day. Since slaughtering facilities charge by weight of the animal, the more animals processed per day, the more money they make. This partially explains why slaughterhouse workers may be careless about cleaning meat from bones and fail to keep the bones from different animals separate from one another. This also prevents the consumer from getting some of the wonderful or specialty cuts that are available from different animals, whether beef, lamb, or pork.

As for pork, many people (including me) aren't that fond of smoked ham. They don't want the bother of smoking it themselves, lack the equipment, or don't want to pay a smokehouse to smoke it. A slaughterhouse or custom-cutting shop will probably offer only one other option: to cut the fresh unsmoked ham in half, removing a center slice from each half (what's known as fresh pork steak), and leave the two ends for fresh pork roasts.

But there are other options. The ham can be seamed out like a round of beef, giving you pork top round, eye of the round, sirloin tip, and bottom pork round. The top round can be sliced into thin pork cutlets (like veal scaloppine, only tastier). The eye of round can be thinly sliced into little pork medallions for a quick cook in a hot skillet. Pork sirloin tip makes a great grilling steak or terrific boneless tied pork roast, and pork bottom round makes a very nice pork oven roast. These are all cuts you should be able to ask for or be given as alternatives to smoking or saw-sliced pork steak and bone-in fresh ham and pork roast.

If you ask the slaughterhouse or custom-cutting shop to do this for you, you're apt to get a puzzled look. They should be happy to do this for you, but as a rule they won't . . . at least not without discussion or compromise. This is why I travel from farm to farm doing custom-cutting work, and it's a good reason to consider learning some butchery skills yourself.

3 Boxes

#2

THE ROYAL BUTCHER LLC
P.O. BOX 223 RANDOLPH, VT 05060 802-728-9901 FAX 802-728-9164

BEEF 1/2

Name .... Date 12-5-12

Business Name .... DOS

Phone .... DOP

Address .... Zip Code

Organic ....yes ....✓.....no

BEEF CUTTING INSTRUCTIONS

FRONT QUARTER				
SKIRT STEAK	yes ......✓	No............	LOT# 773	
CHUCK	steaks ..	Roast ....✓	TAG# C128	
SHOULDER	steaks	Roast ....✓	LOT#	
RIB	steaks ..✓	Roast	TAG#	
SHORT RIBS	yes ......✓	No	LOT#	
STEW MEATS	yes ....✓	No	TAG#	
SHANKS	yes ....✓	No		
BRISKET	yes ....✓	No		

HIND QUARTER				
FLANK STEAK	yes ..✓	NO	~~HEART~~	
TOP ROUND	steaks ....✓	Roasts	~~LIVER~~	
BOTTOM	steaks	Roasts ..✓	~~OXTAILS~~	
EYE	steaks	Roasts ..✓	~~TONGUE~~	
SIRLOIN TIP	steaks ....✓	Roasts	(DOG BNS ~~YES~~)	
SIRLOIN	steaks ......✓	Roasts		
T-BONES	yes ......✓	No		
TENDERLOIN	steaks ..✓	Roasts ..../		

STEAK THICKNESS ......1 1/4 ... PACKAGE SIZE ......2......
ROAST # ... 3-4 ....
HAMBURGER # ......1........ TOTAL WEIGHT 216 ...LBS

Labels ....

You agree that we are processing a chemical free animal ....
Customer Signature

PICK UP: MONDAY – THURSDAY 7:00 AM TO 2:00 PM
 FRIDAY – 7:00 AM TO 12:00 PM

One slaughtering facility's cut sheet. These sheets vary from facility to facility.

VERMONT: THE ROYAL BUTCHER

The Royal Butcher is a small meat-processing facility located just outside the Vermont town of Randolph. Its owner, Royal Larocque, bought the defunct old slaughterhouse eight years ago and has since brought it back to productive life. From the start he had two goals in mind: to support local farmers by focusing exclusively on them, and to change public perception about slaughterhouses.

The facility is small and well run, with a staff of 15 handling the various tasks of killing, cutting, wrapping, taking orders, and so on. It is Animal Welfare Approved, NOFA (Northeast Organic Farming Association) certified, USDA inspected, and works with a HACCP plan.

About 30 to 45 beasts go through the facility each week. As for what animals, Royal lists its capabilities as beef, veal, swine, sheep, lamb, goat, beefalo, and yak

Royal Larocque.

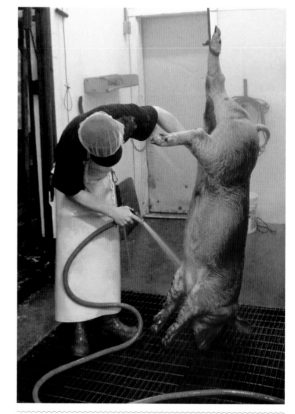

Washing a freshly slaughtered hog.

(yes, yak!). Obviously, this is a very different kind of operation from the highly industrialized processing plants that kill many hundred animals a day. But the basic processes are the same, although more is done by hand at the Royal Butcher.

While many slaughterhouses don't like deviation from their standard cut sheet, the Royal Butcher welcomes it. Staff members take time to go over the cut sheet with each client before the animal is killed, discussing options that will best suit the purposes of the client. A farmer whose client base consists mainly of chefs and restaurants will have different preferences from a farmer whose beast is being cut for his own freezer.

Below are two cut sheets from the Royal Butcher.

The sheet to the left is for two whole beef carcasses, weighing 1,000 pounds total. The customer has included extra specifications. He wants all the offal, a hanger steak, dog bones, a boneless whole rib, plus other special cuts. This is a farmer who raises beef for a select customer base of private individuals. His cut sheet reflects their preferences. His order will be delivered in 14 boxes.

There are cut sheets for every animal the facility processes. To the right is one for a pig. This farmer has asked that his ham and bacon be smoked. Since Royal doesn't offer smoking services, the ham and bacon will be segregated, unfrozen, for the client to pick up and take to a smoking facility. The client has also chosen the breakfast sausage option (the Royal Butcher offers three kinds of sausage: breakfast, sweet, and hot). So as you can see, customers are valued here.

Those customers range from people whose one animal will feed themselves and their family, to individual farmers selling retail and wholesale cuts and primals directly to the ultimate consumers, to larger producers selling wholesale to franchises like Whole Foods (which conducts its own audits of the facility on a regular basis).

Small but top quality: This is the mantra of the facility, and it shows in the friendliness of the staff and its attention to the needs of its customers.

Cut sheet for two whole beef carcasses.

Cut sheet for one pig sow.

Getting What You Deserve

"I brought it to the slaughterhouse. Will I get all of it back?" This is a question I get asked all the time. Frankly, I'm not sure you can be certain you're getting all of your animal back from a slaughterhouse or custom-cutting shop. In defense of slaughterhouses that process the entire animal (kill to carcass to cuts), I believe that in most instances where people think they have been shortweighted, the shortchanging has more to do with carelessness and sloppy cutting procedures than it does with unethical practices.

The average loss in cutting up a side of beef is often claimed to be 30 to 55 percent. I don't agree. A properly cut side of beef should have no more than 25 to 28 percent waste. In fact, I've often gotten that down to 22 percent. Let me explain. If the slaughterhouse butcher doesn't clean the meat off the bones with care, then a small percentage gets thrown into the rendering barrel with the bones. This is loss of product. Trimming too much fat from finished cuts also creates waste.

To me, saving the bones is critical to getting higher yield from an animal. Remember that bones have many profitable uses, such as soup stock, dog bones, and more, yet at many slaughterhouses it's standard procedure to discard them. When you (the informed client) ask for them, the slaughterhouse just picks a bunch of bones from the rendering barrel and gives them to you. These are probably not the bones from your animal. This is a very good reason to learn the anatomy of the animal you are having slaughtered. You'll know just what types and quantities of bones you should expect to receive.

I'm aware of custom-cutting shops that help themselves to their customers' meat. I would advise caution if you're dealing with a custom-cutting shop that also has a retail outlet, whether on-site or at another location. It's too tempting for someone to sneak a few good cuts from each animal being processed for a customer to sell in the shop's retail outlet. I call it the old "one for you and one for me" scam. Total weight received isn't the only issue. There's also the question of whether the meat you're being given is truly the meat from the carcass you purchased. Again, this is why it is so important to educate yourself. When I ran a custom-cutting shop, I refused to cut anyone's animal unless they were present to observe the process—not only so that they could feel secure about my work, but also to allow them input as the animal was processed.

Chapter Five

The Way the Meat Industry Works

In the late 1800s, people bought food at small specialized stores. Grocers sold nonperishable items, canned goods, and so on. Greengrocers sold fresh produce, and butchers sold . . . well, you know what.

The first tiny twinklings of the shift to supermarkets came with the Great Atlantic and Pacific Tea Company. Founded in 1870 as a mail-order tea business, by 1920 the A&P chain had grown to more than 4,500 stores, and over the next decade meat departments were added to some of them.

Here I am at a typical supermarket meat department. I'm smiling on the outside, but actually thinking about what goes on out of view.

Other players soon followed, led by Clarence Saunders's Piggly Wiggly stores, established in Memphis in 1916 and widely credited with introducing America to self-service shopping.

Real growth came in the 1920s, when chain stores began to flex their muscles. Still, most of these stores were fairly small, with no meat departments.

Gradually, stores became larger, offering more kinds of products, and by the 1950s we were living in the age of supermarkets—down this aisle, steak! down that aisle, toilet paper! Large discount chain stores emerged soon after as competition. Today's environment includes many players: "upscale" supermarkets, warehouse stores, superstores, and so on.

Why I am relating this little history of big food stores? To tell you that tooth-and-nail competition is the driver of this business. In the world of staple products like food, pennies can often make the difference between profit and loss. And these days, meat is becoming more and more of a staple product. Or at the very least, a standardized one, if supermarkets had their druthers.

Why this focus on the "big boys" of food? Because industrialized retail chains like supermarkets, superstores, high-end food marts, and low-end multiproduct box stores have become status quo in the meat-buying lives of almost all North Americans. Unless you're that rare individual who buys meat only from trusted local sources, I urge you to read this chapter, because I believe that everybody should understand how meat gets to their table.

HOW SUPERMARKETS WORK

Supermarkets sell products at low prices by cutting margins and relying on volume to deliver profit. How do they do this? A number of ways:

- Selling certain products—normally staple foods—at a loss (loss leaders), counting on the attractiveness of the deal to bring customers into the store. Once the customers are there, they're sure to buy something else.
- Reducing labor costs through a self-service approach—you do the work yourself (including, these days, often checking yourself out).
- Taking advantage of economies of scale by owning their own warehouses and distribution centers, which are generally located in the largest city in the area.

I may be telling you stuff you already know, but keep it all in mind as you continue reading.

Here we are in the warm and customer-friendly aisle of your local supermarket.

It's a Real Meat Market Out There

To give you some context, it's helpful to understand how the meat business works. And believe me, it's *big* business.

There are four major meat processors/distributors in the United States (as well as a number of minor players). In 2013, the top four were:

- Tyson, with about 25 percent of the market share;
- Cargill, with about 21 percent;
- JBS, with 18.5 percent;
- National Beef Packing, with about 10.5 percent

Each week, these companies slaughter and process hundreds of thousands of pounds of beef and other meats, shipping them out to big redistributors and supermarket chain warehouses. Shipments go out in refrigerated trucks holding 40,000 pounds of meat at a time. Supermarket chain warehouses send the meat on to individual stores in the chain. Redistributors send it on to multiple end users, which could be smaller meat markets, restaurants, hotels, and so on.

Now let's consider what this means for consumers, in terms of the quality of meat products offered by supermarkets, meat markets, and many restaurants. You'll recall from chapter 3 that two primary factors that determine beef quality are fat marbling and aging.

Aging is measured from the date of slaughter, because as soon as an animal is killed the aging process begins. In the old days (not really that long ago), the aging norm was 21 days, measured as follows:

- Two days aging in a chilling room at the processing plant.
- Two days aging in a chilled truck during transportation to the distributor.
- Approximately 17 days aging in the distributor's/supermarket's central chilling facility before shipping to ultimate customers.

But then market realities stepped into the picture, as they so often do, and the aging period was reduced to about 14 days. Why? Economics. Meat processors demand their money COD or within seven days. This means that when a redistributor or store chain receives one of those 40,000-pound truckloads of meat from a processor, they have to write a big check on (or almost on) the spot. But since these "middlemen" have to further age the meat for another 17 days while making weekly shipments out to their end users, they need to stock three weeks' worth of meat in their chill rooms—a total of more than 100,000 pounds of beef. It's a huge investment.

Are you getting a hint of where I'm going with this? Why not reduce the cost of sitting inventory by cutting a week off the aging process? Who'll notice?

And there, gentle readers, is the answer to your oft-asked question, "How come I can't get a good steak at a supermarket?" It's not just the quality grade, it's the aging. The difference in tenderness between meat aged 14 days and meat aged 21 days is almost 25 percent. Ponder on that.

A Box of Beef, Please

In Europe, it's still possible to buy a whole or half carcass to be cut up by expert butchers. It's extremely rare in North America, where large chain supermarkets demand a more efficient (read: cheaper) method.

Most beef now is sold to stores and restaurants as boxed beef. This includes most smaller chains as well as individually owned markets. You'd be hard-pressed to find a whole carcass anymore (unless you patronize a traditional butcher, ahem).

In the past 30 years of my career, I have not seen any hanging meats such as sides of pork or beef (with the exception of an occasional whole lamb carcass) in a supermarket meat department. Today the meat is delivered to supermarket warehouses already broken down into primals, subprimals, and in some cases vacuum-packed retail cuts. All packed and labeled neatly in sturdy cardboard boxes. Thus the term *boxed beef*. It eliminates the

A just-opened carton of boxed boneless rib eyes. Each vacuum package will be opened and further divided into cuts for sale to store customers.

need for specialized butchers, since pretty much all that's required to turn boxed beef into retail cuts is a band saw and minimal knowledge.

Meat industry consultant Bob Oros, who teaches food service professionals about the meat sector, explains how boxed beef came about.

Years ago, meatpackers sold beef by the whole carcass. Over the years, customers wanted to be able to buy certain parts of the beef carcass without buying the whole thing. To make things simple, meat packers divided the carcass into sections: rib, loin, chuck, sirloin, round, shank, brisket, plate, and flank—known as primal cuts. Once meat is cut into these large primals, it can be further cut into smaller, more usable pieces called subprimal cuts. Today, most of the items that food service distributors buy are subprimal cuts.

An example of this would be a beef tenderloin. The primal is the loin. Within the loin is the tenderloin (the subprimal). A distributor who wants to sell tenderloins to his restaurant customers doesn't want the whole loin; he just wants tenderloins. So meatpackers will extract the tenderloin from the whole loin and vacuum-package it. They will put five individually vacuum-packaged tenderloins in a box, creating a box of tenderloins.

Buyers of boxed beef still have to further cut the subprimals into individual portions for retail sale (or for serving in a restaurant). Since not everyone wants to do this, some distributors will buy subprimals, remove the vacuum packaging, cut the subprimal into final portions, then reseal these portions in vacuum packaging and pack the portions into 10-pound boxes. This is called **portion-cut beef**. Often, the individual portions are frozen for use as required.

Bravo for Boxed Beef?

Whether boxed beef is something to cheer about depends on who you ask. From the supermarket or large food service company's perspective, it's great. Boxed beef is easy to handle and ship, and easy to store (nice standard boxes of meat all stacked up). Since meat inside the boxes is vacuum-packed, it's not exposed to oxygen, so bacteria cannot grow. This increases shelf life from a week to about a month and a half. And while vacuum packaging inhibits bacterial growth, it doesn't stop the aging process. The meat continues to age inside the box. Because the meat is aging in its own juices, this is called **wet-aging**.

So are there problems with boxed beef? It can be a problem in some retail establishments, which is why it's important to know who you're dealing with. Vacuum-packaged meat needs to be used within four to five days after its vacuum is broken. Beef that shows any signs of spoilage should be discarded rather than used or put into a freezer.

Then there are the leakers. A **leaker** is a package of vacuum-sealed meat that has been punctured. Leakers must be used right away (or, ideally, thrown away). The thing is, there's no way you can know whether or not your meat came from a leaker. The store sure won't tell you.

Plus, there's the matter of how long that box has been around waiting for somebody to eat it. Boxed beef can easily be up to three weeks old before a

---

Cole's Notes

Of all beef served in restaurants, 90 percent comes from only five cuts. Here's how it breaks down:

- Inside round = 40 percent
- Rib eye = 25 percent
- New York strip = 10 percent
- Top butt = 10 percent
- Tenders = 5 percent

Note that we're starting to see changes in this—for example, things like teres major (also called petite tender) or flat iron steaks are appearing more frequently on menus.

supermarket actually receives it. But it could be six to eight weeks old. Why? Because in cases where the warehouse is overstocked with aging merchandise, it sometimes offers a "pallet deal" to the store. Did the warehouse know the boxed beef was old? Yes. Did the store know? Yes. Did the customer know? No.

You can usually tell that a box of meat is old by the persistence of odor. When a boxed beef's vacuum-packed package is first opened, there's a sour smell caused by the type of bacteria that dominate when oxygen is not present. This is normal and will disappear in about 20 minutes. The older the meat is, the longer it usually takes for the odor to disappear. If the odor is still present after 45 minutes and you were to eat that meat, you'd find it has an off taste (and it might even make you sick). But again, if I hadn't told you about this, how would you know? The store staff certainly won't tell you that the meat they're selling is old.

Blooming Meat (No, I'm Not a Florist)

Studies have shown that color is very important to consumers when they select meat. For beef the ideal color is bright cherry red; for lamb, dark cherry red; for pork, grayish pink; and for veal, pale pink. These colors are achieved by allowing the fresh meat to "bloom."

Vacuum-packaged beef is a little pale when the packaging is first opened. This is due to the lack of oxygen. Once the meat is exposed to air for about 15 minutes, it will bloom into a cherry-red color.

The term *bloom* has a specific meaning in the meat sector. All meat contains myoglobin, a protein that stores oxygen. Myoglobin contains an iron atom that can bind with oxygen or water, oxidizing in a chemical reaction that affects the color of the protein. In a just-slaughtered animal, the myoglobin is purple, and thus the animal's meat looks—you guessed it—dark purplish. Freshly slaughtered meat usually remains this color because it's quickly packaged in airtight, dark containers and is not exposed to light or oxygen.

Within half an hour of exposure to oxygen and light, meat blooms; its myoglobin becomes

Green Meat

Don't worry. The term *green meat* has nothing to do with color. All it means is that the meat hasn't aged enough to be considered acceptable. So here, *green* means "young," not "rotten."

❦

The Brevity of Bloom in Ground Meat

Why does fresh-ground meat turn brown in the middle? When meat is ground, the tissues of the meat are oxygenated and turn red . . . they *bloom*. Over time the oxygen is lost and the meat loses its bloom, which happens more quickly in the middle of the package. But the meat is still safe. Nevertheless, please use all ground meat within one or two days of purchase. This happens less if ground meat is packed loosely.

oxygenated as oxygen is absorbed by the meat and binds to the iron atom, forming oxymyoglobin, which is red. This causes the meat to turn from purple to a bright cherry red or pink—just what we're looking for in our about-to-be-dinner.

The exact shade of red is determined by the amount of myoglobin in the meat, which depends on the animal's diet, age, sex, and species, as well as how much exercise it had. Meat from older animals or from muscles strengthened through exercise tends to be darker. Thus beef—which has a higher concentration of myoglobin than pork, lamb, or chicken—is bright cherry red; veal from a milk-fed penned calf is pale pink.

Markup and Pricing

One thing many customers wonder about and complain about is the price of meat. Why does it cost as much as it does? Why does it cost more in one store than another? Is a higher-priced cut worth buying?

The answer isn't always straightforward. There are some legitimate issues that govern pricing, and some pricing practices that aren't legitimate. But to start with, let's look at some of the normal business factors that go into the setting of meat prices. To do this, I need to define a few retail terms.

First, markup. Markup is a percentage of the cost of a product added to its cost to determine a final selling price. Let's suppose that a product cost the storekeeper $1.00. The storekeeper adds 10 percent of the product cost of $1.00 to get a selling price of $1.10. That's a 10 percent markup. If the storekeeper wants a 20 percent markup, he'd add 20 percent of the base cost of $1.00 and sell the final product for $1.20. Got it?

Markup is not that relevant in my sector; instead, we work with profit margins, expressed as gross profit margin and net profit margin. Gross profit margin is the profit on each item before any costs or business expenses are paid. After paying all expenses (which include both immediate costs like packaging as well as ongoing business expenses like rent and salaries), what's left is called the net profit margin. This is my "take-home" profit on each item. It's obvious that if the gross profit margin is set too low, the net profit will be too low, the store will never be profitable, and the hapless storekeeper will soon be out of business.

Meat department markups are different from meat department gross profit margins and vary across the country as well as from town to town based on population and average incomes in the region. In the case of meat, which needs to be trimmed of fat and other inedible items such as connective tissue and gristle, there's a further step, which involves a cutting test. Using the example of a large subprimal called a New York strip, I begin by weighing the whole piece to determine the cost of that particular whole strip, using the fattest or "wastiest" one I can find. Then I cut the whole piece into individual steaks, trim them, and weigh them to see how many pounds of meat I have left. From this I can figure my gross profit margin. How?

The arithmetic is simple. It's based on 100. Suppose I've determined that—factoring in all my costs and business expenses—I need to make a 42 percent gross profit margin on every piece of meat

Gassing Up

The blooming period is short (which is why meats lose their color quickly), but it can be prolonged by avoiding light, keeping temperature low, and using a method of packaging called modified atmosphere packaging (MAP). The MAP method seals food in a package containing a mixture of gases in proportions designed to retard decay by slowing oxidation and microbial growth. For example, for raw red meats in sealed thermofoam packages (supermarket meats), the recommended mix is 70 percent oxygen, 30 percent CO_2. For cooked, cured, and processed meats such as bacon and hot dogs, it's typically 30 percent CO_2 and 70 percent nitrogen.

A self-service meat case. Generally, if a store has both types of meat cases, the quality of the meat in a service case is expected to be superior to that in the self-service case. Unfortunately, this isn't always so.

I sell. I subtract 42 from 100 and get 58. Then all I need to do is to divide the *cost* of my product by .58 to give me a 42 percent gross profit margin. This works for every desired gross margin. If I'm seeking a gross profit margin of 30 percent, I divide the cost of the product by .70.

Let's suppose my cost for the whole strip is $80. I simply divide $80 by .58 and come up with $137.93. That New York strip cut into trimmed New York strip steaks needs to bring me $137.93 for me to achieve my 42 percent gross profit margin. Going further, if the trimmed steaks end up weighing 7 pounds, I also need to divide $137.93 by 7 to figure out what *price per pound* to charge. In this case, that would be $19.70 per pound.

So now—by dividing my total cost of the strip of $80 by 7, I can find the actual cost of the trimmed steaks. So $80 divided by 7 = $11.43 per pound cost of trimmed steaks. To make this a bit clearer, take our final trimmed cost of $11.43 per pound, divide by .58—and guess what? I come up with $19.70 per pound to achieve a 42 percent gross profit margin.

The target profit margin can vary depending on the level of service required, the location of the shop, and your labor costs, as well as many other factors. Large supermarket chains generally offer two types of meat departments. The most common is a **self-service meat case** where everything has been pre-cut and packaged in clear plastic film for customers to pick up themselves. In my region, this type of meat department usually operates on a 28 to 32 percent margin.

The second type is a **service meat case** where meat is displayed in a glass-fronted meat case and a store employee personally serves you. There aren't many of these left where I live. This type of meat department will normally operate on a 38 to 42 percent margin because of the added labor and service involved. You should normally receive higher-quality cutting in this type of meat department (notice I didn't say higher-quality *product*).

Supermarket chains that offer both types of meat cases may well be carrying exactly the same brands, grades, and types of meats in both, but charging higher prices for those in the service meat case. This is because it takes a more experienced meat-cutter to present the product well in a full-service meat counter. The full-service meat departments I've managed always maintained a 52 to 58 percent margin, mainly due to the wide variety of value-added items that I created (because value-added products are pre-prepared and ready to cook, they generally command an even higher margin). Basically, we did the prep work for the customers and they paid for our labor.

So even though the normal fresh meat (steaks, chops, roasts, burgers, et cetera) may be yielding a margin of 38 to 42 percent, the value-added items may be yielding 60 to 75 percent—bringing the average of the whole meat department up—which keeps the store profitable and in business. I feel totally justified charging these prices providing that the product is of superior quality and freshness and that customer service is impeccable.

Margins will vary depending on location. A store in a wealthy town with high-priced real estate will cost the shop owner more in rent or mortgage than he or she would pay in a less affluent town, so expenses are going to be higher, and thus margins have to be set higher. Plus, people in an affluent neighborhood aren't as price-conscious as people in a less affluent neighborhood.

IS YOUR BUTCHER BEING TRUE TO YOU?

Get ready to read about some business practices the meat sector doesn't want you to know about. Some of them may seem far-fetched, but I can assure you that all are true. How do I know? Because I've been there. In my many years of working in meat departments, I've seen some shocking examples of very poor practices, both ethically speaking and

sanitation-wise. These are things I think every shopper should be aware of, both to avoid getting ripped off and to avoid possibly getting sick.

Why some meat department managers think it's okay to cheat their customers this way, I don't understand. Your best defense against their kind of ripoffs is to educate yourself. Don't allow dishonest practitioners to thrive on your ignorance.

I'm giving you the benefit of my inside knowledge here. I refuse to work for large supermarket chains any longer because of their deceptive practices and lack of quality. They survive only because of deceptive advertising and labeling and—most of all—because of customer ignorance.

Not So Special as It Seems

The next time you're wandering past the meat department of your supermarket wearing a sweater because of the Arctic chill, take a closer look at what's being offered. Chances are you'll see meat offerings with labels like REDUCED! or SPECIAL TODAY!! or MANAGER'S SPECIAL!!! What exactly does this mean? Did the manager personally endorse that "Manager's Special"?

During my years working in large supermarkets and chain stores, I learned one important fact: The department manager's authority is surprisingly small. A meat manager's authority is generally limited to ordering products for his or her department, directing his or her staff or reprimanding them for infractions, sitting in on an interview of a potential employee, and making sure that the mandated policy of the company is followed.

Pricing of products, weekly advertised specials, and the determination of what products will be sold are handled by the corporate office. I am very wary of any fresh food with a red label of any kind on it implying that it is a special deal. If a piece of meat is labeled MANAGER'S SPECIAL, it has nothing to do with the individual manager. The department stocks rolls of labels for different circumstances—SPECIAL TODAY! . . . FRESH TURKEY! . . . MANAGER'S SPECIAL! . . .

Is this really special? Or just old? *istockphoto*.

I AM WOMAN, HEAR ME ROAR! (no, wait, not that one)—so when a piece of meat is nearing the end of its shelf life, you can bet that it's suddenly the Manager's Special. 'Cause if they can't sell it fast, they have to throw it out. Probably tomorrow.

Now, suppose you buy a package of chicken parts with one of those labels on it, intending to cook it for dinner, but then you decide not to cook tonight. That chicken might end up spoiled totally if it's not frozen immediately.

Is Close-Dating a Couples Thing?

In the world of meatdom, the term ***close-dated*** refers to a product (typically vacuum-packed primal or subprimal cuts) shipped from the warehouse/wholesaler to the retailer with an imminent use-by or sell-by date. This doesn't necessarily mean the product is bad or spoiled; it simply means that it will have little or no shelf life left once it is opened, and maybe only one or two weeks left unopened.

In the case of beef, if the meat is left in the vac-packed or Cryovac package and the seal is intact, it may well be good for another week, or even two.

If They Won't Let You Smell, Don't Let Them Sell

It's not unusual to see meat items like lamb or veal shanks sliced for osso bucco and offered in vac-pak or frozen packs of two. Very convenient. What you can't tell from the packaging is whether those shanks have been thawed and refrozen several times before packaging (it happens). Sometimes the individual packages aren't dated . . . the correct date was on the original box in which that meat arrived at the store. Often, items that are less popular (and thus take longer to sell) are stored in the freezer. Upon request, the butcher will grab one package, weigh it for the customer, and put the label showing the weight and date on the package. But the date shown will be that day's sell- or use-by date—even though the package of meat may have been sitting in the freezer for six months to a year.

Folks have told me that they've purchased these types of convenience packages, brought them home, opened the package, and found that they smelled to high heaven. Sometimes it's just a little gassy smell; other times they smell really rancid—so much so that even rinsing in cold water didn't help.

If you're considering pre-packaged shanks, be sure to ask the butcher to *open them up first* so you can smell them before you buy. This may save you a trip back to the store. Remember, if you even question the freshness of meat, don't eat it!

But as soon as the vacuum pack is opened and the meat is exposed to air, it must be used or frozen right away; otherwise it will spoil rapidly.

I know one supermarket chain that forces its meat departments to accept close-dated or outdated products from the chain's warehouse—even if those products haven't been ordered. Another chain I know of shipped pallets of close-dated or outdated meats as often as once a week to their meat departments. Great way for the company to clean out old stock that it can't unload anywhere else. It charged the meat department anywhere from 40 cents to one dollar per pound and instructed the meat department to double that price for retail sale. For example, outdated or nearly outdated chicken at a cost of 40 cents a pound would be retailed for 80 cents. Rib eye or New York strip costed at a dollar per pound would sell for two bucks a pound.

But that's not all. Since the meat managers knew that items like rib eye, New York strip, and tenderloin sell well, they often offered these close-dated or outdated cuts to the public at $4.99 or $5.00 a pound (especially if their regular price was $8.99 or $9.99 per pound). Great deal, right? This is how some managers achieve the profit margins their head offices demand of them.

I know retail stores that buy only close-dated products and sell them as fresh. Some freeze the close-dated product, thaw it out as they need it, and sell it as fresh. Here in Vermont (as far as I last knew), such a product must be labeled PREVIOUSLY FROZEN. These stores do not label it as such. It's a practice I detest.

If the meat you're buying is much cheaper than most of the competition in your area, if it doesn't maintain the bright red color it has when first opened for at least 24 hours in your refrigerator, and if it has a gassy odor when it's opened, chances are it is a close-dated meat product (near the very end of its shelf life). It is certainly old and could very well make you sick. So if you're buying New York strip, rib-eye steak, or tenderloin and the price seems too good to be true, *it is*. If you smell a rat, it's a rat.

Marinated Meat

I worry about marinated meat products in a large supermarket, because I know that many meat markets marinate their old stuff to give it more shelf life. This is especially true of chicken. If a store is going to sell marinated items, the meat should be marinated *at its freshest*. And the tray should be dated.

There are two dates to look for: a sell-by date or a "when-packaged" date—this latter means it's up to you to determine whether it's as fresh as you'd like. Most states require stores to label meat with one of

the above. Which one is the store's call. If there's no date, ask when the meat was marinated and make sure that the store is willing to guarantee the product.

Taring Along

By law, meat scales in retail stores must be inspected periodically by state officials for accuracy and tare weights. This is not an option; it's obligatory. The total tare weight is the weight of the exterior packaging—usually Styrofoam and plastic wrap—along with the weight of that little white soaker pad in the bottom of the package you find when you remove the meat from the packaging (soaker pads absorb extra blood and juices that seep out of the meat as it sits in its package waiting to be sold).

Each meat producer has its own price lookup code, which is programmed into the meat scale along with the tare weight. Price lookup codes, commonly called **PLU codes/numbers**, are identification numbers affixed to food products in stores to make checkout and inventory control easier and faster. You've all seen them, although you may not have noticed them.

For example: Ground chuck in a family pack might be assigned a PLU code of 208, ground chuck in a 2-pound package a PLU code of 207, and so on. (Tray sizes are usually expressed in numbers plus the letter *S* for "shallow tray" or *D* for "deep tray.")

Using the example of the PLU 208 ground chuck family pack, when the clerk puts the product on the scale she punches in its PLU number—208—and the scale automatically allows for the tare weight of the packaging as it figures the total weight of product and total price to be charged. So setting the tares accurately for all the various types of packaging used in a meat department is an important part of programming a scale. In Vermont, state inspectors from the Department of Weights and Measures will randomly weigh packages of prepared meats to make sure the establishment is factoring in the proper tare. You, the consumer, are only supposed to pay for the contents of the package (the tasty meat), not the weight of the package (all that chewy white Styrofoam).

After inspecting the scales, the weights and measures official certifies them—providing they're accurate—with a seal of approval. The seal should be stuck to the side of the scale and will indicate the last date of inspection. Stores cannot duck inspections, which are random. If the scale is not accurate, it will be condemned until repaired.

So don't be afraid to ask to see the seal on a store's scale. Why do I recommend this? Because there are meat departments where they don't bother to factor the tare on packaged meats. Sure, the tare on a small package of meat might only be an ounce, but paying for a few hundred ounces of basically nothing over a year can add up. And considering that stores may sell thousands of pounds of prepackaged meats every year, well, leaving out the tare can put a lot of *your* dollars into *their* pockets.

A practice still common today is wrapping up freezer orders for retail customers in freezer paper, offering to double-wrap, and then weighing the product paper and all, without taking the tare. So the customer pays for all that paper. With meat prices as high as they are today, this can amount to quite a sum of money.

Trimming Tricks

When I trim steaks, I like to leave ⅛ inch of fat on the edges (unless my customer requests more fat left on). Why? Because I believe that when you buy a piece of meat you should get as much eating out of it as possible. Although some stores have trim standards (⅛-inch trim or ¼-inch trim), many employees or meat managers ignore the standard and increase the trim, leaving more fat on just to rake in more of your money. Ask yourself: Do you really want that extra fat, or are you just going to be leaving it on your plate? 'Cause you're paying for it.

Some places use what I call an **angle trim**—they trim the fat at the edge of a steak at an angle. Viewed from one side, the steak looks like it has been nicely

Tare It Up

Tare weight is the weight of an *empty* container, without its product. Thus, gross weight (total weight) = net weight (the weight of the goods) + tare weight (the weight of the container). Most scales have a way to set the zero mark to measure only the contents of a container without measuring the weight of the container itself. The word *tare* comes from an early French word meaning "wastage in goods."

Notice the "tare" indication at lower left. *istockphoto.*

Leg of Lamb Secret

If you buy a leg of lamb in a supermarket, ask if it's "oven-ready." If it's not, it shouldn't be in the case. Leg of lamb ought to be weighed *after* it's been made oven-ready, not before. I've seen many markets put big fat old lamb legs out for sale, untrimmed. When a customer asks for leg of lamb, they'll weigh it, then say, "Would you like me to make that oven-ready for you?" A very sly way of making certain you pay for a lot of waste. Your response should be, "Certainly. Thank you. Oh, and would you please weigh it *after* you've trimmed it?"

trimmed, with just ¼ inch of fat left. But when you turn it over, you'll see that the other side may have as much as an inch of outside fat. I call this a nice angle to squeeze more money out of the customer.

The Layered Look

A particularly lean oven roast *needs* some fat; otherwise the roast will dry out. However, stores sometimes go overboard with things like prime rib roast at holiday time. A whole rib usually has a nice covering of fat—doesn't need more. But I've seen prime rib roasts with thin sheets of extra fat tied around them, just to add weight. It's almost impossible for a customer to tell this has been done, and it's something to be wary of.

Tubed Beef Trouble

Do you believe that your supermarket grinds its own meat on site? Much of the time it doesn't. Most ground beef offered in large supermarket chains and some smaller stores is what is known as tubed or coarse ground beef, ready for one additional grinding by the store to the correct consistency for the particular product (hamburger, sausage, what have you) they're making. It's termed "tubed" because that's exactly how it comes; as 10-pound tubes of vacuum-packed ground beef (meat ground one time through a coarse grinder plate and vac-packed). The store then just opens a tube and grinds it again.

Tubes are labeled 93, 90, 85, 80, and 73 percent lean and are generally shipped in 50- or 60-pound boxes by the meat packer. Each box is labeled with the name of the producer, as well as its total weight and a lot number.

That lot number is extremely important because it refers to the specific batch of ground meat that the box of ground beef came from, so that in the case of any foodborne illness you can trace the contaminated product back to its source.

Here in Vermont, each retail meat establishment selling ground beef by fat content must maintain a **ground beef log**, which must record the name of the

A Disreputable Profession

Craig Rogers of Border Springs Farm has this impression of the meat business: "The meat world has been the most dishonest profession since the beginning of time. I've seen everything from truckers who picked up my animals and replaced them with cheaper ones, improperly trued scales, fires at livestock auctions to drive away new buyers, and speculators who play the auctions like Vegas slots. This has been the history of the meat world. One of the reasons auctions are so regulated is because of how easy it is to cheat."

Although most of my own experience has been limited to the retail end of the meat sector, I agree with Craig. The meat business is and has been one of the most dishonest entities since its beginning, especially in the United States. Even though I and many other butchers believe in honesty and integrity, the meat industry has for years been plagued by many more who believed in just the opposite.

company from which the tube was purchased, the date that it was further ground, the fat content, and, most important, the lot number.

The law also requires that the meat department conduct a fat test using a sample of the batch of coarse-ground meat product. This is done with a device that allows you to cook a measured amount of ground beef and catch the drippings in a small clear tube. As the drippings cool, fat separates from other juices and a small dial with an arrow on it allows you to get an accurate measurement of the fat content of the ground beef.

These are all excellent rules to ensure the safety of ground beef . . . except for one thing. They're almost never followed. I can't tell you how many times I've seen meat-cutters, service clerks, or even department managers grind beef without conducting a fat test. Or worse, place ground beef from one batch of grinding into the premeasured equipment, let it cook without looking at the results, and leave that bit of ground meat in the fat tester for days. In other words, go through the motions without doing the work. Rather repulsive.

In many stores I've worked in, the ground beef log was rarely filled out or kept up to date. Then *surprise!* . . . the manager finds out there's a meat inspector on the premises or that someone from the corporate office has dropped in. Like a bad movie, everything suddenly speeds up. The manager grabs the meat log or has one of the cutters or clerks take it to a discreet place to fill out. Not accurately, of course, since generally the boxes the meat came in have already been discarded. So the meat log gets filled in with random, made-up lot numbers. The risk is further compounded by the fact that a single 10-pound tube of ground meat does not represent one animal. Far from it. One tube of ground beef might conceivably contain beef from many different animals in many different states.

I have one question: In the event of an *E. coli* outbreak, how could the ground beef from a meat department like this one be accurately traced? It would simply not be possible. This is one of the reasons that when there's an *E. coli* outbreak, millions of pounds of ground meat are often recalled.

I will not buy ground beef from any store that sources ground beef this way. If you shop in supermarkets, the safest possible way to purchase ground meat is to buy a piece of muscle meat like a roast or steak and have it ground *in front of you*. And before the butcher grinds it, you might ask when the grinder was last sanitized. Read on to find out why . . .

Bait and Switch

This practice is very common in supermarkets that sell pre-prepared ground meat—which is most supermarkets, actually. As more people learn about tubed ground beef, they're opting to have meat ground to order. Makes good sense. So they choose a nice roast from the meat display and ask to have it ground for them. So far so good, *but*—I've often

seen the meat-cutter say "Sure," take the selected roast out of the meat display, go through those white doors into the back room, set the roast aside, pass some tubed ground meat through the meat grinder, wrap it up, and give it to the customer: "Your ground roast, madam." Then after the customer has left, the roast that the customer picked out is put back into the display case. This really pisses me off. Should piss you off, too.

If you ask to watch a roast being ground and the store employee says something like, "We don't do that, but the label from your original roast will be on your package of ground meat so you will know I ground your piece of meat"—don't believe it. It's all too easy to remove the price tag from the roast package and paste it onto a package of tubed ground meat. This is fraud.

An even more disturbing bait-and-switch practice involves poultry. A customer picks a whole chicken from the display case with a nice fresh date on it and requests that it be cut up by the butcher. The butcher or meat department manager picks a chicken package with a nearly expired date, cuts up that old chicken, wraps it, and pastes on the tag from the fresh chicken. The customer gets nearly expired meat, which—in the case of poultry—is a dangerous practice. The original fresh chicken goes back into the display case as soon as the customer walks away from the department.

Bait and Switch—Part Two

When I worked in big meat markets, I'd often see customers pick out a primal cut and ask the butcher to cut it up for them (and yes, there are large meat markets—even supermarkets—that will sell you a primal). The reason—of course—was that by starting with a primal cut buyers would get much of their meat at a lower price per pound. Which makes sense, except . . .

The butcher would weigh the primal cut for the customer, then take it to the back of the store for cutting. At which point, he'd switch out the original primal cut for a smaller primal cut, thus cheating the customer out of several pounds of meat.

But I Already Gave at the Office!

If you're buying a primal cut, the butcher will often offer to cut it up for you. He or she will rarely mention the trimmings (all red meat) that would make great ground meat or stewing meat. If you're not vigilant, you probably won't get them. You've paid for a product, but are not getting all of it . . . a very good way to "donate" a fair amount of useful meat to the butcher shop.

Meat Loaf Mischief and More

I know a butcher who sells whole New York strips and offers to cut and freezer-wrap them for his customers. He tells them they'll get about 12 steaks, 1 inch each, out of the strip. In reality, a strip usually yields 13 to 14 such steaks. He cuts the strip, but removes the nicest steak by sliding it aside and putting a towel over it until the customer leaves. Then he puts that nice steak in his display case to sell. Like a sidewalk shell game.

Many customers ask for beef, lamb, and pork ground together because this mixture makes a delicious meat loaf. Never buy this type of meat loaf mix unless you watch it going through the grinder. I also know meat-cutters who smile and say "Sure thing," then go into the back out of view and grind up some beef with a bit of fat, bringing it out as—*ta da!*—a wonderful three-meat mixture. And they charge extra for the nonexistent lamb and pork. Why would butchers do this? Because they didn't have lamb or pork available, they were too lazy to get the pork and lamb, or they just love to rip people off.

I've always maintained that you can sell anybody anything—once. But if you want people to come back and support you on a regular basis, you *must*

treat them like royalty—with respect and honesty. Customers will pay for quality and service—there will always be a market for these values.

I Have This Grinding Feeling

Another common practice is to take a tube or two of 90 percent lean tubed ground meat, put it through the grinder, and then package it up, labeling half of it 90 percent lean ground beef and the other half ground sirloin for 30 cents a pound more. None of it was sirloin to begin with.

This is also done by markets that don't use or sell tubed ground beef at all but use only muscle meat for their ground meats. For example, they may use whole beef chucks or shoulder clods for their ground chuck. This is fine, but then they remove some of the leaner parts, grind them, and sell the result as ground sirloin at ground sirloin prices. One store I recently helped out for a few weeks does just that.

Yuck! That Saw!

Equipment sanitation in large chain supermarket meat departments is a major concern of mine. The issue is closely tied to regulation and inspection. Meat inspection and sanitation regulations vary from state to state and are overlain by federal rules and guidelines. In my state, you can get copies of the regulations from the Department of Agriculture, which governs the sale of domestic meats and retail meat departments. I'm not going to get into the details of the regulations, but I *can* pass on some of my experiences concerning the subject of sanitation.

I occasionally help out retail meat establishments that have lost their butcher to illness or injury. Recently, I walked into one that appalled me so much that I contacted the state meat inspector; it was obvious that the store's management was negligent, incompetent, or worse. The inspector closed the department down. Why? Well, the meat saw hadn't been cleaned in such a long time that the inside was packed with meat sawdust covered in mold. The meat grinder hadn't been cleaned for weeks, and the meat cooler—which was also used as the cutting room—was just as filthy. That department was not allowed to reopen until a crew of employees had completely cleaned and sanitized the entire meat cooler and equipment, and the department was reinspected. But guess what? The same hour that the meat department reopened, the health inspector arrived to shut down the deli department.

There's no way for customers to know whether such appalling conditions exist where they shop, because they can't peer into the innards of the equipment, or walk through the meat cooler doors.

This situation is not as uncommon as you'd think. In 1997, I remember a meat inspector closing down the entire meat department of a large supermarket that had opened less than a year before—essentially a brand-new store. That supermarket had the entire store staff cleaning the department until the inspector came back in the afternoon and allowed them to reopen.

When poor sanitation happens in supermarket chains, it's often due to low morale and motivation on the part of workers who are underpaid, overworked, and underappreciated.

Obviously, all supermarkets aren't like this. I have only worked for four or five large chains. Still, let me leave you with a delightful—and true—visual image . . .

In one chain, I was sent to another state to cover a vacationing meat-cutter in one of their supermarkets. I found the saw, the meat scrap barrels, and the grated floor drain covered in maggots. 'Nuff said, I think.

Is There Any Value in Value-Added?

For the most part, value-added products that are sold ready to take home and pop in the oven are a great way to enhance profit for a *service* meat department (notice I stress *service* meat

department!). They're a good thing . . . providing that they're fresh, extremely high quality, and made on-site—hopefully with the butcher's own recipes.

From the customer's perspective, value-added offers lots of benefits: convenience, new recipes, and variety. Now let's look at the benefits to the meat market. A small retail meat market may operate on a 40 percent gross profit margin on fresh meats such as roasts, chops, steaks, and ground meat. That 40 percent gross profit margin has to cover operating expenses such as labor, electric, refrigeration repairs, maintenance, license fees, taxes, the cost of the meat, and so on. Oh, and profit. It quickly becomes obvious that you need to do considerable volume to cover all your costs and make a livable wage yourself.

But by creating special products like shish kebabs, stuffed chicken breasts and pork chops, chicken roulades, fresh hot stews, soups, gourmet burgers, or fresh homemade sausage, a store can easily command 60 to 70 percent in gross margin for these new, value-added products. The store has added value to its basic commodity—meat—by turning it into ready-to-go dinners. This, in turn, raises the overall gross profit margin of the entire meat department, often bringing it to an average of as high as 58 percent.

But—as you've guessed—there's a dark side to value-added. I have worked for many retail meat markets and supermarket chain stores that haven't the slightest idea what *fresh* or *quality* means. Value-added to them essentially means "How can I get rid of this old product?"

Once, working as a floater (a meat-cutter who travels from store to store for a chain supermarket covering days off and vacations), I encountered a meat wrapper who had found a few packages of out-of-date boneless chicken breasts and was preparing to stuff them. When I asked her why she was stuffing outdated chicken breasts instead of fresh ones, her reply was (and I'm quoting) "That's what we've always done."

Another store only adds what they call "value" to outdated or nearly spoiled products via a box of stuffing, garlic marinade, or hot and spicy seasoning to hide any hit of bad odor. And I know of supermarket chains that bring in their "value-added products" frozen and sell them as fresh in their meat cases.

All too often I've watched staff smile and nod when a customer says, "Oh, this place has the best meat," when behind the scenes frozen meat is being thawed and sold as fresh, the marinated chicken breast is a week old, and today's deli lunch special is from last week's outdated meat.

Value-added has two sides: a good side and a bad side. If it's created with the freshest possible product, prepared from scratch with the highest-quality ingredients and great and tasty recipes, and used as a tool to provide a service to the customer, it's a good thing. But if value-added is strictly a tool to get rid of old product, then it's not only a bad thing, but a very dangerous thing.

WHERE DOES THIS LEAVE US?

I still work at a few meat departments from time to time. One small store I helped out recently still does all the things that I have complained about for 48 years . . .

They redate outdated products. They label previously frozen meat as fresh. They label pre-ground burger as fresh in-store ground. They label select as choice. They put Certified Angus labels on meat that isn't Angus at all. They even sell venison from Arizona as local Vermont venison. In the deli, they make soups, stews, and casseroles from spoiled meat and offer them for sale in the prepared foods section.

Faced with horror stories like these and the others in this chapter, you ask: *Is there any way to verify the wholesomeness of the meat I buy?*

I'm very sorry to say this, but no, not unless you actually work in the meat market you patronize. It all comes down to three important principles—know your butcher, know your meat market, know their practices.

I firmly believe there will always be a demand for quality and customer service, even though these elements seem to be disappearing as large chains take over the meat business. The norm these days seems to be that a customer is just a number coming through the door, and if customer number 20 won't shop here anymore, well, here comes number 21. I remember a time when this attitude would have put you out of business.

The corporate management of retail supermarket chains certainly wouldn't sanction the practices I've described in this chapter, but meat department managers are under such pressure to meet profit margins mandated by their corporate overlords that guidelines slip. And it has to be stated that many meat department managers lack the knowledge to run a profitable department without such tactics.

Real change will only come when consumers demand to know how their meat is raised and how it is handled. You can be one of those consumers.

Chapter Six

Processing Your Own Meat

Welcome to the heart of this book: in-depth information on how to process your own meat. Processing a side of pork or beef or a whole lamb is satisfying on many fronts. It's not just bragging rights when you have friends over for dinner—*Hey, that steak you're eating is my personal handiwork!*— but also that nice warm feeling knowing exactly where your meat came from, how it was cut, and how fresh it is. You'll feel secure knowing that you and your family are eating top-quality meat, humanely raised by a local farmer. An abundance of options will be open to you: grass-fed, grain-finished, specific breed, organic, antibiotic-free, and so on. You will finally know the provenance of your meat!

The most practical reason to butcher your own meat is the money you'll save on your annual meat bill. Since you'll be buying a much larger piece of meat—a whole carcass, side, or quarter of the animal— you'll get the benefit of volume savings. A large, unprocessed piece of meat costs less than it would after it's been divided into retail portions. In the case of meat, the whole *is* cheaper than the sum of its parts.

Only a century ago, a large part of the population raised their own animals for meat. They slaughtered, cut, processed, and preserved that meat for their own consumption throughout the year. Today with modern freezers, processing your own meat makes even more sense! It's not an old-fashioned thing; it's totally modern.

Yes, it's fun learning to butcher. And being one!

WHO SHOULD CONSIDER THIS?

Learning how to cut meat isn't just for apprentice butchers. Plenty of people can benefit from these skills, from foodies to home cooks to chefs to farmers.

Home Cooks, Meat-Lovers, and Foodies

If you're a dedicated home cook, you know how important it is to ensure the quality of the food your family consumes. Since buying large cuts of meat involves developing relationships with animal farmers, you're now in the driver's seat. You can source exactly the kind of meat you want (humanely raised, please) and match your eating and taste preferences to farm resources (heritage pork? grass-fed beef? specific breed?). Thus, to a great extent, you can control the nutrient quality and wholesomeness of the meat you eat. You'll be lowering your annual food bill, and you'll also be freed from the vise grip of the limited offerings (and sometimes iffy practices) of large supermarkets and chain food stores.

For those of you seeking a hobby that doesn't involve wool and needles, meat cutting's great. It can be a fun family project, as well as a great educational experience for the entire family or a group of neighbors and friends. It really is . . . my classes are full of all kinds of people, all ages, all walks of life, having a great time learning to do this.

Even if you only do this once or twice, consider the learning experience. What you'll learn will stay with you forever, making you a much more informed meat shopper. Knowledge is power.

Here my son Todd (yes, also a butcher) is teaching a student how to use a meat saw correctly.

Chefs and Restaurant Owners

For those in the restaurant trade, cutting your own meat isn't just an effective way of offering more variety and delectable dishes to your diners; it guarantees top-quality product (assuming you commit to sourcing the highest-quality meat), supports local farmers, and increases profit margins.

Perhaps you've been buying already portioned table-ready cuts of meat, as many restaurants do. This is convenient, but more expensive than it needs to be. That's because you're paying someone else to do work that could easily be done right in the kitchen of your restaurant.

If your restaurant is small, I'm not advocating that you purchase an entire side of beef. But you could buy a whole rib eye or a whole New York strip and cut the individual steaks, instead of buying pre-cut individual steaks. Leftover trimmings can be used in soups, stews, or ground meat. And *you* determine the size and thickness of each cut.

You could also consider purchasing a primal cut such as a whole loin of beef, which would yield New York strip, tenderloin/filet mignon, sirloin steak, and some trim items such as kebab meat, stews, and ground meat. The bones could be used for soup stock. Purchasing whole or half pigs is another option. A pig carcass is much easier to cut and process than a side of beef. The incredible variety possible from a whole or half pig would give a restaurant a great selection of options: pork belly, fresh pork steak, pork cutlets, pork roast . . . not to mention standard offerings like pork chops and pork tenderloin. Meat doesn't get any fresher or higher quality than a locally raised pig.

Farmers

If you're a small farmer raising meat animals, you should consider learning to process your own meat. Yes, you'll need to have your animals slaughtered at a USDA- or state-inspected plant (keeping in mind

Cole's Notes

The style of whole-animal butchery I demonstrate in the teaching CD that accompanies this book may be somewhat different from what you'll find in other books. It's based on a traditional Vermont premise: Use as much of the animal as possible. Not only does this honor the animal's life, but it also greatly increases the yield of usable meat. This is the style I learned in the 1970s and '80s from the traditional country butchers I apprenticed with.

And a comment from Joe Monthey (who's been apprenticing with me for about a year): "I've seen Cole butcher a whole pig and end up with only a cereal bowl's worth of waste!"

that USDA-passed meat can be sold out of state, while state-inspected cannot), but you will certainly achieve higher return on the investment of raising that animal by eliminating the middleman.

You might start by offering whole sides, quarters, or primals for direct sale to consumers. Or you might go farther and process your meat into retail cuts for sale via local markets or country stores. This happens a lot here in Vermont, and customers quickly recognize and seek out their favorite farm meats. A whole new market.

LET'S DO THE NUMBERS

I generally use beef as my exemplar when I want to explain the dollar savings of buying meat in bulk. So let's compare buying a side of beef versus going to the meat market or supermarket for individual cuts.

In my neck of the woods, a side of beef costs from $3.89 to $4.80 a pound hanging weight, which is the weight of a side of beef as it hangs in a meat cooler after slaughtering. A whole side of beef will range in weight from 250 to 300 pounds. Using a side weight of 250 pounds at $4.25 per pound, I can calculate the total cost of the side of beef *before cutting*:

$$250 \text{ lbs.} \times \$4.25 \text{ per lb.} = \$1{,}062.50$$

If the side is properly processed with every part utilized as it should be, I'd expect a 28 percent loss, give or take, in the cutting process, leaving you with approximately 180 pounds of usable meat. That 28 percent loss covers things such as extra fat, sinews, and so on that you will discard.

Next, divide the cost of the uncut meat by the pounds of usable meat:

$$\$1{,}062.50 \div 180 \text{ lbs.} = \$5.90 \text{ per lb.}$$

This cost per pound applies to every cut of meat that will come out of that side of beef: tenderloin, flank steak, strip steaks, New York strips, chuck roast, and so on. When was the last time you bought tenderloin at under $6 a pound? To put these savings into perspective, consider that for local, antibiotic- and hormone-free, humanely raised beef, you'd probably pay at least $6.99 to $7.99 per pound just for ground beef!

In my location, New York strip steak from local beef runs approximately $18 to $22 per pound, as do rib-eye steaks. Chuck roast and shoulder pot roast or steaks are $7 to $8 a pound. Top round, eye of the round, and bottom round run from $8.99 to $9.99 per pound. You get my drift: You're going to save a *lot* of money.

To clarify further, let's take a look at some typical retail cuts, priced in line with most supermarkets and meat markets. The following chart represents one side of beef, processed into typical retail cuts.

But you—brilliant reader—paid $1,062.50 for the side of beef you cut yourself. You just saved $617.70! And if your family is large enough to use a whole beef (two sides) in a year, your annual savings would double to $1,235.40.

CUT AND NUMBER OF POUNDS	RETAIL PRICE PER POUND (typical meat market)	TOTAL RETAIL PRICE
Rib: 15 pounds	$18.00	$270.00
Strip: 11 pounds	$18.00	$198.00
Tenderloin: 5 pounds	$26.00	$130.00
Chuck: 28 pounds	$7.00	$196.00
Shoulder: 20 pounds	$7.00	$140.00
Top Round: 13 pounds	$8.99	$116.87
Bottom Round: 9 pounds	$8.00	$72.00
Eye Round: 5 pounds	$8.99	$44.95
Sirloin Tip: 10 pounds	$8.99	$89.90
Flank Steak: 1.5 pounds	$12.00	$18.00
Short Ribs: 10 pounds	$7.99	$79.90
Shanks: 9 pounds	$6.99	$62.91
Ground Beef: 33.5 pounds	$6.99	$234.17
Bones: 11 pounds	$2.50	$27.50
TOTAL RETAIL PRICE, ALL CUTS LISTED		**$1,680.20**

WHICH ANIMAL?

If you're a beginner, I recommend starting with half a pig. Pigs are the easiest and most straightforward animals to cut. Plus, there's sausage! I often lead sausage-making parties here in Vermont. Sausage making with all of the pork trimmings is so much fun . . . and who doesn't like sausage?

Leaving sausage aside, you'll also get much more variety processing your own pig, because you'll be able to get many cuts that simply aren't offered in supermarkets anymore. Things such as shoulder pork chops, boneless Boston butt pork roast, fresh pork hocks, pork belly, fresh ham steaks, and more.

If you don't eat pork—*oy vey, tref!* (look this up in a Yiddish dictionary)—then you can start with a beef primal or subprimal. See the enclosed CD to decide which primal/subprimal you prefer, then just follow along.

What Size?

Consider the size of your family first. If there are just two of you, you're probably not going to eat an entire beef in six months, even if you're on the "Caveman Diet."

A family of four to six could certainly consider a side of beef. Much depends on your living situation. Do you live in a house, or a condo/apartment? If you live in a small city apartment, you're probably not going to be shlepping a side of beef up the elevator (although it's been done, and imagine how your neighbors will admire you).

Assuming a side of beef in the elevator is unlikely, consider a quarter beef animal or a whole beef primal cut, such as a loin, a boneless top butt for sirloin steak, or a whole rib for rib roast or rib-eye-Delmonico steaks. You can easily handle a side of pork or whole lamb—quite manageable, even in a small elevator.

A good size for a pig is around 200 pounds. For lamb, I like 40 to 60 pounds. Beef I like to see at about 500 to 550 pounds. But again, the size you seek depends on what you want to do . . . and how large your freezer is . . . and whether or not you'll be sharing the animal with others.

You can save a bit of money by purchasing primals to cut yourself. If your specific taste is (for example) top round roast or steak, just get a whole top round. If you prefer chuck roast, then buy just a whole chuck to start with, or a whole rib or strip or short loin. These are small enough to process on the kitchen table, and the process won't seem nearly as daunting to you for your first time doing this. With a small kitchen grinder or grinder attachment for your food processor, you can turn all the trimmings into your own safe ground beef

Cole's Notes

According to all the old-timers I used to work with, there used to be three styles of cutting meat in the United States: Chicago style, California style, and Texas style. They told me that California style was mostly cut boneless, that Texas style was always bone-in, and that Chicago style was a combination of the two. How true this was I am not sure, but I do know that there are different styles and techniques of cutting meat all over the country just as there are in countries other than the United States.

(refer to the precautionary tales about supermarket ground beef in chapter 5).

I would recommend starting with a boneless primal cut, then gradually working up to a bone-in primal if you find that you want to continue to buy in bulk and process your own meat at home. And whatever you buy—be it a primal, a quarter, a side, or a whole carcass—make sure it that has aged the proper number of days. This is very important (for more about aging, read "Meat Quality" in chapter 3.)

SOURCING

Permit me to offer a personal mantra: For the safest, freshest beef, *Buy local and buy large*. You may think this is impossible, because you live in the middle of a high-density city. How on earth can you find local meat? Or you may think you can't afford it, because locally sourced beef costs more. Which it does, if you buy it pre-processed into table-ready cuts.

I'm sure you can think of lots of other reasons for resisting my mantra. Well, here's my unequivocal answer to all of your concerns: Yes, and no.

You live in a city? *No*, it's not more difficult to find a source of local meat (in this case, *local* will mean within a reasonable radius of your city). Go online and type in "buying a side of beef in [city name]" and watch what happens. Or visit a farmers' market or specialty meat shop and ask them where they get their meat. I bet it probably won't take more than a day—if that—to identify several local sources.

Too awkward? *No*, because you probably won't be buying half a steer and trying to carry it up three flights of steps. Instead, you can buy one primal cut, which is much easier to carry and to work with in a small space. Don't be daunted. This is much easier—even for city dwellers—than you might imagine. (Get that mental image of you carrying a side of beef up three flights of stairs out of your head. Stop it.)

It costs more? *Yes*, but there are many ways to purchase local meat economically.

For instance, you can go in with a few friends and purchase half an animal. That way, each of you can select the primal cuts you prefer. Hey, some couples play bridge; others cut up a side of beef. Sharing in a cow is even becoming a trend. It's called cow-pooling.

I recently met a Midtown Manhattan couple who had purchased a whole steer from a farmer when it was a calf. The farmer raised and finished it to their specs, and was about to slaughter it and deliver primals to them in their Manhattan penthouse. They were planning a butchering party, and bought several copies of my *Gourmet Butcher* DVD for their guests. They even asked me to autograph the DVDs! My ego is still reeling.

But . . . the same Manhattanites also mentioned that their cow was a lovely Jersey. Took all my self-control to keep from pointing out that Jerseys are dairy cows, and that "their" farmer was laughing all the way to his bank at having fobbed off an unwanted male calf on some unsuspecting city slickers.

Finding a farmer/rancher/butcher you can work with and trust is much easier than you think—even if you live in Manhattan. I advise folks just starting out to get references from as many people as they can about the source of the meat. Far too many

people I know have bought meat they weren't happy with because they didn't do their homework and trusted blindly. Never trust blindly.

Think about it. If you were planning to do business with somebody—say, a plumber, carpenter, or insurance broker—wouldn't you check him or her out first? I know I would. And here we're talking about a product that actually goes *into your body*. So do your homework—get online, or ask around. I guarantee you'll discover that all meat, and all meat purveyors, are not the same.

Most farmers raising meat for sale have websites, so this is a great place to start. They should welcome your questions; if they don't, run away. Any reputable farmer enjoys sharing his or her knowledge and enthusiasm with others. Most farmer websites will tell you what breed(s) they're raising, describe the raising method, include testimonials, and be chock-full of great photos. So do a little web surfing, make a short list, then start calling. Can't find a farmer website right off the bat? Google "meat CSA [city name]" or check out EatWild.com.

Another option is to solicit recommendations from people who source and process their own meat. Or you can visit an organic grocer who sells meats and copy farmers' names from the labels on the meat packages. Sneaky but effective.

Where the meat was raised is not a big deal. It's much more important to know the reputation of your farmer or rancher. Some raisers actually offer tastings (like wine, only chewier). Read the rest of this book and—armed with knowledge—ask questions. You'll be rewarded in the end, and the farmer should respect you for your thoroughness.

Sample Before You Buy

In my opinion, you should never blindly buy a side of beef or a whole lamb or pig without sampling the product. We're talking a big piece of meat and a healthy chunk of change; you don't want a freezer full of meat that you can't or won't eat while you throw good money away in the process.

I've found that most of the farmers I have dealt with are more than happy to give potential clients a sample to take home and try. By "try" I mean get a sample, cook it, and eat it before you commit to a carcass. And don't just go by the sample; ask for references, too. One thing people will be honest about is the meat they eat. If they think it's below par, they'll say so.

You could also go to a farmers' market and meet farmers who are selling their products there. If you do, ask lots of questions—read the list of questions under "Doing Your Homework" in chapter 4. Be in charge of your own meat destiny!

This book is a good start in helping you understand the ins and outs of meat production, but it can't introduce you to the meat raiser whose animal you'll be purchasing. There's a ton of information out there, so look at as much of it as you can. Then weigh it all. Why? . . . because some of it is good, accurate information and some may not be. Compare farmer/raiser websites and see how folks present themselves and their products. Be choosy.

Finding a good source of meat will make all the difference in determining whether your first experience in meat processing will be great—or a total disaster. This book is a great guide, and you can do more in-depth research, too. Contact your state's grass-fed growers' association or your local Extension Service office.

GETTING IT FROM THERE TO HERE

Slaughterhouses wrap carcasses for their customers, but different slaughterhouses do it in different ways. The best I've found is Royal Butcher here in Vermont, which wraps the carcass first in cheesecloth, then in bubble wrap, and finally in a plastic bag.

If it's fall or winter and the temperature is below 60°F (16°C) *and* if you're transporting the carcass a short distance (no more than a one-hour trip)—paper-wrapped is fine. But cheesecloth and bubble wrap is better. Keep in mind that many places will deliver for a small fee, and delivery may be included in the cost of the animal. Verify first—ask.

An animal carcass should not be transported on a hot day unless it's in a refrigerated truck. Nor should it be processed on a hot day, unless you can do it in a cooled environment. This is why we folks up in the Northeast process animals generally in the cooler (some would say frigid) months of fall, winter, and early spring.

GETTING WHAT YOU PAID FOR

Remember that *you* are in control. If you purchase an entire carcass, you—not the farmer—are the owner of the animal. Thus, *your* wishes should guide the slaughtering facility in how the animal will be cut. This means that you don't have to accept what the farmer or the slaughterhouse offers.

If you read chapter 4, you'll be familiar with cut sheets. Ask the farmer to discuss the slaughterhouse cut sheet with you *before* the animal is slaughtered. This is your opportunity to pre-plan. If you enjoy liver and onions, make sure that the offal is not discarded. If you have a dog, you might want the bones. Yes, you'll be doing the cutting, but you want to make sure that the carcass is processed to your specs before it gets to you.

After slaughter, when the carcass is delivered back to the farmer to be turned over to you, the processor should provide the actual carcass weight and total delivered weight along with the actual cutting loss.

It's in both the farmer's and ultimate consumer's best interest to check the actual weight against the final list and question large discrepancies. Many people never question that final list. But you may have to, in order to get back your entire animal, not just what they've decided to give you. You paid for it.

THE CUTTING AREA

Many folks choose to cut in their kitchens. Some prefer the garage. Either is fine.

I have cut meat in some pretty bizarre conditions—out in a pasture on a couple of sawhorses with a plywood top, in barns with sawdust on the floor, and in unheated greenhouses in sub-zero weather. So no need to get too picky. But wherever you process, *clean is key*. Keep in mind that in an unrefrigerated room, bacteria will nearly double on cutting surfaces and tools *every hour*. So it is extremely important to clean and sanitize surfaces and tools often during the cutting process.

I clean with antibacterial detergent or soap. And I sanitize with a solution of water and chlorine bleach: 10 parts water to 1 part bleach. Ideally, you should clean and sanitize *once an hour*. This will minimize the growth of harmful bacteria as well as clean and sanitize anything that comes in contact with the meat—including your hands.

You will need access to a reliable supply of hot water. The floor should be clean and free of dust, as should the walls and ceiling. To help make cleanup easier, it is also wise to place a tarp of some kind—like a disposable plastic sheet or plastic tarp that can be washed and reused—on the floor around the area where you will be cutting. It'll make life a lot easier.

You must have a good cleanable, sturdy cutting surface. This can be simply a solid wooden table or a specialty nylon-surfaced meat-cutting table (don't invest in one of these until you've gained some experience). Any sturdy table will do as long as it has a clean—and cleanable—surface. Some

folks simply place a large cutting board on top of a table (to prevent scratching its surface).

The ideal height for your table will depend on how short or tall you are. Most professional cutting benches are about 34 to 35 inches high (of course, if you're as tall as Lebron James, you might want something higher). So let's just say you want a comfortable height so your back won't be killing you after a few hours of work. A good length is 6 feet and a good width is 2½ to 3 feet.

You will need a separate and very clean surface for wrapping the meat. This can be a table, a countertop, or a piece of plywood supported on two sawhorses. You'll also need a place to hold tubs or bins for the trimmings you'll use for sausage or ground meat. The important table is your cutting table—no uneven legs, please.

Waste cans (30-gallon size) lined with plastic garbage bags work well for collecting unusable scraps. You may need two if you're doing a large cutting project. You should also have two food-grade plastic bins or tubs (known in the trade as lugs or luggers) for all your usable trimmings.

You will need a supply of paper towels and a few, preferably white, terry-cloth towels to clean your hands as you go. Also a good firm nylon scrub brush for cleanup. You should have an 18-inch roll of freezer paper for wrapping the meat.

The temperature should be cool (ideally, between 40 and 45°F, or 4–7°C) and the area must have good lighting . . . you're working with knives and saws here. If you're doing this in your kitchen, either lower your thermostat and open the windows (if the weather's cool), or turn up your air conditioner. If all else fails, try to work quickly (but safely) and have a helper beside you wrapping each piece as it's cut and placing it into the freezer. Obviously, I don't recommend cutting on a hot day if you don't have air-conditioning.

And there should be relatively fast access to the freezer so that the product can start freezing as soon as possible. There should be a sink with hot and cold running water to clean tools often as well as to wash your hands.

Our setup is in a garage—the concrete floor is easy to clean!

Food-grade luggers—you'll need these.

If you are going to invest money in a facility to do this on a regular basis, then I recommend that the ceiling, walls, and floors be covered in a smooth washable surface. It would be beneficial to have a small cold room. This could be just a small insulated room, cooled with an adapted air-conditioner unit if you are processing for home use.

HOW MUCH TIME?

How long the cutting process will take from start to finish depends on you as well as the animal or primal you're learning on. In a recent lamb class I instructed, we worked with six whole lambs on several long tables placed end-to-end. Each of the five students cut up one lamb. I processed the sixth lamb, giving step-by-step instructions to the class as they followed along working on their own lamb. It took us about six hours.

Learning from a book or a video will take longer at first, because you need to be certain you understand each step of the process before you attempt it. As you use this book and CD set, I strongly suggest you study each butchering section in the CD several times before starting to cut. When you're satisfied that you understand most of the process, it's time to plunge in. Take ample time to understand each segment before starting.

I'd plan on spending the better portion of a day—8 to 10 hours for a beginner—to process a beef quarter, half pig, or whole lamb. If you're working with a smaller beef primal, it should take from three to six hours (depending on which primal you're working with).

I've worked with some people who are "naturals" and with others who have to work hard to understand aspects that seem simple to me. I had one student—a physical therapist who was familiar with human bone structure—bone out a leg of lamb perfectly on his first attempt.

I advise all my students not to worry if their steaks aren't perfectly cut. It's meat, not an art project. I have never had a student fail to master the art of butchering meat. It may take some people longer than others, but they all succeed. And so will you.

One local family I know set up a cold room and a cutting room with power meat saw and commercial meat grinder to process their own beef. They said that it used to take them a week of evenings and weekends to process a whole beef animal in the 600-pound range, using only a meat chart. They knew they weren't getting the best cuts or utilizing the animal properly, but they did it. I give them so much credit for even attempting to do this on their own, without instruction. These are terrific people who won't mind me using their names. Congratulations to both of you, Dan and Cecile Bouregard from Franklin, Vermont.

BUTCHERING WARDROBE, FOR NON-FASHIONISTAS

I always wear a three-piece suit and tie when I cut meat because it's important to look fashionable at all times, particularly when covered in grease.

But for my students, I recommend old jeans and an old shirt . . . clothes you don't mind getting dirty. You *will* get a lot of blood and fat on them. A large apron will minimize this, and give you that nice professional, butchery look. Of course you'll want to dress warmly if you'll be cutting in a cold room or unheated space in colder climates.

It's important to wear clothing that isn't too loose or baggy such as long loose sleeves or anything that can get caught in a piece of power equipment. This includes long hair—guys as well as gals. Either put your hair up or wear a hat or hairnet. One of the first recommendations I give

Cole's Notes

If there is a covering of fat on your beef, don't throw it away. This is good stuff. Instead, slice some of the fat into sheets ⅛ inch thick. Cut sheets of waxed paper approximately the same size as the fat sheets. Then, one by one, place each sheet of fat between two pieces of waxed paper. Continue layering until you're done, wrap everything up, and freeze. Use the sheets of fat to cover lean oven roasts during cooking. You can tie one on with twine. It helps prevent very lean roasts from drying out. Another way to maximize yield from your beef. Now you're thinking like a butcher!

This helps prevent any major cuts or injuries to the hand holding the meat. It's a good idea to wear a plastic glove under either type of cutting glove.

Wearing a knife holder around your waist is also a good idea, because it trains you to put your knife back into the holder when you're not using it. This eliminates the risk of a cut from knives lying on the meat bench or cutting surface. I've often seen people reach to pick up a piece of meat, only to get cut by a knife lying under or beside the meat.

A good pair of durable work boots or heavy shoes is a must in case you drop a knife, cleaver, heavy bone, or even a whole primal cut; you'd be surprised how often this happens (even to yours truly). Wouldn't want any broken toes or punctured feet. You might consider steel-toed work boots—they're not very expensive, and with appropriate accessorization become a fashion-forward statement.

It's extremely important to remove all jewelry—rings, watches, and so on—not just for sanitary reasons but for safety reasons, especially if you'll be using any power equipment.

beginners is to wear a metal mesh or Kevlar butcher's glove on the *opposite hand* from your knife-wielding hand. If you're a rightie, you wear the glove on your *left* hand; if you're a leftie, you wear the glove on your *right* hand.

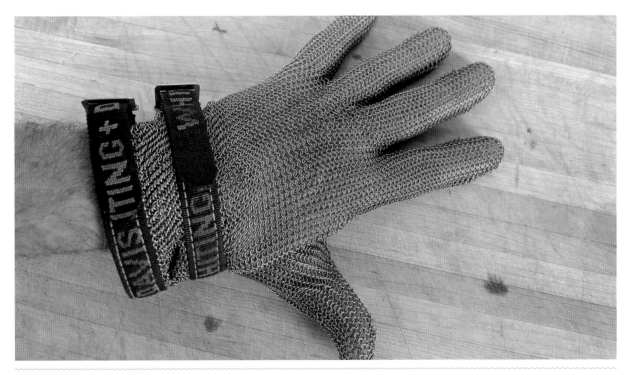

Cutting gloves are available in both Kevlar and metal mesh.

Butchering Day Checklist

Here's a summary of everything I recommend to my students to have on hand on butchering day. With the exception of a commercial power saw, most of these items are available at a restaurant supply store or online (see the resources section at the back of this book). Keep in mind that power tools like power meat saws and grinders can be found used on Craigslist or eBay quite reasonably.

Tools

- One 10-inch steak or butcher's knife, called a scimitar.
- One 6-inch curved flexible boning knife.
- One 6-inch curved semi-flexible boning knife.
- One 6-inch firm straight boning knife.
- One 25-inch hand *meat* saw (please don't use your dad's old carpentry saw).
- One 3- to 5-pound meat cleaver (optional).
- One hard rubber-headed mallet (to tap the cleaver when chopping through bone, also optional).
- One butcher's meat needle.
- One fine to medium steel.
- An oilstone (to sharpen your knives). I use a three-sided one with three stones graded coarse, medium, and fine (see the picture on page 115).
- One stainless-steel or plastic bone duster or meat scraper (used to remove bits of bone dust or meat particles from the surface of freshly cut meat).
- One electric meat grinder.

Accessories

- First-aid kit—this is not optional!
- Kevlar or metal mesh cutting glove.
- Plastic gloves (in boxes at your pharmacy).

- A couple of food-grade plastic tubs (called lugs or luggers); find these online or at a restaurant or butcher supply store.
- One roll of white butcher's twine.
- One or two white terry-cloth towels (or similar) to wipe your hands on.
- One table scraper.
- One floor scraper (especially if you're not using a tarp).
- A nylon brush to scrub the cutting surface when you're cleaning up.
- A stiff-bristled push floor brush.
- A 30-gallon plastic garbage bin and liners.
- Freezer paper and freezer tape.
- A plastic tarp (which will reduce the need for floor scrubbing).

Attire

- Work boots (best are steel-toed).
- Large apron.
- Sturdy work clothes you don't care about too much.
- Knife holder (optional).

Bottom row, left to right: hand meat saw, meat needle, large steak knife or scimitar, medium steak knife/scimitar, small steak knife/scimitar, straight boning knife, curved semi-flexible boning knife, flat steel or hone, steel, cleaver. Top row, left to right: table scraper, bone duster, butcher's twine, and freezer paper on dispenser/cutter.

YOUR TOOL IS YOUR BEST FRIEND

Without the right tool, you are nothing. Tools are the foundation of this craft, which bears not a little resemblance to sculpting. Knowing which tools to use, how to grip them, how to use them accurately and safely, and how to care for them are the bedrock of butchery.

Especially knives. Back in the day, respect for the knife was ingrained into every professional butcher, and you never, *never*, disrespected another person's tools. Every butcher would mark his or her knives with a personal identifier like a notch or an X or initials on the handle. Those knives were respected by all other butchers, who would *never* touch them.

Not so today. Lately, some of the places I've worked in are staffed by people who just don't get it. Some of these ignorant folks think nothing of picking up one of my quality butcher knives to cut up cardboard or slice a plastic band around a box. In the early days of my career, doing something like that was a good way to get your arm broken—a lesson I learned very early on.

Your New Best Friend—The Knife

These days, lots of chefs and trendy butchers buy knives that cost hundreds, or even more. But believe it or not, it isn't necessary to spend a fortune on knives. In the early days of my career, a lot of us used Chicago Cutlery with concave blades. It's been years since I've seen a professional version of those knives, but I truly loved them.

Then it was Forschner. Forschner still exists under the brand name Victorinox (which also makes Swiss Army knives), and this is the brand I use. My knives take a pounding. Nonetheless, I still get five years from a Victorinox 10-inch steak knife or scimitar (as they are also called). My 6-inch boning knives give me several years of good service. Victorinox is a reasonably priced knife that holds an edge very well, as long as it's sharpened on an oilstone and not a high-speed grinder (sharpening instructions are given later in this chapter). Dexter is another brand that I like. Nevertheless, I still use and prefer Victorinox.

How many knives you need depends on how often you'll be processing, as well as how many animals you will be processing. I process daily from September to April and am absolutely fine with one scimitar and three boning knives. This is quite sufficient for my needs and should be for yours.

A scimitar (*left*) works well for big slicing cuts through meat. A boning knife (*right*) has a narrower and often flexible blade designed for cutting meat away from bones.

The scimitar is for cutting layers of boneless meat or large steaks. They come in lengths of 10, 12, and 14 inches. Larger is not necessarily better; select the one that feels right for you.

Boning knives are intended for removing meat from bones. They are typically long and narrow in the blade, running from 6 to 8 inches, and usually come in firm, semi-flexible, and flexible. The flexible boning knife bends, making it easier to follow the contour of bones, so you get a much cleaner bone with less waste. I use a firm boning knife for part of the cutting. Even though there are specific knives for this job called breaking knives, I find them a bit too long for my technique of breaking.

Then there is the meat cleaver (a tool historically associated with butchery). The blade is rectangular and is broader, thicker, and heavier than a knife. Cleavers are used for chopping through bone and cartilage. I prefer a good heavy one that is well

balanced. A note here: You may not really require a cleaver, and frankly, until you're adept with your knives I'd recommend against one. Too easy to cut off a body part.

The scimitar or steak knife can be purchased in the $30 range, and boning knives can be purchased in the $20 range. I haven't priced cleavers recently. Mine was made for me by my grandson—how happy am I?

If you want your knives to last and perform, you need to care for them and store them properly. A well-maintained and properly stored knife will give you years of reliable service. What you *don't* want is your knife rolling around loose in a metal or wooden box or a kitchen drawer.

There are many ways to store knives: a magnetic strip on the wall, a nylon or Kevlar sheath or scabbard, or even a homemade cardboard sheath. Usually, a wooden knife block won't accommodate the size of a large butcher's knife. There are stainless-steel and plastic knife holders designed to hang on the wall or on the end of your cutting table as well. You should be able to find these items, as well as good Victorinox knives, at a restaurant supply store and online (see the resources section at the back of this book for a starter list of suppliers).

How to Sharpen a Knife

You'll need two tools: an oilstone and a steel. Oh, oops, and your knife.

I use a three-sided oilstone with a reservoir incorporated into the bottom for mineral oil. It's actually a multi-tool, with three stones mounted on it. You simply grasp the handles on each end, then lift the tool and turn it to expose the stone you want to use for sharpening. The three stones are graded coarse, medium, and fine. You will need all three surfaces as you work to sharpen a knife.

The bevel is the very narrow edge on the angled cutting surface of a knife. Before you can begin sharpening, it's essential to locate the bevel on your blade.

A cleaver's heavy blade is designed to cut through bone and cartilage.

This is a three-sided oilstone. And yes, that *is* duct tape holding the base together. It's old but it won't quit—kinda like me.

This is the bevel: the tiny line of shiny metal that runs along the edge of the knife. If you look very closely at it, you will see its angle. This is actually what does the cutting.

Step 1: Coarse Stone

Begin sharpening on the coarse stone, keeping in mind that laying the knife flat against the stone will not sharpen it. You must *angle* the knife so that the stone contacts the very sharpest part of the knife from its edge to the end of the bevel.

I count my strokes. I start with 10 or 12 in one direction on one side of the knife, then I repeat the process on the other side of the knife with *exactly* the same number of strokes. I use the entire surface of the stone, dragging the blade from the right-hand end of the stone to the left-hand end for one side of the knife. Then, to sharpen the other side, I push the knife from left to right. After 10 or 12 strokes on *each side* of the knife, I drop down to 8 to 10 strokes on each side, then 6 to 8, then 4 to 6, and finally 1 to 2.

Step 2: Medium Stone

Next, I switch to my medium stone and repeat the same sequence of strokes as in step 1.

Step 3: Fine Stone

Then I switch to the fine stone and repeat the sharpening process for a third time.

Step 4: Testing the Blade

I then clean the knife with hot soapy water. Now it's time to check the blade. Being careful not to run my hand or fingers across the edge of the blade, I run the flat of two or three fingers along the edge from the thick back part of the blade to the tip to see if I can feel a catch or very tiny hitch on the edge of the newly sharpened blade. This is a very subtle thing, but important.

Step 5: Testing the Other Side

Then I turn the knife to the other side and do the same thing. If I feel the least little hitch or roughness along one side, that is the side I need to run down the steel. *Why?* you ask. Because a just-sharpened knife edge is so thin that it often curls slightly to one side. To make straight and even cuts, the edge must be perfectly even without any curl.

Step 6: Straightening the Edge with the Steel

The steel is designed to straighten the edge of the knife, not to sharpen it. Just glide your knife along the steel, front and back.

My final test is to use the finger test again. If there's no catch on either side, my knife should be sharp—assuming I began with the right bevel. I usually test the final result with a piece of paper. I hold the paper up in front of me and see whether I can slice a piece of it off, or get partway through the sheet in one clean strike, without feeling it catch like the knife has a burr in the blade. And don't worry too much about this, because you *will* get it.

Keeping your knife sharp as long as you can is important. Often the knife may feel dull after a bit of use, but unless you've hit bone or used too much pressure and hit the table, you should only have to run it over the steel a few times to straighten the edge again.

Using Your Knife

I know you're all sure you know how to use a knife. You're wrong. Please read the following section carefully and reread it before you take that brand-new scimitar out for a spin.

How to Hold a Knife

Most commercial meat-cutting knives have a handle with a fitted grip (in other words, the handle is shaped to fit your hand and isn't just an oval shape). Your grip should be firm, and your hand should wrap around *just* the handle with your thumb pointed toward the tip of the knife (the overhand grip). Make sure that neither your hand nor any fingers touch the blade. Always grip the knife handle around the top, with your fingers wrapping around the underside of the handle.

Never grip the knife with a stabbing grip, especially when pulling the knife toward you. I have a nice stab scar from doing this back when I was age 17.

Step 1.

Step 2.

Step 3.

Step 4.

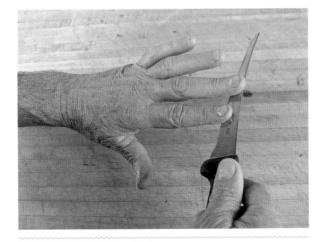

Step 5.

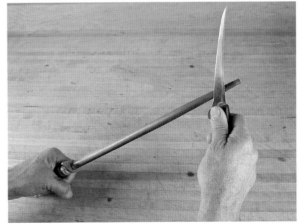

Step 6.

Cole's Notes

A steel is a long straight-shafted instrument with a handle. The steel shaft is round with fine or coarse lines running its length. A steel is used to straighten the edge of a knife, because the sharp edge of a knife tends to curl to one side or the other during use. The steel centers the edge—that is, it makes the cutting edge straight. When you use it, you are steeling your knife.

Developing the Butcher's Mind

Once there was a butcher who lived in China, in the state of Wei. His fame was so great that the duke of Wei heard about it. Thinking that this butcher should work for him, he devised a test to see for himself how good the butcher was.

The duke invited the butcher to demonstrate before an audience of nobles. The butcher began by killing and hanging an animal in the traditional manner. This did not make much of an impression.

But then the butcher took his cleaver, and with a few strokes so fast that the nobles could not follow the cleaver's movement, he cut the carcass into 12 pieces almost exactly the same size. Everyone was astonished. How had he done this so quickly?

The butcher explained that he had spent years studying the skeletons of cattle, and carried a mental map of every joint. "I am working with my mind as well as my body. I do not cut the bone itself, but in the space that exists between all the animal's joints."

This story is the source for the Chinese idiom *you ren you yu*. It's used to describe someone who does a job with skill and ease.

If you'd like to see more examples (lots of them!) of how to hold a knife while cutting meat, simply fire up the CD and start browsing. And an important note here: You will notice a few photographs where I am holding the knife in a stabbing grip; this was done for the photographer in order to make the process more visible. With years of experience, you can gain the skill to hold a knife safely using this grip. However, the *years* of experience and practice are essential. So until you become a highly experienced butcher, for your own safety, don't use the stabbing grip.

How to Cut

When using a steak knife or scimitar, strive for long smooth strokes. The aim is to get through the piece of meat with as few strokes as possible. Never "saw"; a sawing motion will leave little frayed bits of meat behind (we call them curtains or rags).

Once you've started a cut, *don't lift the knife*. Keep the knife in contact with the meat and—pushing forward then backward in as few strokes as possible—try to keep the blade in a vertical position to produce a nice clean, even cut. This skill comes with practice. Don't worry, you'll get it.

Remember to let up on the pressure as you near the cutting table surface. You want to avoid forceful contact of the knife with the table surface, because that dulls your knife. All you want to cut is the meat, not the cutting surface.

You should hold a boning knife the same way as a butcher knife—with the overhanded grip, not a stabbing grip. Normally, when you're using a boning knife you'll either be trimming a piece of meat or removing the bone. It's important to stay as close to the bone as possible in order not to ruin a good piece of meat. A flexible boning knife will make it easier to follow the contour of the bone. On the CD, you'll find many sequences that coach you through the process of removing a bone, such as the shoulder blade or femur, from a piece of meat. Other sequences show special techniques such as butterflying to create a large, thin piece of meat and separating bones at a joint.

Curling your fingers under the handle of a knife (*left*) is the safe way to hold it. Curling your fingers over the upper side of the handle (*right*) is dangerous.

The Most Common Mistakes

One of the most common cutting mistakes is using the wrong grip with too much force, which causes the knife to slip and cut into the next-closest thing—you.

Another common mistake is using a dull knife. A dull knife simply doesn't bite into the meat like a sharp one, meaning that the likelihood of it slipping and catching you in the finger, hand, arm, or some other part of your body is much greater. A cut from a dull knife is almost always a ragged cut, which hurts much more than a cut from a sharp knife. A sharp knife makes a much cleaner cut (not that I'm recommending you test this). Clearly, any cut is something to avoid, so handle all knives with caution and common sense.

What to Do if You Cut Yourself

Scream and faint. No, wait . . . there's no need to panic. You're fine, because:

1. Your properly sharpened knife gave you a clean cut, and
2. Your first-aid kit is *right at hand*.

It seems obvious that if you're going to be using sharp tools, you'd have a properly stocked first-aid kit with you. I'm always surprised when folks don't.

So you have a cut. Now what? If it's a fairly minor cut, first clean it with antibacterial soap and warm water. Once it's clean, apply some antibacterial cream and wrap the cut with a Band-Aid or bandage. If the cut is severe, you may have to apply a tourniquet and seek medical attention.

Something I strongly recommend is to have another person with you as you work. If you're going to be cutting your own meat and using knives and saws, you should always have someone with you in case you do get a nasty cut or any other injury. Never attempt to cut your own meat without another *adult human* present. No, your cat does not count.

Do You Need a Meat Saw?

Yes. However, you don't need to go all in on a power meat saw; you can start with a handsaw. In fact, I recommend this. Note that a hand meat saw is a unique tool. Do not substitute that old carpentry saw in your basement! I recommend a 25-inch hand meat saw. This is long enough to handle most large pieces of meat and bone. I recommend saws made by Dreizack.

If you really get into meat cutting, I know you'll lust after a power meat saw. Not only does it make

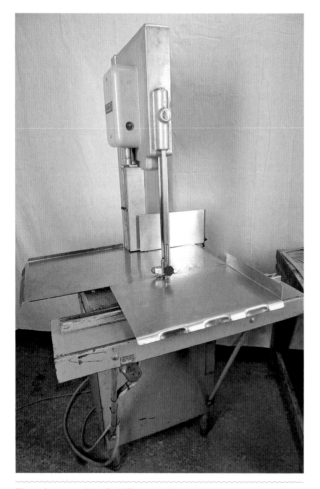

There's a world of difference between a hand meat saw and an electric meat saw. You'll see how both are used in the CD.

the work easier, it makes the whole process faster. The best power meat saw is made by Hobart. It's the most durable saw that I have ever used—perfectly engineered for easy disassembly and cleaning.

I've also used power saws made by Biro, Butcher Boy, and Toledo, but none can compare to a Hobart (and the same goes for all the other Hobart meat-processing equipment I have used).

Never buy a power meat saw from a catalog that sells tools for home use. I did once, and experienced the horrors of having to assemble a power saw that not only worked backward but was so poorly designed that it was extremely dangerous to work with. Home tool catalogs are for home tools, not for meat-cutting equipment.

A new Hobart power saw will cost thousands of dollars. If you're only cutting a few animals per year, it wouldn't be a practical investment. This is why I recommend that you go online to Craigslist or eBay to look for a used one. You'll often find them at very reasonable prices. Mine was probably an $8,000 saw new, but I purchased it for $500 in great working order. I've seen them as low as $200 (possibly needing a little tuning up) and up to $1,500. I am talking used here. Keep in mind that the older Hobarts still perform as well as new ones, and that parts are easily found.

Transporting the power meat saw from farm to farm was a challenge to me at first. The saw fits easily in the back of a pickup truck, but it's extremely heavy, so it takes some serious man- or womanpower. There are several ways to overcome the challenge. One is to have an electric winch; another is a small trailer with ramps and a dolly. I've used a trailer, but finally had wheels and handles welded to the saw, which works quite well.

Saws Versus Knives

Yes, power meat saws make the work more efficient and, to some degree, easier. But power saws are not ideal. A power meat saw tends to tear the meat rather than cut it, pulling out stringy connective tissue and creating bone dust. These residues must be scraped off with a scraper tool called a bone duster, which at the same time removes some of the meat juices from the cut. (Some processing plants leave these residues on the meat, but that's a big no-no.)

If you put a knife-cut pork chop and a power-saw-cut pork chop side by side, you'll see a visible difference. The saw-cut chop will have a dull ragged look; the knife-cut chop will glisten as the juices settle on top. There's actually a picture of this in the pork section of the attached CD.

The best way to cut chops (and similar bone-in cuts) is to first cut the chop to the bone from the outer fat side with a steak knife or scimitar, then put the loin on the saw and finish by cutting just the bone. Not only does this create a pretty-looking chop, but the chop won't need scraping, it will retain all its juices, and its shelf life will improve. So yes, power saws are great, but only when they're used properly.

THE PROCESS

The attached CD will help you visualize the butchering process. As you start to cut, you'll be removing large pieces of meat from the carcass for further division. When you're working with these (they'll be either large primals or subprimals), simply move pieces you're not cutting to the side or back of the table to hold them until you're ready.

As you finish cuts, you have two options: either place them into your food-grade tubs (luggers) and refrigerate them until you're ready to wrap, or have a helper with you who will wrap each piece as it's cut and place it into the freezer.

Remember to clean your knives often as you cut—every 30 minutes or so—and wipe your cutting surface with sanitizer and a clean cloth or towel once an hour. If you decide to take a break for lunch, it's a good idea to refrigerate the meat that is out beforehand. It's also recommended that you clean and sanitize your tools, bench, and equipment as well before taking a lunch break.

Beware the Boneless Blade!

Power meat saw manufacturers offer blades for boneless cutting (as well as bone-in cutting). *Never* use a boneless blade! Why? A boneless blade produces irregular cuts, giving the meat a ragged appearance. Worse, it heats the meat. Meat processed with a boneless blade does not have a long shelf life because of the heat generated by the blade. I also find the boneless blade much more dangerous to use.

As you process, you should also be separating out the trimmings into your two luggers. You want to separate the unusable scraps from valuable trimmings. If you've processed the animal properly, retaining such things as the bones for soup stock, then the unusable scrap pile will be minimal. Dispose of it by putting it into your regular garbage or—preferably—by composting, which adds value and will make your garden and houseplants happy. Those that aren't too fatty can be used for pet food. Your usable scraps or trimmings should be ground for hamburger. See? No waste.

WRAP LIKE A BUTCHER!

You're done! Congratulations. Now all you need to do is to wrap what you've cut (although, if it's a big job like a side of beef, I recommend wrapping as you go).

Make sure you have enough paper on hand before you start cutting. If you are wrapping small quantities of meat, you can get by with the 75-foot rolls of Reynolds freezer paper that can be found in most supermarkets. Reynolds freezer paper comes with pictorial instructions on the side of the box, which will serve you well although it's

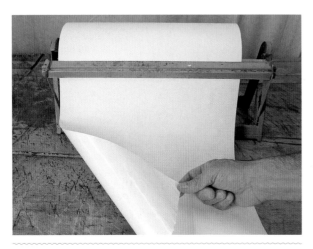

Notice the difference between the coated and uncoated sides of this freezer paper. Meat should be placed on the shiny, coated side when wrapping.

not my preferred way of wrapping. My preference comes from my days working in small markets where we wrapped meats in either butcher paper or plastic-coated freezer paper. And just to set you straight, butcher paper and freezer paper are not the same thing. Butcher paper is plain uncoated paper. Freezer paper is paper with one side coated with plastic or a waxy material, which allows it to resist moisture and keeps the paper from sticking to the meat.

Freezer paper (occasionally called locker paper) can be purchased from restaurant supply stores. Most grocery wholesalers also carry it—you may have to ask a grocery store to order it for you. It will run $60 to $80 for a 1,000-foot roll.

You'll be wrapping with freezer paper. *Always* place meat on the plastic-coated (shiny) side when wrapping. The outside of the package should be the dull-coated side—this is where you will label the package, describing what it contains and any other pertinent information (including date of slaughter, processing, wrapping).

Freezer paper comes in green, white, and brown. I don't know whether there's any difference between colors, but the light green one seems to be less brittle and more flexible. This is the one I prefer.

It takes practice to master the skill of wrapping meat properly. Here's how to wrap like a butcher.

Step 1

Cut off a piece of paper large enough to accommodate the piece of meat (or package of bacon, as shown here) you intend to wrap. Lay the paper on the table with the *plastic (shiny) side up*. If you are wrapping a large piece of meat and don't think the paper is wide enough, cut two pieces the same length and lay them side by side, overlapping one over the other by 6 to 8 inches, as shown here. Place the meat *diagonally* in the center of the paper (corner-to-corner).

Step 2

Lift one corner of the paper that is close to you up and over the piece of meat. Using your fingers, tuck the corner of the paper under the side of the meat farthest from you, and gently pull the meat toward you, making the paper tight.

Step 3

Fold the adjacent corner of the paper toward the middle, making sure the edge is even with the edge of the meat.

Step 4

Do the same with the opposite corner of the paper.

Step 5

Roll the meat toward the remaining corner, being sure to keep the package nice and tight. As you roll, the paper will narrow a bit.

Step 6

Once you come to the end, make sure your package is tight and apply your freezer tape. I always wrap tape completely around the package so that the tape sticks to itself on the other side of the package. I also wrap the package with tape in the opposite direction. Use tape liberally—the purpose is to seal the ends of the package well enough to keep all air out.

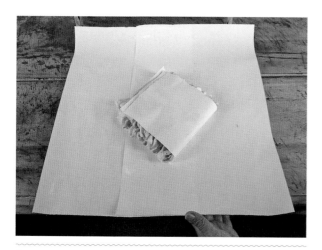

Step 1.

Step 2.

Step 3.

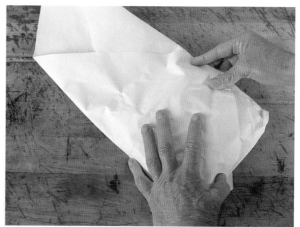

Step 4.

Step 5.

Step 6.

If you're wrapping multiple steaks and you want to stack one on top of another, cut strips of freezer paper to fit the steaks and fold in half, making sure that the plastic side is always what touches the meat. Place one piece of folded freezer paper between each steak, then wrap the entire stack. The interleaved freezer paper between each steak will make breaking apart the frozen packages much easier when you remove them from the freezer—just one or two taps of the package on the counter, and the steaks will separate.

Properly wrapped and frozen meats will keep up to a year in a freezer. In fact, I have had frozen meat that has been in my freezer for up to two years that was still a delight to eat. The keys are to wrap properly to minimize the possibility of any air in the package, and to freeze meat as quickly as possible after wrapping.

STORAGE ISSUES

Storage space? If you're serious about doing this, I would advise that you get a small to medium chest freezer and rotate the meat for the first few days to ensure that all areas of each piece freeze evenly (because it's the contact with the cold that's doing the freezing). After cutting, a side of beef will take up about 10 cubic feet of freezer space.

I prefer simple fresh meat or fresh meat that has been properly flash-frozen. The exception is corned beef, which I brine myself. So when I cut my own beef, pork, or lamb, I like it all fresh. Where pork is concerned, that means not cured or smoked. I like pork steak, pork cutlets, roast, chops, and fresh bacon or side pork. All of these cuts, as well as all the beef and lamb cuts, only require me to properly wrap, properly freeze, and properly thaw to enjoy all the meats I like to eat.

There are other packaging options for those who like cured or smoked meats. There's a wealth of information out there on the subject of curing and smoking meats, something that is too involved for me to get into in this book. Canning is another effective way to preserve some meats after processing, but it requires more technical knowledge and can be unsafe if proper procedures aren't followed. For long-term storage, freezing is about the best way to store meat.

Setting Up and Packing a Freezer

The first thing to do is to choose a site for your freezer. Position it on level ground well away from heat sources like stoves, washers, dryers, or direct sunlight (even coming through a window). There should be about 2 to 3 inches of space all around it to allow for air circulation, and of course you'll want it near an electric outlet. It's not a good idea to use extension cords with freezers, so make sure it's close enough to the outlet for its cord to reach easily. You should also make sure that the freezer's lid can open and close without rubbing against or hitting anything.

As for freezer temperature, 32°F (0°C) is the official freezing point, but a meat freezer should run at 0°F (-18°C) or colder. I recommend turning on the freezer the day *before* packing it.

You want your meat to freeze as fast as possible. To achieve this, you need to place the meat packages into the freezer with enough air circulation around them to get them frozen fast.

The way I do this is to pack the meat packages in layers, with each meat layer separated by a thin layer made of narrow wood slats or strapping cut just a little shorter than the length of the inside of the freezer. (Strapping is a narrow board about 2 inches wide and ½ to ¾ inch thick.)

Start by laying four or five lengths of strapping on the bottom of the empty freezer. This prevents the meat from sitting directly on the freezer bottom (and probably sticking to it!). Put in a layer of meat, then a layer of strapping, another layer of meat and another layer of strapping, and so forth, until all of your meat is in the freezer.

The strapping allows airflow around each layer of meat. This helps the meat freeze quickly, before the packages begin to bleed. Packing a freezer full of meat without something between each layer will cause the center of the meat not to freeze for a long time.

After packing the freezer, lower the temperature to the coldest setting for one week. Then you can set it from -10° to 0°F (-23 to -18°C) and be safe. Don't forget to check your freezer often to make sure that everything stays frozen.

How to keep your freezer organized? One family I know uses a different-colored freezer paper each year, keeping a chart so that they know (for instance) that everything from 2012 is packed in brown paper and everything from 2013 is in green paper, and so on. It's always a good idea to put the date on each package as well.

Some people have two freezers and rotate the meat from one to another every few months. This gives them the opportunity to check everything from the center to the bottom of the freezer to make sure it is all well frozen, or to identify any torn or damaged packages that should be rewrapped.

What to Do if the Power Goes Off

If you're unlucky enough to experience a power failure, don't panic. Frozen meat will stay frozen for at least 24 hours as long as you don't open the freezer door. Keep it closed! If your freezer is not full (important point), cover it with a thick layer of blankets.

If the power failure looks like it might last awhile, fill the freezer with ice (those packages you can buy at the corner store or garage) and keep replacing the ice every 24 hours.

When the power comes back on, examine the meat packages. If the packages are not all wet and bloody, you should be fine. If some of the packages exhibit these conditions, and the freezer hasn't been without power *for more than two days*, they can probably be *rewrapped* and placed back in the freezer. If the freezer has been out *for over two days*,

check each package for smell and color. Bad odor and brown color—get rid of it. It will be a combination of smell and color that will determine if you can use it or throw it out.

CLEANING UP: 10 STEPS

You've finished cutting and it's time to clean up. Here are the steps:

1. Scrape all tables and cutting surfaces with a flat scraper to remove any obvious heavy fat and grease residue.
2. Sweep all loose meat scraps and fat from the floor. Scrape and remove all fat, scraps, and bone sawdust from your meat equipment.
3. Spray all tables, cutting surfaces, meat equipment, and the floor with *cold* water (you don't want to cook any of the meat or other organic residues).
4. Spray all tables, cutting surfaces, and meat equipment with an antibacterial detergent and let it "set" for a short period of time. While I'm waiting, I wash my meat trays, dishes, and equipment parts in a three-bay sink, which in order from left to right requires soap, rinse, and sanitizer—so wash, rinse, sanitize, and stack to air-dry. Since you probably don't have a three-bay sink, you will have to do this in two or three steps.
5. Go back to the tables, cutting surfaces, and meat equipment you sprayed with detergent and spray again.
6. Using a stiff brush, scrub all surfaces, table legs, equipment surfaces, and (if necessary) interiors that have come into contact with the meat or its residues.
7. Scrub the floor with a stiff push broom, unless it's been covered *completely* with a tarp.
8. Rinse with hot water, getting every nook and cranny on the tables, cutting surfaces, and equipment.

9. Remove excess water from tables and cutting surfaces with a table squeegee and from the floor with a floor squeegee.

10. Spray with the sanitizer and let air-dry.

Yes, it's a lot of work. But it won't be if you've worked cleanly and carefully and haven't sprayed meat bits all around. Try not to be a slob—you'll cut down on cleanup.

Chapter Seven

A Side of Beef

I'm going to be your tour guide on a voyage through the world of beef. It's a big, confusing place, conjuring up all sorts of mental images: steak on a plate, cows (not really the right word, but the one most folks use for beef animals), small-*p* political discussions about the good or bad of hormones and antibiotics, passionate voices of animal advocates, slick marketing pitches for "certified" this and that kinds of beef, conflicting health claims, and so much more.

In the next several dozen pages, I'm going to do my best to disentangle all the issues involved with beef. I'll introduce you to the many different ways it's raised, present both sides of the battle about the use of antibiotics and hormones in beef production, and give you a handle on beef-raising terminology. I'll explain the way beef is quality-graded and how it's done, and take you on a mini-tour of a side of beef, naming the cuts and explaining how best to cook them. I'll introduce you to the fascinating history of beef cattle: where they came from, how they were domesticated, and where they're going—with a neat list of today's meat breeds to help you identify what your beef farmer is offering you. You'll be amazed at the diversity.

As for the actual butchery, you'll find a step-by-step guide to cutting up a side of beef in the attached CD. Those of you who bought this book for its do-it-yourself information should begin by reading this chapter carefully, then insert the CD into your computer and go through the course.

Finally, a note about the people quoted in this chapter. They represent the very tip of the iceberg. I spoke to many, many more: agricultural scientists, geneticists, health experts, epidemiologists, historians, breeders, raisers, animal advocates, and on and on (their names are listed at the back of this book). Those of you who wish to pursue the topics discussed in more detail should refer to the references section at the end of this book, where you'll find a partial list of the scientific journals and books I consulted as I researched the thornier subjects in this chapter. All this is to say that I have tried to present the most verifiable, factual, and up-to-date information available at the time of writing.

The magnificent head of a great bull. *Photo © Ronald Goderie, TaurOs Programme.*

Heme Iron

There are two forms of iron in foods: non-heme iron in plant foods and heme iron in meat, poultry, and fish. Of the two, heme iron is the one better absorbed by our bodies.

Subcutaneous fat is deposited on the *outside* of the carcass, while intramuscular fat or marbling is deposited *inside* the muscle.

WHY EAT BEEF?

You're reading a book about culinary butchery, so you're probably not averse to eating meat, right? And since you've read chapter 2 of this book, you already know that our human heritage is deeply tied to meat eating. We are what we are today because our ancestors ate meat—much of which was beef.

Beef is a good food. It's one of the best sources of protein, which our bodies need to maintain muscle mass as well as a healthy immune system. Beef is an excellent source of B vitamins, including niacin, vitamin B$_6$, and vitamin B$_{12}$, as well as all the essential amino acids. Beef is also an excellent source of zinc, plus heme iron. Plus it tastes wonderful.

How Do You Like Your Beef?

So beef is a great source of protein and iron, but when it comes to why we love beef so much, fat may be the real motivator. We all love fat. It's hard-wired into human brains. (Yay, *bacon*!) This is probably because we can't live without it. Fat helps maintain body temperature, provides energy, and improves the taste and texture of all kinds of foods. Fat provides essential fatty acids (remember that for later). In addition, vitamins A, D, E, and K—called fat-soluble vitamins—rely on fat for absorption into your body. Nutritional deficiencies related to a fat-free diet have been linked to arthritis, cardiovascular disease, PMS, and headaches. So obsessing too much about lean meat may be a little . . . obsessive.

But besides the fact that we love fat and it's good for us (in moderation), it also plays an important role in developing overall flavor in an animal's carcass. That's because first, the outer layer of fat protects the meat as it ages, preventing it from drying out or spoiling. (I've noticed that when I'm cooking meat with very little fat on it, I end up with not much flavor and not a lot of tenderness.) And second, it's the fat contained *within* the actual meat that gives it much of that mouth-feel and taste we all love so much.

There are four kinds of fat in cattle: **abdominal, subcutaneous, intermuscular,** and **intramuscular.**

- Abdominal fat protects vital organs and is a handy source of energy the animal can tap into if its food intake falls below what's needed for maintenance (as in, staying alive). When a steer is slaughtered, the processing plant pulls the abdominal fat from the carcass and blends it with lean meat to create various ground beef products (90:10, 80:20, 70:30 lean-to-fat percentages).
- Subcutaneous fat lies between the skin and underlying muscle, and acts both as an energy reserve and insulation. Most subcutaneous fat is trimmed from the carcass during processing and used—like abdominal fat—for ground beef products.

- Intermuscular fat is deposited between muscles, primarily in the chuck, brisket, and round. As the industry increasingly moves to value-added products, some of this fat is removed, thus generating new products. For example, a tender muscle called the infraspinatus found in the shoulder was once included in blade roasts. Now processing plants separate it and market it as a flat iron (or top blade) steak.
- Intramuscular fat is deposited within muscles between muscle fibers. Commonly known as marbling, it's really only important in what are called the "middle meats"—the rib and loin, collectively forming the longissimus dorsi muscle. Because marbling is most apparent in the rib and loin subprimals, rib and loin cuts are priced by quality grade.

The degree (amount) of marbling is a primary determinant of quality grade and hence value. Research results clearly demonstrate a correlation between degree of marbling and eating quality (tenderness, juiciness, and flavor). Quality grades (in order of descending marbling content) are prime, choice, select, and standard (we'll get to them shortly). Marbling appears as white fat flecks or streaks visible in lean pieces of meat.

Since so much of the beef eating experience depends on marbling, both the US and Canadian governments grade meat primarily according to the amount of marbling.

YOU *CAN* ALWAYS GET WHAT YOU WANT!

At least, in beef. And what do you want? Lots and lots of consumer panels, studies, polls, and more have asked consumers exactly what they want in their beef. And they answered:

- Bright red color. People choose beef on the basis of color.
- White fat color. Not yellow, thank you very much.
- Consistently firm muscle texture (not mushy).
- Finally, marbling consistent with the USDA standards.

Hardly a surprise. But keep the above in mind as you read on.

Understanding the Conversion of Live Animal into Meat

Meat grading is the procedure in which carcasses are segregated based on quality/palatability (tenderness, juiciness, and flavor) and yield (how many closely trimmed edible cuts can be obtained from a carcass). Do not confuse meat grading with inspection. Meat inspection is *mandatory* and deals with issues related to human and animal health.

Quality grading is *voluntary*.

Quality Grade

This denotes the expected palatibility (yumminess) of the meat. The United States Department of Agriculture's beef grading system has been designed to favor marbling, and has a number of different grades (starting with the highest): prime, choice, select, standard, and several below-standard grades, including commercial, utility, cutter, and canner. Grading is done by a certified meat grader after the carcass has been chilled for a minimum of 12 hours. The grader evaluates many things, including maturity, meat color, fat color, carcass muscling, fat coverage and texture, meat texture, and marbling level (more of this below).

In Canada, the system is slightly different; Canadian grading uses only four of the USDA marbling levels. Grading is handled by the Canadian Beef Grading Agency, a private, not-for-profit corporation accredited by the Canadian Food Inspection Agency. Canadian meat quality grades are (from

BEEF QUALITY GRADES
MARBLING SCORES

Canada				USA
Prime		Abundant		Prime
Prime		Moderately Abundant		Prime
Prime		Slightly Abundant		Prime
AAA		Moderate		Choice
AAA		Modest		Choice
AAA		Small		Choice
AA		Slight		Select
A		Trace		
				Standard and below

Beef grades in the United States and Canada.

highest to lowest), Canada Prime, Canada AAA, Canada AA, Canada A, plus others of lower quality, generally not sold in better stores.

The chart above summarizes the terminology of Canada and the United States.

Canada Prime and USDA prime are virtually identical. Canada AAA is the same as USDA choice, and Canada AA is the same as USDA select. Canada A is somewhat above USDA standard (according to the Canadian Beef Grading Agency, Canada A has no exact USDA equivalent).

Prime has the highest marbling, so it costs more—both in shops and at restaurants. You will rarely—if ever—find it in your supermarket. The grade generally sold in retail stores is choice, although select may also be offered as a less expensive option.

In the United States, prime, choice, select, and standard are grades more generally seen in younger cattle while commercial, utility, cutter, and canner are generally older cattle; these lower-quality grades are not sold as beef cuts, but as material used in ground products and cheaper steaks for family restaurants (note this last comment!).

How is quality grade determined? Beef carcasses are assigned a quality grade by a certified meat grader, who assesses the animal's maturity (a specific term defined on page 132), the color and texture of the lean meat, and the degree of marbling within the lean meat.

Marbling matters most and is rated by a marbling score, using officially approved photographs issued by the grading agency (see the following section). The grader will then roll or stamp the appropriate quality grade on the carcass with food-grade (edible) vegetable dye using a tool that looks like a paint roller. Carcasses that do not have enough marbling for a select grade will not be stamped. Thus, they are designated as "no roll," receive a discount price, and are marketed appropriately (in lower-priced beef markets). The best way to think of a "no roll" is that it is an inspected piece of meat that has not been quality-graded.

Marbling Score

The amount of marbling in a carcass is evaluated in the rib eye between the 12th and 13th ribs. The more marbling, the higher the quality grade of the beef. Marbling is expressed in increments from most to least: abundant, moderately abundant, slightly abundant, moderate, modest, small, slight, and devoid. Within each there are additional subdivisions. (And I am losing my marbles attempting to explain all this.) Generally—and to keep beef graders from going nuts—marbling scores are evaluated in tenths within each increment of marbling.

Now, I just know that one of you out there is going to take a look at these marbling photos and write me an irate letter, asking, "Hey, where's the photo of 'abundant'?" It's not here, and this is why, directly from the lips of Stephen Cave, deputy director of the USDA's Grading and Verification Division:

The reason the USDA doesn't publish official photos of Abundant marbling (or the even higher grades Very Abundant and Extremely Abundant), is that industry meat graders don't sort Prime carcasses according to marbling content.

To summarize the relationship between quality grades and marbling scores:

- Abundant or above marbling = prime
- Small to moderate marbling = choice

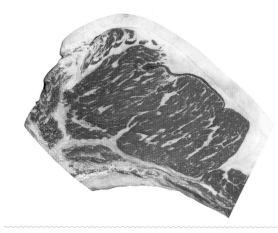

Moderately abundant.

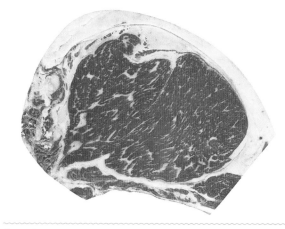

Slightly abundant.

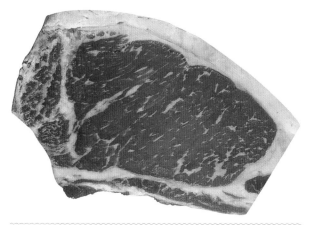

Moderate.

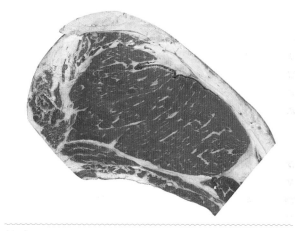

Modest.

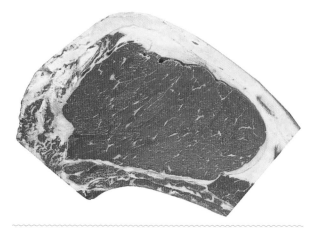

Small.

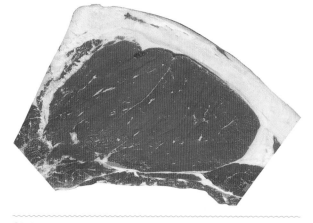

Slight.

Meat graders use these official photographs to help determine the grade of meat. *Courtesy The Beef Checkoff.*

Chine Bone

Beef and pork carcasses are generally split into right and left sides after slaughter. The central vertebral section (central spinal column) that is now exposed is called the chine bone.

- Slight marbling = select
- Slight or below marbling = no roll

Physiological Maturity

Maturity in beef production refers not to the actual (chronological) age of the animal at slaughter, but to its *physiological* stage or verifiable physical condition. In other words, maturity has nothing to do with age (kinda like people).

To determine the animal's maturity, graders look at the condition of the animal's bones and cartilage as well as the color and texture of the rib-eye muscle. More attention is given to bone and cartilage than color and texture, since these latter elements can be affected by other things like stress (for a fuller explanation of the effect of stress on meat, see chapter 4). Like humans, young animals have more flexible bones—more cartilage (that's why kids bounce). As they get older, the cartilage ossifies (turns into bone).

This is particularly apparent in the spinal column. For this reason, the condition of the bones is determined from the animal's split chine bone:

A = Red, porous, soft
B = Slightly red, slightly soft
C = Tinged with red, slightly hard
D = Rather white, moderately hard
E = White, nonporous, extremely hard

As for Canada, to be graded Canada A, Canada AA, Canada AAA, or Canada Prime, the animal must be judged "youthful" on the basis of skeletal development in the split carcass.

Sex

The sex of an animal will affect meat color and palatability. A mature intact male animal (a bull) won't taste the way you want your beef to taste. As the Canadian Beef Grading Agency puts it: "Pronounced masculinity in animals affects meat colour and palatability." So, since better meat comes from younger animals, you can understand why young male animals are castrated.

Carcass Weight (or Hot Carcass Weight)

This refers to the unchilled weight in pounds of an animal immediately after slaughter and after the hide, head, intestinal tract, and internal organs have been removed. The meat grader usually writes it on a tag or stamps it on the carcass. The amount of kidney, pelvic, and heart (KPH) fat is evaluated subjectively as a percentage of the carcass weight (this usually will be from 2 to 4 percent of carcass weight). The area of the rib eye is determined by measuring the size (in inches, using a dot grid) of the rib-eye muscle at the 12th rib.

Yield Grade

This is an estimate of the yield of closely trimmed, boneless retail cuts that can be obtained from a beef carcass; in other words, the ratio of meat to bone (or, *How much eating can I get from that steer?*). When beef is graded and stamped for quality, it must also be graded and stamped for yield. In the United States, beef yield grades are 1, 2, 3, 4, and 5. Yield grade 1 is the leanest and highest in yield or cutability; yield 5 is the fattest and lowest in yield or cutability.

- **Yield grade 1:** The carcass is covered with a thin layer of external fat over the loin and rib; there are slight deposits of fat in the flank, kidney, pelvic, and heart regions. Usually, there is a very thin layer of fat over the outside of the round and over the chuck.

- **Yield grade 2:** The carcass is almost completely covered with external fat, but lean is very visible through the fat over the outside of the round, chuck, and neck. Usually, there is a slightly thin layer of fat over the inside round, loin, and rib, with a slightly thick layer of fat over the rump and sirloin.
- **Yield grade 3:** The carcass is usually completely covered with external fat; lean is plainly visible through the fat only on the lower part of the outside of the round and neck. Usually, there is a slightly thick layer of fat over the rump and sirloin. Also, there are usually slightly larger deposits of fat in the flank, udder, kidney, pelvic, and heart regions.
- **Yield grade 4:** The carcass is usually completely covered with external fat, except that muscle is visible in the shank, outside the flank and plate regions. Usually, there is a moderately thick layer of external fat over the inside of the round, loin, and rib, along with a thick layer of fat over the rump and sirloin. There are usually large deposits of fat in the flank, kidney, pelvic, and heart regions.
- **Yield grade 5:** Generally, the carcass is covered with a thick layer of fat on all external surfaces. Extensive fat is found in the brisket, kidney, pelvic, and heart regions.

In Canada, if a carcass qualifies for Canada Prime or any of the Canada A grades, a calculation of lean yield is also made, using a special ruler placed on the surface exposed between the 12th and 13th ribs of the carcass. The aim is to predict the amount of lean (muscle) meat in the carcass. There are three Canadian yield grades:

- Canada 1 (Y1): 59 percent estimated yield
- Canada 2 (Y2): 54–58 percent estimated yield
- Canada 3 (Y3): 53 percent or less estimated yield

Fat Thickness, Color, and Texture

Fat thickness refers to the amount of *external* fat on a carcass, and is used in determining the yield grade.

What Most Stores Carry

Most beef sold in retail stores is USDA choice, yield 2 or 3. Certain retail chains also sell a "house brand," which may not be USDA-graded, but which meets the retailer's own specifications of marbling and yield grade.

Since much external fat will be trimmed off the carcass, a higher fat thickness means a lower yield grade. Fat coverage, color, and texture are also important factors in determining meat quality. Consistent fat cover enables the carcass to cool effectively, and firm fat is also a designator of a good carcass. Most consumers also prefer that the fat be white, rather than yellow (in fact, Canada's regulations state that Canada A, AA, AAA, and Prime may not include carcasses carrying fat with a yellowish tinge).

Rib-Eye Area

This is the total area of the rib-eye muscle at the 12th rib, expressed in square inches, and used in the determination of yield grade.

Internal Fat

The amount of internal fat is expressed as a percentage of the hot carcass weight, and also used in determining the yield grade. Meat producers measure backfat (another term for subcutaneous fat) to get an idea of an animal's total body fat, because the two are related. The degree of backfat on a beef animal generally correlates to the degree of total body fat. Generally, an animal with ¼ to ½ inch of backfat will yield a carcass with reasonable marbling. Still, there is a law of diminishing returns: Too much back fat means that so much fat will need to be cut off after slaughter, the overall yield of meat will suffer.

Carcass Muscling

This relates to the general build, outline, and shape of the carcass, which contributes to the determination of grade as well as the percentage of meat to bone. A well-muscled side with good length will give a higher yield of high-priced cuts than will a shorter side of comparable muscling.

Let's Talk Carcass

Hot carcass weight is usually about 63.5 percent of live weight (this is considered industry standard). An average live beef animal weighs from 1,200 to 1,350 pounds, with a hanging hot carcass weight of about 750 to 900 pounds.

Hot carcass weight does not include organs like the tongue or liver, both of which are, of course, edible and delicious.

Each carcass side (a "side of beef") is—obviously—50 percent of the hanging hot carcass weight. When a carcass side is split into carcass quarters, the forequarter represents about 55 percent of the hot carcass side weight, while the hindquarter approximately 45 percent.

Help! I'm Shrinking!

When a beef animal is slaughtered, it immediately starts to shrink. Over the first 48 hours following slaughter, it will lose 1 to 2 percent of carcass weight (depends on things like fat thickness, as well as the cooler's temperature and humidity). If the carcass is hung in a cooler for 14 days, it will lose another 4 to 6 percent. If it is dry-aged for up to 14 to 24 days, it loses even more weight. These days, not many carcasses are dry-aged. The majority of meat-processing facilities wet-age for up to 14 days.

What, exactly, *is* aging? As meat ages, the connective tissue within the muscle breaks down—this makes the meat more tender. It generally takes between 14 and 21 days for this to happen. Without any aging, the meat is extremely tough and has little flavor. Aging greatly enhances the flavor and adds to the tenderness.

Dry-aged beef is stored in a temperature- and humidity-controlled environment for 20 or more days. The process requires a near-freezing hanging room, and is only suitable for high-quality grades of meat, so it's quite expensive and is rarely done outside certain butcher shops and high-end steak houses. Dry-aging creates superb meat by reducing moisture in the muscle—thus concentrating flavor—and by giving enzymes in the beef time to break down connective tissue, thus increasing tenderness.

Wet-aged beef is aged in a vacuum-sealed bag. This is the way most beef is aged in the United States. Wet-aging is popular because it takes less time and no weight is lost in the process.

Each process produces a slightly different flavor. Dry-aged beef is often described as having a nuttier, more "roasted" taste, while wet-aged beef is described as milder.

There are three reasons why aging is so important: taste, uniformity, and blood.

1. **Taste:** The animal should hang for more than seven days (the length of time will depend on the depth of the fat layer on the carcass) . . . the more fat, the longer you can hang it. And the more taste it will have. Prime dry-aged beef usually hangs more than 21 days.

2. **Uniformity:** When a freshly slaughtered animal is hung, it will take on the uniformity it needs in order to be accurately cut. In other words, the carcass will take on the correct shape so that the butcher's cuts are accurate.

3. **Blood** . . . *or in this case, the lack of it:* The animal will continue to bleed as long as it's warm. Once it's firmed up during hanging, it will lose some moisture, which in turn will reduce the bloody mess when you're cutting. I recently had a chat with an acquaintance who told me he had decided to butcher a cow himself, but that the results had been "disappointing." I asked what the problem was, and he replied that the meat was "too bloody." From that description, I knew he hadn't aged the carcass before he cut it up.

Unmaking Connections

Connective tissue includes tendons, ligaments, and other supporting and connecting structures of the body. Connective tissue in meat is a long, stiff protein that's quite tough; that's why cuts with a lot of connective tissue are best cooked slowly in moist heat. The amount and composition of connective tissue within muscle will vary with the position of the muscle in the animal as well as with its age.

Tenderness and Aging

Generally speaking,

- 6 days improves tenderness 50 percent,
- 10 days improves tenderness 62 percent,
- 18 days improves tenderness 78 percent,
- 24 days improves tenderness 88 percent,
- 28 days yields small additional improvement.

A Cattle Diet Digest

- **Roughage:** Fibrous indigestible material in vegetable foodstuffs. Also, coarse, fibrous fodder.
- **Cellulose:** The structural component of cell walls in green plants (including wood).
- **Starch:** A carbohydrate found in the seeds, fruits, tubers, and roots of plants, especially corn, potatoes, wheat, and rice.
- **Grain:** A small, dry seed of a cereal grass like wheat or corn. Also used in the plural to mean the entire crop.

Additional moisture and trimming losses happen as carcasses are broken down into retail cuts and ground beef. When fat and bone are removed, the weight of take-home product from a carcass decreases. The breed of cattle, as well as cutting and trimming procedures, can also have a dramatic effect on retail yield—another reason to get to know your breeds.

Bone represents from 15 to 19 percent of a carcass. A general rule of thumb when buying carcass beef is to expect about 25 percent in bone and trim loss, then about a 25 percent yield in steaks, a 25 percent yield in roasts, and a 25 percent yield in ground beef.

It's hard to be precise in calculating exactly what you'll get in terms of bone, fat, steaks, roasts, et cetera from a given carcass. Each animal will be different. To help you get a handle on all of this, there's a Beef Cutout Data chart on page 136. It comes to us courtesy of Professor Rick Machen of the Texas AgriLife Research and Extension Service, and shows how one side of a 723-pound choice, 3.7 yield grade carcass breaks down. Professor Machen points out that this carcass (with .90 inch of backfat) is fatter than industry standard, so the fat percentage is a bit higher than average. Different steers will break out—obviously—differently from this one.

HOW CATTLE WORK

Now that you know more about fat and muscle and how their balance affects beef quality, let's look in depth at how cattle turn the raw materials they eat, such as grass and grain, into fat and muscle. Why? Because what a beef animal eats affects the taste of its meat.

Cattle are complicated, digestion-wise, because they have four stomachs. To be more accurate, their stomach is actually a complex four-compartment gizmo made up of the rumen, the reticulum, the omasum, and the abomasum. (Gesundheit.)

ANSC 437 BEEF CUTOUT DATA
Texas A&M University, April 2012

Group:

Cutting Assistant:

Steer Name:	ANSC 437	Actual Fat Thick.	.90
		PYG:	4.0
Mat.:	A	Rib-Eye Area:	12.2
Marb.:	Small 20	%KPH:	2.0
QG:	Choice-	Yield Grade:	3.7

Live Wt.:	1114.0
Carcass Wt.:	723.0
Dressng%:	64.9
Side Wt.:	349.1

IMPS	SUBPRIMAL	WEIGHT	PERCENTAGE OF SIDE WT.	PRICE/LB.	VALUE
	Forequarter				
120	Brisket, bnls, deckle-off	8.9	2.55%	$2.04	$18.16
114A	Shoulder clod	18.8	5.39%	$2.03	$38.16
116B	Chuck (mock) tender	2.9	0.83%	$2.15	$6.24
115D	Pectoral muscle	1.5	0.43%	$2.42	$3.63
130	Chuck short ribs	3.6	1.03%	$1.88	$6.77
116A	Chuck roll, bnls (3X3; 2 cerv. vert.)	13.3	3.81%	$2.21	$29.39
121D	Skirt steak, inside	2.3	0.66%	$3.49	$8.03
121C	Skirt steak (diaphragm), outside	1.0	0.29%	$7.52	$7.52
109B	Rib cap and wedge meat	2.5	0.72%	$2.42	$6.05
124	Back ribs	2.7	0.77%	$1.27	$3.43
112A	Rib-eye roll	13.5	3.87%	$6.36	$85.86
	Hindquarter				
189A	Full tenderloin, (side muscle on, defatted)	6.0	1.72%	$9.18	$55.08
180A	Strip loin, short-cut, bnls (0 X 1)	11.0	3.15%	$5.22	$57.42
184	Top sirloin butt	12.8	3.67%	$3.12	$39.94
185A	Bottom sirloin butt, flap (denuded)	3.4	0.97%	$4.36	$14.82
185B	Bottom sirloin butt, ball tip (denuded)	0.8	0.23%	$2.25	$1.80
185C	Bottom sirloin butt, tri-tip (denuded)	2.4	0.69%	$4.48	$10.75
167A	Knuckle (peeled)	11.3	3.24%	$2.27	$25.65
168	Top (inside) round	19.0	5.44%	$2.13	$40.47
171C	Eye of round	5.2	1.49%	$2.26	$11.75
171B	Bottom round flat	13.1	3.75%	$2.09	$27.38
193	Flank steak	1.9	0.54%	$4.87	$9.25
	Trimmings/Waste				
	Special trimmings	2.0	0.57%	$2.42	$4.84
	Regular trim (80%) FQ 50.2 HQ 17.6	67.8	19.42%	$1.77	$120.01
	Bone FQ 32.2 HQ 23.7	55.9	16.01%	$0.08	$4.47
	Fat FQ 19.5 HQ 44.7	64.2	18.39%	$0.28	$17.98
	Kidney	0.8	0.23%	$0.32	$0.26
	Total side	348.6	99.86%		$655.10
	Total carcass (side X 2)	697.2			$1,310.20
	Drop credit/cwt			$13.26	$147.72
	Total value of drop credit and carcass				$1,457.91
	Minus price paid for live animal (feedlot)			$119	$1,325.66
	Net gain or loss				$132.25

Eatin' here . . . cattle eat by grasping a clump of forage (and whatever might be hidden inside it), ripping it off, chewing just long enough to coat it with saliva and turn it into a ball, then swallowing it whole. *istockphoto.*

Hardware Disease

When cattle eat, they hoover up large mouthfuls of whatever's in front of them without noticing if anything doesn't belong in that mouthful of grass. Things like nails, bottle caps, and old beer cans sometimes end up inside them, leading to something called "hardware disease." The foreign object gets stuck in the reticulum, the second compartment of the cow's stomach. This can eventually kill the animal, as contractions force the object through the reticulum and into the heart. Believe it or not, several companies sell magnets for insertion into the cow's reticulum, designed to hold metallic objects in place and reduce the chance that they will perforate the stomach wall. And just to help your education along, the official name for this disease is bovine traumatic reticuloperitonitis.

This anatomy gives cattle (and similarly stomached animals such as sheep) unique attributes. In a nutshell, they are perfect machines for eating stuff that one-stomached animals (like pigs and humans) cannot—especially roughage (high-fiber cellulosic plants like grass).

The process begins with getting food into the mouth. Simple, just chew thoroughly and swallow, right? Not quite. Cattle have fewer teeth than other animals, with eight incisors on the bottom front, six molars at the back top and bottom, and no teeth at all on the top front of their mouth, which instead consists of a hard leathery pad. This dental pad, plus a relatively immobile upper lip, means that cows mostly use their tongues to get food. Cattle eat rapidly, swallowing their feed without chewing it sufficiently. So how do they digest? Read on . . .

You've just found out that cattle don't chew much before swallowing. This characteristic makes them "ruminants"—named for (you guessed it) the rumen, or compartment number one.

The **rumen** is the largest compartment in a bovine stomach, and it's huge—a mature rumen can handle over 55 gallons of material. Inside live billions of happy bacteria and microorganisms designed to digest large amounts of plant material—from grass to hay, corn, grains, cornstalks, and so on. When the rumen is full, the animal lies down and regurgitates the contents to re-chew. This is easy for them, because their esophagus functions bidirectionally (I'm so glad mine doesn't).

This process of rumination or "chewing the cud" is where swallowed feed is forced back to the mouth for further chewing and mixing with saliva. This cud is then swallowed again and passed into the reticulum, which is sort of the second compartment. I say "sort of" because animal scientists consider the reticulum and rumen to be one organ called—wait for it—the reticulorumen.

Rumination

A cow may spend as much as 40 percent of each day ruminating (how many hours do you?).

The **reticulum** is made up of tissues arranged in a honeycomb pattern. Swallowed feed moves freely between the rumen and the reticulum, whose main function is to collect smaller particles of feed and move them on to the next compartment (the omasum), while retaining larger particles for further digestion in the rumen. Some cattle people have another name for the reticulum; they call it the "hardware stomach." That's because this is where stray rusty nails, beer cans, and wire come to rest, nestled into the honeycomb structure of the organ. You'd think cows would know better.

On to the third compartment: the **omasum.** This is a globe-shaped structure full of tissues arranged like pages in a book (weirder and weirder). The folds increase surface area, which improves the efficiency of the omasum's function—the absorption of water and nutrients from digestive contents. The omasum acts like a filter, releasing water and keeping plant residues inside so that bacteria can keep digesting them.

And finally, the **abomasum.** It's also known as the "true stomach," because it's the closest in function to our own stomachs. This is where acids and enzymes break down the cow's feed, after which it's on to the small intestine (we're out of the stomach now), where nutrients are absorbed by the cow's body or eliminated. Cow poo.

What this all means is that the cow's digestive system allows it to consume plant materials such as seeds, shells, and stems that would otherwise be of no economic use to farmers. Which brings me to the debate over feeding cows grain.

First, some clarification . . . natural forages are a high-*fiber* feed. By natural forage, I'm referring mainly to pasture plants like grasses and legumes, plus hay (dried pasture), silage (fermented pasture), and so on. Grains—wheat, corn, and the like—are a high-*starch* feed.

You may have heard or read that cattle aren't "designed" to eat grain—that they simply cannot digest it. This is not accurate. What *is* accurate is that, yes, ruminants are primarily set up to be fiber digesters, but they will naturally take advantage of starchy grains when they're available. If you put cattle onto pasture where the grasses have matured and produced seedheads (read: grain), they'll enjoy eating the seeds, and will do so naturally. In fact, they'll eat the seeds first. Tasty, I guess.

How does this tally with claims that cattle cannot digest grains? It doesn't. Some of the billions of microbes inside a cow's rumen are better at digesting fibrous grasses; others are better at digesting starchy grains. The cow's population of rumen microorganisms is adaptable, with "starch digesters" and "fiber digesters" that shift to deal with the type of feed available.

However, while the microbes in cattle rumen *can* digest a wide variety of feeds, they are extremely sensitive to sudden changes in diet. Ruminants like cattle are very different from mono-gastric creatures like people. I can eat a salad today and a steak tomorrow. But a cow cannot—its rumen microbes won't be able to cope with the sudden change.

Keep this in mind as you read on . . .

FEEDING AND FINISHING: FACT AND FICTION

Cattle raising is a complex business—much more so than for any other meat animal. There are so many ways of raising beef animals, and so many passionate proponents of each method, that the whole subject is kinda like a maze of conflicting opinions. But it's a very interesting maze, full of fascinating paths.

Before we start our journey, I want you to keep one fact in mind. All cattle are raised on pasture or rangeland, eating forage (grass and other plants), *for most of their lives*. Some remain on open pasture until slaughter; others (called "conventionally raised" cattle) spend the last few months of their lives in penned systems called feedlots, where they are "finished."

In beef production lingo, *finish* refers to the degree of body fat on the animal, and *finishing* is the final process of fattening the animal prior to slaughter.

However, that's just the basics; the terminology used in the beef production industry gets pretty dense. The federal government has set specific standards for terms such as *grass-fed* and *organic*, but it gets pretty iffy out there.

Some terms are simply multiple names for the same process, while others can be rather deceiving.

Let's take a quick gander at a list:

- Grass-fed
- Grass-fed and grass-finished
- Grass-fed and grain-finished
- Pasture-raised
- Pastured
- Feedlot
- Feedyard
- Grain-finished
- Conventional
- Traditional
- Industrial
- Factory-farmed

And there are various combinations of the above, plus *natural* and *organic*. These are all terms for the *pre-slaughter management of the animal*. One method is not necessarily better than another. Their differences lie in qualities like taste, amount and color of fat, relative cost, personal preference, political agenda, and so on.

No matter what the method is called, let's get one thing straight. All cattle are raised on grass (well, to be exact, forage) up to 4 to 10 months of age while they are nursing, at which point the calf is weaned

and the process of finishing begins. Here, methods diverge. So when we talk about "grass-fed," "conventional," or "organic," what we're referring to is generally how that animal has been finished in the final three to six months of its life before slaughter.

Grass-Fed

I'm starting with **grass-fed** because I'll bet it's a term you've been hearing a lot. Denoting (you might assume) an animal that has eaten only grass right up until the time it is killed. Right or wrong? Both. And then some.

Right: According to the USDA standard, *grass-fed* is defined as an animal that (with the exception of its mother's milk) has consumed only forage ("any edible herbaceous plant material that can be grazed or harvested for feeding")—with the *exception* of grain. Grass-fed animals must have continuous access to pasture during the growing season. When they cannot graze outside, they may be fed crop residues, silage, hay, et cetera, as well as grain plants in their pre-grain (pre-seed) stage. The USDA's GRASS-FED label does not forbid the use of antibiotics, hormones, or pesticides.

Wrong: All *grass-fed* means is that the animal has been eating forage. Could be in a pasture, could be in a barn, could be grass, could be hay or silage. The interpretation of this term—like so many others in the beef industry—depends on who's talking.

As I said earlier, technically speaking, *all* cattle are grass-fed, because all cattle spend most of their lives eating natural forage. The term *grass-fed* has nothing to do with either the age of the animal, its degree of body fat (finish), or even how it was managed or treated. But "grass-fed" certainly has a nice ring, doesn't it?

Grass-Fed and Grass-Finished

Aha, now we're talking. This term does have a precise meaning. A grass-fed and -finished beef animal

Paul List's grass-fed and -finished Lowline Angus cattle spend their entire lives—summer and winter—outside.

is one that has reached its mature size and weight on forage alone and then is slaughtered. For more on this from the perspective of a grass beef farmer, read Paul List's story in chapter 4.

Since grass feeding and finishing must be done on forage alone (without grain supplementation), the nature and quality of the forage mix is paramount. Ideal forages are cool-season annual grasses (grasses that grow best in cool conditions) such as orchard grass, fescue, and timothy, as well as immature (green or pre-seed stage) grains such as wheat, rye, and oats. Warm-season forage such as big bluestem, eastern gamagrass, and indiangrass tend not to have as high a nutrient content as cool-season forage. For this reason, as you move north on the American continent, conditions for grass-fed and -finished beef production get better. That is, until winter, when most farmers must feed their cattle hay or harvested forage. This is what Paul List does.

Grass-fed beef is generally leaner than grain-finished beef, because the energy density of grass is lower than grain. That leanness is displayed throughout the carcass—in the animal's subcutaneous, intermuscular, and intramuscular fat.

No matter which preslaughter production system is used (grass, grain, etc.), subcutaneous fat will be trimmed to a level deemed acceptable to the consumer, generally ¼ inch or less.

Professor Rick Machen of the Texas AgriLife Research and Extension Service works with both grass-fed and traditional producers. Here's his take on grass-finished beef:

Grass-fed beef can vary widely in taste from a very gamey, grassy, metallic flavor to a taste that's indistinguishable from traditionally produced beef. The flavor of grass-fed beef depends on the age of the animal at slaughter and the quality of its diet (forage + supplement).

The American Grassfed Association standards (the most stringent in the business) allow for supplementation of animals on grass using approved supplements (primarily no cereal grains). So grass-fed animals supplemented with something like soybean hulls or distiller's dried grains (coproduct of ethanol production) taste much like grain-fed cattle. On the other hand, cattle fed dormant winter forages and harvested in a leaner condition can have a flavor very different from grain-fed animals. I have eaten everything from young (16–18 months old) cattle to old cows, all of which qualified as "grass-fed." There is a huge difference in eating experience.

But some consumers expect (maybe even prefer) the gamey, grassy flavor of some grass-fed products, and if they do we need to be producing it for them.

Pastured Beef

This is actually what it says it is. Hooray. Pastured beef animals have spent most their lives outside (barring extreme weather), on pasture. That's basically it. Pasture = where the animal lived. Grass-fed = what it ate. However, beef marketers and stores may use the terms interchangeably—another reason for knowing the source of your beef.

Pasture-Finished

This is a somewhat deceiving term. It seems to suggest that the animal has been fed only natural forage until slaughter. But that's not always the case. Pasture-finished animals may be raised on natural forage until they're considered mature and slaughtered, or they may be raised on pasture that is supplemented with additional feed like dried distiller's grains, soyhulls, or pasture cubes (pellets of compacted feed like alfalfa). Often, feed additives are given.

This method is more dependable than grass finishing, producing faster growth and heavier market weights, because the nutrients in grain are consistently available and thus more predictable than natural forage, which varies according to regional, climatic, and seasonal differences. In many ways, this is very similar to the conventional or traditional method.

Traditional or Feedyard/Feedlot

Traditional or conventional beef production is sometimes called industrial or factory-farmed beef (terms deliberately chosen for their negative connotations). A **feedlot** is a fenced or enclosed area where animals are fattened for market. More precisely, it's a livestock management system where grazing animals are confined to an area that produces no feed, and are then fed on stored feeds.

Does Good Forage Just Happen?

Beef farmer Paul List: "Good forage is not just green and lush; it must contain the proper balance of trace minerals and pH level. If the topsoil is lacking in organic matter/fiber, it cannot generate the needed nutrients without fertilizer additives. Managing the land naturally takes more time, but produces much healthier forage and results in much healthier beef."

❦••◗ ◖••❧

No Free Range for Cattle

If you see the label FREE RANGE on a beef product, something's seriously wrong. In the United States and Canada, the only animals to which this term can be applied are poultry. In other words, there's no such thing as a free-range steer. However, change *free* to *open* and you get *open range*, which is the term for a cattle-raising system widely practiced in the western states where cattle are allowed to roam freely across vast rangelands and are often mixed with other herds of cattle. These cattle are most often grazing state or federal rangelands in addition to private ranchlands (hence, the reason for branding).

And by the way, both countries define *free range* as a poultry animal that has "access" to the outside (that's it, folks—that chicken could be outside only five minutes a day to qualify).

Feedlots are sometimes referred to as "drylots." Today, the vast majority (upward of 90 percent) of beef eaten in America represents the feedlot method of farming.

Obviously, a feedlot system is as different as you can get from a grass-finished system. Feedlots are

These cattle are being fed in amounts that are carefully controlled. *Photo by Franzfoto, Wikimedia Commons.*

associated with traditional or conventional beef production. Which sounds like they have always been with us. But in the case of beef production, 'tain't so.

In the early 20th century, there was no feedlot system. Feedlots appeared in the 1950s . . . tied to the expansion of corn production in the United States . . . which in turn is tied to World War II.

Wars require bombs. Bombs require nitrogen—a key component of high explosives like TNT. To supply the demand for explosives during the 1940s, the US government built 10 nitrogen plants—all located in the center of the country. Come peacetime, these plants produced a lot of excess nitrogen, so manufacturing switched from explosives to nitrogen-rich ammonia fertilizer. Which led to the war's final "explosion"—the startlingly fast expansion of corn production. Notice where the US Corn Belt is located? Pretty much the same region as those 10 nitrogen plants.

All this resulted in lots of surplus corn (this was before the era of ethanol). What to do with it? Use it for cattle feed. Suddenly you could feed thousands of cattle in one convenient place. No need to truck them miles to natural feed; just bring the feed to them. Like an incredibly oversized restaurant for

Ten Million Feedlots and Counting

The size of a feedlot is expressed as its total onetime feeding capacity (maximum number of animals that can be fed at the same time). According to the USDA, on August 1, 2010, the number of animals on feedlots with a capacity of 1,000 animals or more was 10.7 million.

cows. Soon, the old ways of cattle raising—pasturing on diversified farms that fed their cattle on forage grown right there—began to decrease as beef production became more specialized. Today, feedlot farming is by far the largest method of beef production in America, with hundreds of feedlots. Small feedlots handle a few hundred animals apiece; some huge operations are capable of handling more than 800,000 animals at one time.

The discussion about grain-fed versus grass-fed beef is interesting. Seems to me that what most people worry about is the generic use of corn. It's not just that the animal was fed GMO corn, but also that corn in general has taken over other crops—resulting in an agricultural system that increases risks associated with monocultures. Diversity is what we all ought to be seeking.

Anyway, when we refer to traditional or conventional beef farming, we're talking about this method of beef production. The feedlot system has other names: Intensive Livestock Operation (ILO), Confined Feeding Operation (a term used in Canada), or Concentrated Animal Feeding Operation (CAFO). As these acronyms suggest, the North American livestock industry is big business. It has to be, to satisfy the appetites of so many people.

As you learned earlier in this chapter, *all* cattle—including feedlot—are raised on pasture from birth until weaning, typically at six to eight months of age. Generally, a traditionally raised beef animal will eat natural forage until it's about 750 to 900 pounds, at which point it's moved to a feedlot, where it's fed a more concentrated diet consisting of 20 percent forage and 80 percent grain (more or less; there are many variations, depending on the grain type and other factors). The aim is to increase slaughter weight and improve overall body fat content. Corn is the most common grain fed to beef cattle, although as you move north, wheat, barley, and grain sorghum are fed. Increasingly, a by-product of ethanol production called distillers' grain (DDG or Distillers Dried Grains) is used.

Feedlot farming has its own vocabulary. Here are some terms you may hear.

Backgrounding is the growing, feeding, and managing of cattle from weaning until they enter a feedlot, done usually on grass. Feedlot operators like backgrounded cattle because they know they're getting a consistent animal for finishing that is weaned and thriving on forages. Backgrounding is a way of controlling the animal's weight gain up to a specific weight (typically, 750 to 900 pounds) . . . the ideal base weight for finishing. Backgrounders typically produce two types of animals:

- **Stocker cattle** are kept on pasture or rangeland to increase their weight and maturity before they're placed in a feedlot.
- **Feeder cattle** are weaned calves ready to be put on pasture or a high-energy food like grain for finishing prior to slaughter.

Some cattle raisers prefer to finish their own cattle on their own feedlots. This is called self-finishing. There's still a feedlot involved, only it's on the farm where the animals were raised. It's an efficient way for diversified farmers to use their own grain for cattle feed.

Are Feedlots Bad?

The terms *industrial* and *factory farm* are obviously intended to produce a negative emotional reaction. But let's keep emotion out of this. While aspects of traditional feedlot cattle raising do resemble assembly-line production, beef production is still conducted outdoors without many of the confinement aspects of pork and poultry—which is both bad and good.

ENVIRONMENTAL ISSUES

Nevertheless, there are serious issues to be considered in traditional cattle methods. Feedlots can create environmental problems. Hundreds—or thousands—of animals in one place create a lot of manure and urine. Runoff can become a real problem, unless some form of containment system is in place. Nitrogen released from the huge amounts of manure and urine becomes ammonia gas in the atmosphere, which can harm ecosystems or combine with other substances in the air and be inhaled by humans. So the environmental issues around feedlots include air pollution as well as ground- and surface-water contamination.

As a result, most feedlots require some type of governmental permit and must have plans in place to deal with the large amount of waste that is generated. This, of course, doesn't mean that infractions never happen or that accidents don't occur. It simply means that there's an increasing public awareness of the environmental impacts of poorly managed feedlots, and growing pressure to address the problems they create. In the United States, the Environmental Protection Agency regulates animal feeding operations. In Canada, authority is shared among all levels of government. Whether or not the various governments are properly addressing the environmental issues of large feeding operations remains a matter of opinion.

FOOD SAFETY ISSUES

You have probably found out—in the most unpleasant way—that what happens in your gut (more accurately, that bunch of bacteria called your microbial population) affects your health. Those gut microbes are a product of genetics, environment, and the food you eat. Same for cows.

Grass-fed and -finished beef is generally marketed as safer for human health than grain-finished beef, with grass-fed advocates pointing to an increased incidence of *E. coli* contamination in feedlot cattle. On the other hand, studies conducted by researchers from various institutions, including one from Purdue University and Zhejiang University in China, have found no difference between the samples of grass-fed and conventionally raised beef they tested. For those of you who want to read the scientific literature on the topic for yourselves, as I have, please see the resources section of this book.

Curious to find out more about cattle digestion (who wouldn't be?), I contacted the lead scientist on this study about *E. coli*. Paul Ebner is associate

Cow manure can actually reveal the quality of the feed; this "cowpat" indicates proper fiber and moisture content. *Photo by Jeff Vanuga, USDA Natural Resources Conservation Service.*

professor in the Department of Animal Sciences at Purdue University. He told me that whether an animal is grass-raised or conventionally raised, it will still face what he called "microbial chaos" once it leaves the farm to be slaughtered. Why? Because it has to travel in an often less-than-pristine truck to the slaughtering facility, where it will be confined in pens with other animals, and so on. This exposes the animal to different bacteria.

To quote Dr. Ebner: "The processing of a grass-fed animal versus a feedlot animal is essentially the same. The difference between the two lies in a different mix of fatty acids like omega-3 and -6, as well as in the fact that grass-fed animals are leaner, with less fat. Most people don't realize that an improperly managed grass-fed farm is just as big a problem as a poorly managed feedlot."

ANIMAL HEALTH ISSUES

Critics claim that feeding a high-grain diet to ruminants creates severe health problems for the animals. This is the claim I've probably heard most often from grass proponents.

Again, it helps to dig a little deeper. Feedlot operators certainly recognize the fact that they cannot suddenly switch a cow from forage to grain.

Professor John Hall is extension beef specialist and superintendent of the Nancy M. Cummings Research Extension and Education Center at the University of Idaho. He says, "If you try to push too much grain into the animal too fast, you can see lesions on the liver. But that's bad for the animal, and people don't want to see that happen."

So when cattle are brought into a feedlot, they go through an adaptation process during which their diet is gradually shifted toward grain—this gives the microbial population inside their rumens time to adjust. Typically, a feedlot will feed about 80 percent grain and oilseed products and 20 percent natural forage such as hay or sileage. This keeps the rumen (and its animal) healthy.

ETHICAL ISSUES

This is more problematic, because it's subjective. Many people point to feedlots as inhumane places where animals are kept penned up in close proximity to one another, and are treated badly. While it's true that cattle are herd animals, genetically programmed to stay close together in a group for protection against predators, the sudden exposure to animals they haven't grown up with does cause stress. One would hope that feedlot operators would be committed to the humane treatment of their animals, not only because the USDA requires it, but also because stressed animals yield poor-quality carcasses. Poor-quality carcasses sell for lower prices. Ergo, less profit. After all, this is big business, and live cattle are the inventory.

On the other hand, many people prefer to know that the animal they're eating spent all its life eating natural forage in a pasture. It's a reasonable position to take, particularly if you're interested in supporting local farmers. For my own part, I always try to purchase my meat from local sources and, in doing so, know that I'm helping the nation's growing grass-fed and grass-finished beef market.

Maybe it's best to simplify the whole thing and concentrate on appearance and taste. Generally, the fat from grass-fed and -finished beef is slightly yellower than that from grain-finished beef. But the

What Does "Humane" Mean?

Professor John Hall on the humane treatment of animals: "One thing we stress in beef quality programs is proper animal handling and care. We all feel the same way: We've got to make sure that the animals are treated properly. And one of things we're currently struggling with in our society is *what does that mean?*"

This young steer will drink from 1 to 2 gallons of water per 100 pounds of body weight a day. Obviously, having fresh, clean water at hand is very important in raising cattle.

difference is quite slight. As for taste, most people think that grass-fed and -finished beef has a distinct "grassy" flavor. It's really a personal choice, based on your taste preference and personal philosophy. Why not do a side-by-side comparison? Buy and cook two steaks: one from a top-quality grass-fed and -finished animal, the other from a conventional steak no higher than choice grade. Then serve your family and score each.

Quite often, grass-fed and -finished farming is a method chosen by smaller farmers. As Dr. Ebner points out, "I think grass-fed beef is a good way for people to get into beef production. It requires a fairly small input, and pays a premium." What he means is that there's very little infrastructure required to get into grass farming. Read Paul List's story in chapter 4; he leases land to graze his cows, which remain outdoors all year—even through the Vermont winter. No barn. No extra feed. Yet his grass-fed and -finished products sell for higher prices than local supermarket meat because the cattle grow much more slowly than feedlot cattle. Grass beef farming is difficult and requires superior animal husbandry skills.

What About Nutrition?

You will read or hear that grass-fed and -finished beef is a rich source of important fatty acids—in particular, omega-3 fatty acids, omega-6 fatty acids, and conjugated linoleic acid, or CLA. How do we know? We measure.

To measure the fatty acids in meat, scientists separate the fat from everything else in the meat sample, then analyze that fat. This is because fatty acids are found—and stored—in fat, not in lean tissue (muscle).

Most analyses show that grass-fed and -finished beef does contain statistically higher levels of omega-3, omega-6, and CLA than grain-fed beef. The key word here is *statistically*, because of how this word is used in science. Time for a scientist . . .

Professor Rick Machen specializes in animal and resource management at Texas Agrilife Research

Cooking Grass-Fed and -Finished Beef

Grass-finished beef is leaner than conventional beef, so it needs to be treated gently.

Here are some cooking tips:

- Thaw the meat in your refrigerator, not in a microwave.
- Don't remove any fat before cooking.
- Cook roasts at a lower temperature than you would use for grain-fed beef.
- Grass-finished beef continues to cook after it's removed from heat; generally, it will need about 30 percent less cooking time. Verify this as you go.
- Since grass-finished beef is low in fat, it's a good idea to coat it with a light oil (like olive oil) to help it brown, enhance the flavor, and prevent drying and sticking.
- For some cuts, it might help to use a mechanical meat tenderizer such as a metal pounder (although personally, I don't like having to pound my beef).
- Make sure the fire or heat is consistent, but not hot enough to burn or scorch the meat.

Fatty Acids

Omega-3 fatty acids are essential nutrients that our bodies can't make by themselves and must get from our food. There are two types of omega-3. The first is found in certain vegetable oils, walnuts, and green vegetables like brussels sprouts; the second comes from fatty fish like salmon. Omega-6 fatty acids are also essential nutrients, but are much more common in our diet. Conjugated linoleic acid (CLA) is a nutrient found primarily in the meat and dairy products of ruminants. CLA may increase lean muscle mass and decrease body fat, and may also be an anti-carcinogenic substance.

Best Sources of Fatty Acids

According to Tufts University, cold-water fish are the highest source of omega-3 fatty acids, although walnuts and flax are also good sources. The current recommendations are to have 7 to 11 grams of omega-3 fatty acids each week.

A lovely baby at John Clark's organic Applecheek Farm in Vermont.

Center, part of the Texas A&M University system and the state's premier research agency in agriculture and life sciences. He's also a member of the animal nutrition section in the Department of Animal Science. Professor Machen works with both grass and conventional producers. According to Machen,

When the amount of omega-3, -6, and CLA is measured in beef, it's expressed as nanograms per gram of fat (a nanogram is one-millionth of a gram). Using CLA as an example, when we measure CLA levels in both grass-finished and grain-finished beef, we find that, yes, the level of CLA is statistically higher in the grass-finished beef. In other words, one number is higher than another; however, the actual difference is so small that there is little or no nutritional difference. Think about "statistically higher" this way: 10 is statistically higher than 1. Ten is also statistically higher than 9.9.

Continuing with our CLA example, when levels are expressed as a function of a typical *portion, the grain-finished beef may actually deliver more CLA, simply because it has more fat than grass-finished beef. Remember, fatty acids are found in fat, not in lean muscle tissue, and grass-fed and -finished beef is leaner than grain-finished beef.*

So what to do? Again, consult your own taste buds. As every scientist I talked to told me (and

it became rather monotonous), when it comes to omega-3 and -6, beef—regardless of how it's produced—is a fairly poor source. If you're after omega-3 and -6, you should be eating wild salmon.

And beta-carotene? Grass-fed and -finished beef is higher in beta-carotene than conventional beef, but again, it's a poor source compared with leafy greens, green vegetables, and carrots.

The Pre-Baby Animal

There's another aspect to consider when you talk about supplementing the grass-only diet of a meat animal with grains. Kimberly Vonnahme is an associate professor of animal sciences at North Dakota State University. She works in an emerging field known as fetal programming, which simply means looking at how the nutrition of a pregnant animal affects its offspring.

When we look at livestock production, we need to consider the entire life cycle of the animal, including the period before birth, because 30 percent to 40 percent of a meat animal's life is spent in its mother's uterus. All livestock producers know that the better care you take of Mom, the healthier her babies. So what she eats during her pregnancy is critical to their development—not just how healthy they will be, but also how their carcasses will grade, how much marbling they'll have, and how much meat they'll yield.

My work focuses on the best food for Mom. And we're starting to understand that to produce the best animals—programmed in the womb for health and good muscling (meat)—we have to make sure the mother receives adequate protein and other key nutrients during her pregnancy.

Every farmer and rancher I know wants to be sustainable. But environments vary, forages vary, and weather constantly changes. Factors like inadequate rainfall can impact the protein content in the forage. We're starting to understand that to make sure our cattle, sheep, and pigs get adequate protein during pregnancy, we need to provide supplements. If an animal is on low-protein pasture, we need to provide Mom with a good protein source like dried distiller's grains, or soy, or a higher-protein forage.

In recent studies by a colleague of mine, when supplementation was given to pregnant cows during their last trimester and their offspring were followed, their female calves had much better reproductive performance in later life, and their male calves produced meat that had better marbling. My point is that you cannot ignore environment. Yes, the genotype of an animal is important (you certainly don't want to raise a Holstein for meat!), but if you don't provide the best environment—including in the womb—that animal will never reach its potential.

Natural and Organic

Are "naturally raised beef" and "certified organic beef" the same thing? Not at all.

I'm So Natural

The USDA defines **natural** beef as a product that (1) must be minimally processed, (2) must not contain artificial ingredients, and (3) must not contain any preservatives. Under these guidelines, most beef from traditional feedlots would be termed natural. As might beef from any other method of production. It varies from producer to producer. To summarize,

in the United States there is no formal certification process for natural beef. In Canada, the Canadian Food Inspection Agency (CFIA) regulates the use of the words *nature, natural, Mother Nature,* and so on in labeling, but there is no "natural" certification program for meat.

Many natural beef producers have programs where their products are branded to help people recognize them. Each has its own specifications. For example:

- "Never-ever" programs (sounds like Peter Pan): This means no antibiotic use at all.
- "Not lately" programs: No antibiotic use within the last 100 days before slaughter.
- Ionophore use: yes or no.
- Growth-promoting hormones: yes or no.
- Feed types.

Obviously, verification and enforcement are a critical component of branded programs. These are generally handled by the branding company's own certifiers.

And I'm So Organic

To produce beef that can be labeled ORGANIC requires a different kind of verification process with very specific standards. In fact, the correct term is *certification*, which must be done by an arm's-length accredited certification organization. These can be found by doing an online search under "accredited organic certification."

Organic beef production requires more time, effort, and documentation than any other production system. The USDA requires that organic meat come from cattle born and raised on certified organic pasture (no chemical fertilizers or pesticides). They must never have been given antibiotics or growth hormones and must be fed only certified organic grasses *and grains* (my italics). Use of genetically modified (GMO) crops is prohibited. And all this must be verified via a production system that keeps records on every animal in the herd, including its breed, vet care, and more. However, under an organic raising system, cattle may be fattened in confinement areas.

Genotype, Phenotype

- **Genotype:** The genetic makeup—*not* the physical appearance—of an individual. Genotype determines the animal's hereditary potential.
- **Phenotype:** The entire characteristics of the animal (anatomy, behavior, appearance, and so on). Phenotype is based on the combination of an animal's genetic makeup and its environment. In other words: Phenotype = Genotype + Environment. Take a look at the breed list at the end of this chapter to learn about some phenotypes.

Beef Brands

Branded beef meets a set of specifications determined by the branding company and is identified by a specific brand (logo or wordmark) on the package, whereas a package of regular (unbranded) beef just has the name of the beef cut on its label. Every branded beef program is unique. In general, there are three categories:

- **Breed-specific:** Restricted to cattle from a specific breed (like "Certified Black Angus").
- **Company-specific:** Can involve any beef breed, but must meet specifically defined criteria such as grade, marbling, feed, and use/non-use of pesticides, antibiotics, and growth hormones.
- **Store-branded:** Some store chains brand their beef under criteria developed exclusively for that store chain.

Much of all this nomenclature is created by the brilliant minds of marketers who have come up with an impressive litany of ways to suggest that their beef is better, healthier, and more nutritious than anyone else's. What, exactly, does *naturally raised* mean? Heck, I'm naturally raised, and so are you.

HORMONES

Along with feeding and finishing methods, there are two more flashpoints in the debate over beef quality: use of hormones and antibiotics.

Language is an interesting thing. You can use words to create emotional responses. Like "hormone-free" beef. *There's no such thing. It's impossible.*

Why? Because all animals—like all human beings—produce natural hormones in their metabolisms. In fact, every multicellular critter (cattle, people, dogs, birds, vegetables, whatever) creates its own natural hormones. If you're worried about hormones in beef, what you need to look for are the terms *no added hormones* or *no hormones administered*.

What we're really talking about here is the administration of a growth hormone to stimulate weight gain in young cattle. There are several reasons for this. People who eat beef prefer (and have come to expect) tender meat. This means eating the meat of younger animals, because the meat of a younger animal is naturally more tender than that of an old animal (obvious, really). For this reason, beef cattle are usually slaughtered at a very young age (from about 18 months to two years). Most male cattle are neutered when they're young (a process that turns bulls into steers); female cattle (heifers) are spayed. Neutered steers and spayed heifers produce fewer natural hormones than older animals (same with humans), so small amounts of certain hormones are given to them to help them grow.

Ear Implants

Most cattle finished in the conventional way are given growth promotants via ear implants. These

Hormones in Food

A hormone is a chemical released in one part of the body that affects another part of the body. Hormones like **estradiol**, **progesterone**, and **testosterone** are naturally present in beef, pork, poultry, milk, eggs, and fish. Plant foods like potatoes and wheat contain significant levels of progesterone, and certain oils and plants like wheat also contain testosterone. Milk products provide about 80 percent of the progesterone, 30 to 40 percent of the testosterone, and 60 to 70 percent of the estrogens in our diet. Meat and fish provide about 5 percent of the progesterone, 20 to 30 percent of the testosterone, and 15 to 20 percent of the estrogens in our diet.

This cowboy is implanting the ear of a young steer (it doesn't hurt!).

are small pellets placed underneath the skin in the middle one-third of the ear—a place where there's no way they could accidentally be included in products intended for human consumption, and no risk of hormone residues entering the meat. The pellets dissolve gradually over 60 to about 120 days, and must be completely depleted before slaughter. Feedlot operators must verify that the pellet is "dead" at least 30 to 45 days before the animal is killed.

After the pellet is implanted, its active ingredients are slowly released into the animal's bloodstream, increasing its blood hormone level just enough to stimulate additional growth in muscle. Implanted animals grow faster, are leaner, and use feed more efficiently. Since the conventional cattle feeding business is defined by very narrow margins affected by seasonal swings in feed costs, the use of implants more than pays for itself in delivering larger, leaner cattle. Note this comment, however, from an implant information sheet: "All growth implants . . . show the greatest improvement in gains and efficiency on higher energy diets." So the type and quality of feed is critical.

Implants increase weight gain by 5 to 23 percent and improve feed efficiency from 3 to 11 percent.

Essentially, the animal grows larger faster, and so can be slaughtered younger. The hormones used are identical (either natural or synthetic versions) to the hormones cattle naturally produce. Today, both the United States and Canada have approved three natural hormones (estradiol, progesterone, and testosterone) and three synthetic hormones (zeranol, trenbolone, and melengesterol) to help cattle feed efficiently, bulk up faster, and develop leaner meat. Note that melengesterol is administered via feed, not implant.

Implants work by speeding muscle growth and reducing fat deposition. Comparing an implanted animal with an animal of the same weight that has not been implanted, the implanted animal will be leaner. But since beef grades are primarily based on the degree of marbling (measured as the amount of internal fat in the rib-eye muscle), implanted cattle will generally need to be fed longer, so that they have time to put on more muscle and marbling.

In the United States, hormone implants are regulated by the Food and Drug Administration. The Food Safety Inspection Service of the USDA tests for *synthetic* hormone residues in meat, but they don't test for natural hormone residues. It can't be done . . . animals produce natural hormones throughout their lives. It's simply not possible to differentiate between hormones occurring naturally, and those from implants. This means that to some extent, the use of growth-promoting hormones cannot be regulated.

Are Hormone Implants Safe? Part 1: The Politics

There are as many opinions about the safety of beef from implanted cattle as there are people.

In the late 1980s, the European Union banned the import of all hormone-treated meat from Canada and the United States, citing dangers to human health. This suggests that some sort of hormone-related risk from US beef caused the European community to press for the beef ban.

Let's shed a little more light on the issue. The EU ban was triggered by an early growth promotant called DES (diethylstilbestrol), one of the first hormones used in feedlots. When DES was proven to cause cancer, the United States banned it in chicken and lamb production in 1959 and in all cattle production in 1979.

The European concern was triggered *later*—in the 1980s—when DES was detected in baby food made with veal. The baby food was made in France from French cows treated with DES, and thus had nothing to do with the United States. The resulting uproar led several European countries to ban the use of all hormones in cattle, which effectively banned most North American beef.

The dispute over the use of hormones in cattle is long running. The United States and Canadian governments maintain that the hormones being used in meat production today are safe. The European Union claims they are not. So why is this issue such a hot political potato?

Many economists and scientists believe that the EU ban was deliberately protectionist, and

that beef was singled out because Europe has historically depended on beef from other countries—particularly the United States and Brazil. Why? Because there's not much good grazing pasture in Europe. To protect the "inefficient" beef markets of the EU, the ban was created and received generous support from European consumers despite the lack of conclusive evidence.

Arbitrating the matter is the World Trade Organization (WTO), which rules on international trade between nations. The process by which rulings are arrived at is complex, involving hearings from both sides of the issue, expert/scientific testimony, drafting of an interim report, review, a final report, and ultimately a ruling, which becomes international law. If one side breaks the terms of the agreement, the penalty generally involves some sort of trade sanction.

In 1997, the World Trade Organization sided with the United States and Canada, saying that the EU's health claims weren't scientifically justified, and that the hormones used were unlikely to pose a hazard to human health, *if* good animal husbandry was practiced. However, the ruling still stands.

To help alleviate the situation (which was hurting both North America and Europe), in early 2012 the European Union offered concessions to partially defuse this ongoing trade war. The deal allowed the EU to retain its ban on imports of hormone-treated beef in return for an increase in its import quota for non-hormone-treated beef from the United States and Canada. In return, the US and Canada lifted import tariffs on a range of European farm produce like cheeses, chocolate, truffles, and other specialty products.

Are Hormone Implants Safe? Part 2: The Science

Science, like opinion, changes. But opinion is subjective and science, ideally, is objective. It's just that as new information or findings emerge, they are incorporated into scientific research, and so what we know today is colored by and rests upon what we knew yesterday.

So for every report that concludes that the use of hormones in beef production poses a threat to children, another report points out that our bodies continually make the exact same hormones that the EU nations have banned in meat, and that we eat eggs and butter, which contain the same hormones in much higher concentrations than in beef.

An experiment by Iowa State University found that growth-promoting hormones decrease the land required to produce a pound of beef by two-thirds, and reduce greenhouse gas emissions nearly 40 percent per pound of beef compared with cattle never implanted with hormones. That seems like a big check in the positive column for hormone use.

However, other studies have focused on the negative environmental impacts of hormone residues in cattle manure. When manure from feedlots runs off into the surrounding environment, these hormones can contaminate surface water and groundwater. Multiple studies have shown that children are particularly sensitive to these kinds of hormones, even at very tiny levels. Even the USDA has commented on the danger of hormone residues flushed into groundwater. But since cattle produce hormones naturally, this may be more of an issue of manure management than implant residues—it's very difficult to isolate which hormone residues are the result of implants and which are not.

I'm not a scientist. I'm a butcher. All I can do is try to present the most factually accurate information out there. One expert I spoke with is Marcia Herman-Giddens, adjunct professor at the University of North Carolina School of Public Health, Chapel Hill. In the late 1990s, she was the lead author of a seminal study on early puberty in girls, in which she expressed her concern about the effect of hormones and other endocrine-affecting chemicals in our environment. Herman-Giddens told me this:

When I wrote that study, there was very little data on hormone implants in cattle, although the doses were not inconsequential. To me, the use of hormone implants is still an open question, and a complex one. Effects, if any, could depend on the amount of meat consumed, interactions with other substances, and possible cumulative effects of similar exposures. I believe the reasons for early puberty involve a multiplicity of causes.

My generation did not grow up in a chemical-infused world. Now it's all around us . . . in plastics like BPA, in herbicides and pesticides, even in dental sealants on children's teeth. We are living in a sea of endocrine-disrupting chemicals. So it's only natural that, depending on lifestyle, some people will encounter more exposure than others. It's not just one thing—it's cumulative.

In my opinion, the use of hormones to stimulate cattle growth is still a significant issue for the public. I believe that what's needed is open, independently funded research to avoid public perception that the way the beef industry is framing this issue suggests that there's something to hide. So for me, the jury's still out.

In a 2009 report written for the Canadian Institute for Environmental Law and Policy, analyst Susan Holtz noted the dramatic increase in drugs used annually in Canadian animal husbandry—particularly in intensive (read: feedlot) operations. She pointed out that much support for the EU ban on growth promotants had to do with European countries' history of embracing traditional values of livestock raising, as well as support for local and artisanal food and the heritage value of agricultural landscapes. In a phone call, she said that while the use of hormone implants may be safe, she is very concerned about the effect of antibiotic use.

And to conclude: The assessment that administering hormones to beef cattle is a safe practice has been endorsed by the Food and Drug Administration of the United States, Health Canada, the Codex Alimentarius Committee of the World Trade Organization, the Food and Agriculture Organization of the United Nations, the World Health Organization, and the European Agriculture Commission.

Help, I'm Eating Hormones!

Since estrogen, progesterone, and testosterone are naturally produced by the bodies of animals, including people and cattle, should you be worried about additional exposure via cattle ear implants? I say no, and here's why.

Let's take a look at estrogen. The normal production of human estrogen per day is about 4,000 nanograms (1 nanogram = 1 billionth gram; 1 gram = ¼₅₄th pound) in boys, a little higher in girls, about 100,000 to 150,000 nanograms in men, from 5 to 15 million nanograms in nonpregnant women, and from 4 to 64 million nanograms (or more) in pregnant women.

A 6-ounce serving of beef from a beef animal that has *not* been implanted contains about 2.6 nanograms; the same serving from an animal that *has* been implanted will contain about 3.8 nanograms of estrogen. Exact amounts will vary, but remain around these mean numbers.

Looking at some other foods:

8 fluid ounces (one glass) of milk = about 30 ng
300 peas = about 330
3 ounces of cabbage = about 2,000 ng
3 ounces of soybean oil = about 1,680,000 ng

And medications: One human birth control pill contains the same amount of estrogen as 105,000 pounds of beef from implanted steers.

By the way, the administration of growth hormones to pigs and chickens is forbidden in both the United States and Canada.

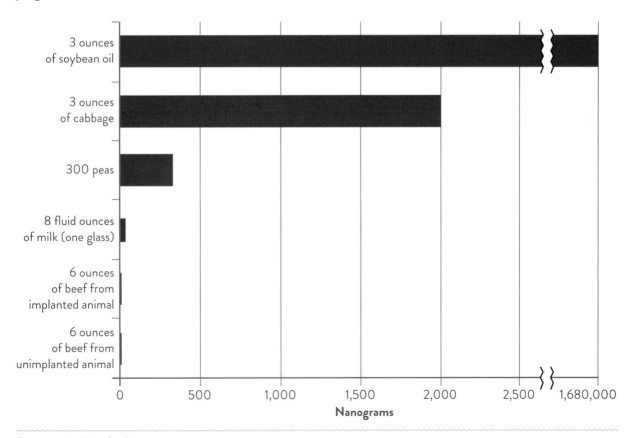

Estrogen levels in food.

ANTIBIOTICS

Giving antibiotics to beef cattle is a very hot-button issue, and rightfully so. The development of antibiotics is one of the major medical breakthroughs of the 20th century, saving the lives of millions. Antibiotics were a medical key to the eradication of diseases that had plagued humankind for thousands of years. It was easy to assume that they would always be part of the modern medical tool kit, ready for deployment against whatever nastiness came our way.

Increasingly, though, resistant strains of bacteria are emerging, confounding doctors and killing patients. How does this happen?

Antibiotic resistance occurs when bacteria change in a way that reduces the effectiveness of the very drugs intended to destroy them. Bacteria are living organisms, and despite the fact that they have no brains they are very fast adaptors. So when you (for example) don't complete your 10-day antibiotic regimen, some bacteria will be left inside you that have been exposed to the antibiotic, but have not yet been killed. And you know the old saying, What doesn't kill us makes us stronger. The resistance is now passed along to the next generation of bacteria.

Half of all staph infections that occur in the United States these days are caused by bacteria that are resistant to penicillin, methicillin, tetracycline, and erythromycin, leaving very few drugs left to fight the good fight.

Scientific studies show that antibiotic misuse is tied to the emergence of resistant bacteria. Problematic uses include the wrong diagnosis by your doctor, the increasing popularity of "antibacterial" household cleaning products, the unnecessary prescribing of antibiotics for illnesses that do not respond to antibiotics, and—many say—the sub-therapeutic use of antibiotics in livestock feed for growth promotion.

Ionophores are a class of antibiotics used in animal feed. They are antimicrobial feed additives that depress "bad" digestive bacteria in a meat animal's gut, increasing digestive efficiency so they grow faster. Ionophores are popular because even a tiny increase in growth rate makes a difference in this highly competitive industry—where pennies are the measure of profit.

According to veterinarian Dr. Gail Hansen of the Pew Charitable Trusts, ionophores aren't an issue, since they're not used in human medicine and don't appear to interact with human antibiotics. However, she points out that "when you remove ionophores from the equation and look only at antibiotics that can be administered to both humans and animals, you'll find that about 70 percent are used in the raising of food animals."

Lobbyists for the conventional meat sector claim that of all antibiotics sold for animal use, only about 15 percent of the total amount are used to promote growth while the remainder are used to target an identified pathogen and treat the disease it causes—uses considered therapeutic by the FDA.

Regarding this statement, Dr. Hansen responds:

It would be nice to have verifiable figures. The FDA allows antibiotics to be used in food animals for the prevention and treatment of disease (together termed therapeutic use) as well as for growth promotion. What's glossed over is a significant gray area in the interpretation of the word prevention. *Levels of antibiotics given for prevention are often the same as levels given for growth promotion. Now, if an antibiotic is used at a level not designed to treat a disease [kill the bacteria], there's a much higher likelihood that resistant bacteria will develop. In human medicine, the use of antibiotics for disease prevention is rarely effective, and is not recommended. I have yet to see solid scientific evidence to show this would be different in any other creature.*

The only tracking of antibiotics in food animals is done by the FDA through the most recent Animal Drug User Fee Act (ADUFA). The system records only the sales of drugs by total weight, with little or no oversight of how those drugs are actually used. Not much use.

So is this issue of concern to scientists? You bet it is, to the tune of literally hundreds of studies investigating links between therapeutic and sub-therapeutic use of antibiotics in meat animals. And to my surprise, the vast majority of those studies have not yet found a clear association between feed-grade antimicrobial use and resistance in humans. That doesn't mean there is no connection; it just means that it's very difficult to determine and has not, to date, been proven conclusively by science.

Part of the challenge comes from the fact that antibiotic substances also occur naturally in nature. In fact, natural antibiotics (those not created by humans) and resistant genes occur in nature in a stunning variety. As basic tools of life's battle against bacteria, they are extremely common and mutate naturally. So resistant bacteria can be found both in conventional and organic production systems—as well as in antibiotic-free production systems.

The World Organization for Animal Health (OIE) represents more than 178 countries. The organization monitors animal disease, shares veterinary information, and develops health standards for the international trade in animals. In March 2013, the OIE held a conference on "the responsible and prudent use of antimicrobial agents for animals." Among its recommendations were the following:

- That data be collected on antimicrobial agents used in food-producing animals (including through medicated feed) in order to create a global database hosted by the OIE.
- That veterinary education include knowledge of antimicrobial resistance, as well as codes of good veterinary practices for the responsible use of antimicrobial agents in animals.
- That good agricultural practices be promoted to minimize the development and spread of antimicrobial resistance.
- That research to improve understanding of current antimicrobial agents be supported, and that

new molecules and other alternatives be developed to replace the use of antimicrobial agents.
- That current use of antimicrobial agents not intended to combat animal diseases be carefully assessed and evaluated.

What I've presented here is just a fraction of my research. The issue of resistant bacteria is phenomenally complex, sitting at the center of a sea of vectors that include veterinary practices, human use and misuse, the drive for profit in the meat sector, and our overreliance on substances we have invested with almost magical properties. Hey, if you're not in an operating theater, you probably don't need antibacterial soap!

Overall, I've concluded that although there may be some sort of relationship between the use of antibiotics in food animals and the increase in multi-drug-resistant bacteria, before a confirming statement can be accepted the discussion must incorporate the effects of the overuse of antibiotics in all areas, not just animal husbandry. So until science can deliver a fully researched, factually based, and robust decision, the jury is still (mostly) out. And yes, this *is* a nuanced statement.

The nub of the issue is not the use of antimicrobials for *true* therapeutic reasons—to treat a sick animal—but their use for nonexistent problems such as "preventing disease." And I'm also concerned with the meat sector's reliance on feed additive manufacturers to tell them what level constitutes safe "sub-therapeutic" use.

If you are concerned about antibiotics in meat, seek out meat that carries a certified organic label, or question the farmer you are buying your meat from.

So Where, Exactly, Does This Leave Us?

Follow your conscience and personal preferences. Follow your taste buds. Find and patronize an ethical producer, who raises her or his animals

The beauty of a well-raised, healthy animal: a Highland steer on the island of Mull, Scotland.

Name That Cut

Before 1973, there were over 3,000 different names for basic beef, pork, and lamb meat cuts. In 1973, nomenclature was standardized, and the number reduced to about 300 names.

gently on the land and makes certain that they are slaughtered in a humane fashion.

First things first—make certain that the animal was slaughtered humanely. I hate cruelty to animals and so should every farmer, chef, and consumer. But this goes beyond issues of animal welfare. There is a direct relationship between the way an animal is slaughtered and the taste and quality of its meat.

You have invested your best efforts in finding the most humane and ethical farmer. You have verified the breed and its characteristics. You have checked out how it was slaughtered, and found out how and for how long it was aged after slaughter.

Do this research and you'll be rewarded with an excellent product—one that's probably far superior in quality and taste to supermarket beef. Plus, you'll have kept your food dollars in your local community.

I'll give the last word to animal science professor Kimberly Vonnahme:

When we think about how animals are raised, we need to keep in mind that animals are adaptable. And I'm basically a capitalist—I'm all for anybody who wants to make it in agriculture. My issue is when people don't tell the truth, or put the other product down. There's no way we can feed all the people on the planet without all forms of livestock production—how else are we going to provide high-protein products that people need and plants can't fully provide?

BEEF CUTS FROM TAIL TO NOSE

No matter whether you're a farmer or homesteader who raises your own beef cattle, or a chef who cooks and serves beef as a regular part of the menu, or a beef-lover who buys a side of beef from a local farmer to cut yourself, or a consumer who believes (and I stress *believes*) that your budget only allows you to buy supermarket meats, you ought to know where those beef cuts come from. Knowledge is power, and a little knowledge helps a lot in your final enjoyment of the meat you've invested in.

At my workshops, people always ask me about tenderness. Well, scientifically speaking, beef tenderness is determined by genetics (breed), how the animal was raised and aged, and how the meat was cooked (can't always blame that leather-like steak on the animal—maybe it was the chef).

But tenderness also varies from one cut of beef to another in the same carcass. It's obvious that not all parts of a beef carcass will be equally tender. As discussed in chapter 3, there's a very simple way to understand carcass tenderness. It involves two basic principles. The first relates to activity and diet, and the second relates to location. Let's begin a detailed examination of a carcass, and how to apply these two principles to various cuts of beef will become clear.

BOSTON AND CAMBRIDGE:
SAVENOR'S BUTCHER & MARKET

If you live anywhere near Boston, you'll know Savenor's. It's been top of the heap ever since it was founded by Abraham and Dora Savenor in 1939. The couple handed down the then grocery store to their son Jack, who expanded into meats, exotic game, fresh seafood, and specialty foods. Some of you may remember Jack Savenor as Julia Child's favorite butcher.

Today, Savenor's is in the hands of the third generation. Ron Savenor runs the operation, which has further expanded to two retail stores and a wholesale operation. Ron says:

Sometimes I feel like the last of a dying breed. This industry has changed so much over the past 25 to

30 years that it's amazing. Twenty-five years ago there was no such thing as boxed beef. Everything came in refrigerated trucks as hanging carcasses or primal cuts. Then we'd cut it down into retail cuts. Our butchers were skilled craftspeople . . . they had to be. A butcher knew every muscle in the animal: how to cut it and how to cook it.

Although they don't know it, most people are eating boxed beef, because that's how it comes into supermarkets and most stores. I'm not a fan, to put it mildly. I simply don't think it tastes great, and I don't like the idea of meat sitting in its own juice. What you want to do is gradually reduce *the moisture, which breaks down the enzymes in the*

Courtesy Ron Savenor.

Ron Savenor.

meat and enhances flavor. This is called dry-aging, and it's all we sell at Savenor's.

I am extremely picky about what we sell. Our beef is USDA-certified prime or choice. Recently, I've been working to find the very best local grass-fed or pastured meat from farmers I know. I don't leave this to chance; all Savenor's farmers sign an agreement setting out the criteria we insist on: raising methods, slaughtering, quality, and so on. My requirements are very specific—exactly how that animal must be cut. I want everything— the cheeks, the tongues, the kidneys. I'm trying to get as direct from the farm to the table as I can, and I want to develop a reputation as someone who supports local farmers. But the end product must be excellent—that's all that matters. How does it taste?

Meat is great stuff, and people come to Savenor's expecting the best. Our customer base is centered on the metro Boston area, where Savenor's has become the place you go to on the weekend for a special outing. We do a little online business, but holidays only. And our wholesale business—chefs and restaurants—is Boston only.

You're only as good as your staff, and finding qualified butchers is a real challenge. It's painful . . . we advertise, look for someone with passion. But generally, we have to train them ourselves. Most of our butchers used to be chefs or kitchen workers, or just people who were fascinated by the art of butchery.

Why do I do this? It's fun to sell the best food money can buy. People come in and love you for it. I've met some of the coolest people here—people who are passionate about food. I just love it.

Before we start the tail-to-head discussion of a side of beef, I have to be honest: Butchery terms can be rather confusing, even for butchers. Some cuts of beef can be called by more than one name, and some terms, such as *sirloin*, can be used in multiple ways. I'm truly sorry, but that's just the way it is.

And one more point (because I know you'll want to call me about it): Like life, butchery has options. You can go one way or another, but not both. Or, to use a simpler analogy, give a cashier a dollar and ask for change. You may get 10 dimes, or you may get 4 quarters, or you may get 2 quarters, 4 dimes, and 2 nickels. My point is that there are several ways to cut up that dollar.

Same goes for carcasses. A particular chunk of meat may make a nice roast, but it might make equally nice steaks. You can have either one or the other, just not both at the same time. So as you read through the next section, keep this in mind, especially when you get to the loin. And if you get confused, just consult the attached CD, which explains the different loin options with pictures . . . and contains detailed photo sequences that will demonstrate what I explain below.

Meet the Back End: The Round

Any cut of meat with the word *round* in it originates from the very back end of the animal: the buttocks, the rump. This part of the animal does a considerable amount of work moving the animal around. Which means more exercise, which means—you got it—muscles. So you can reasonably expect cuts involving the back end to be tougher. And they are.

Parts of the Round

The round can be further subdivided into the top round, bottom round, eye of the round, and knuckle or round tip—sometimes also called the sirloin tip (although it has nothing to do with sirloin).

Top round does well as oven roast or rotisserie roast, but should be cooked rare and sliced thinly. This will increase its tenderness. It also does pretty well as London broil, but again, cooked rare and sliced thinly. Top round is also well suited for steak tartare or kibbeh (also spelled kibbe)—a raw meat and bulgur dish popular in

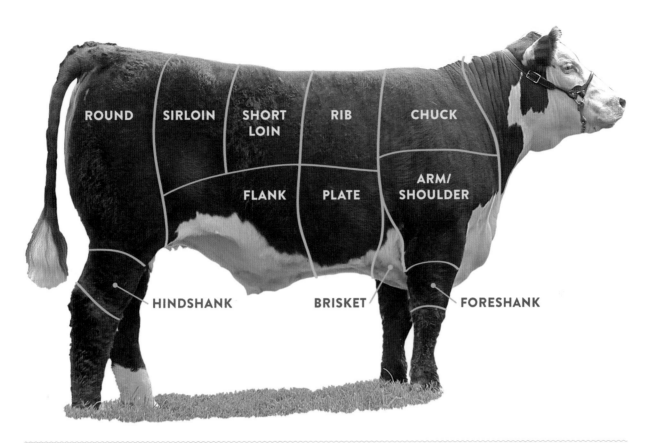

ROUND   SIRLOIN   SHORT LOIN   RIB   CHUCK

FLANK   PLATE   ARM/SHOULDER

HINDSHANK   BRISKET   FORESHANK

Courtesy American Hereford Association

the Middle East). It's also good sliced thin and pounded for rouladen (a German meat dish involving wrapping a thin slice of meat around a stuffing, then braising it) or braciole (an Italian dish similar to rouladen).

Bottom round is best suited for slow moisture cooking or braising, because it's a bit tougher and stringier. It also works well as beef jerky. Gets those jaws working.

Eye of the round has never impressed me. I think it makes a lousy steak. In the old days, we used it for cube steak. If you slice it really thin and pound the heck out of it between sheets of waxed paper, it does make a good sandwich steak.

The knuckle/round tip does fairly well as grilling steak. However, I would recommend marinating it first, then cooking only to rare or medium rare. It also makes a great rotisserie roast.

Thanksgiving Beef

I don't want to spend all day with a turkey (other than my relatives), so instead, this year it's going to be *roast beef*. Easiest thing in the world to make.

1. Take a standing rib roast of from three to six ribs, depending on a) how many people you're feeding or b) how many yummy leftovers you want.
2. Place into 350°F (177°C) oven.
3. Cook for 22 to 25 minutes per pound.
4. Take out of the oven and rest for 5 to 10 minutes.

Ta da! Done.

Steik It to Me

The word *steak* comes from *steik* (roast meat) and *steikja* (to roast on a spit). So what about all those fancy steak names like *filet mignon*?

- **Chateaubriand** is a thick steak created for François-René de Chateaubriand, who was an ambassador under Napoleon and secretary of state to Louis XVIII.
- **London broil** is not a cut, but a recipe that involves broiling marinated flank steak or other cuts such as top round or shoulder steak (in the United States), then cutting it across the grain into thin strips. London broil is completely unknown in London, England.
- **Filet mignon** comes from the French words *filet*, meaning "boneless meat," and *mignon*, meaning "small."
- The **porterhouse** steak was served at coach stops, or porterhouses, in the 19th century. T-bone and porterhouse are almost identical. Both contain a T-shaped bone with meat on each side. The larger section of meat by itself is a strip steak; the smaller is a tenderloin steak. Porterhouse steaks include *more* tenderloin; T-bone steaks *less*. How much of each? Well, this is actually defined by the USDA, but it involves measuring (and it's a little embarrassing to be seen measuring your meat in a restaurant—better just to trust them). Because there's more tenderloin in a porterhouse steak, it usually costs more than a T-bone.
- And **sirloin** comes from the old French word *sur-longe* (upper part of the loin). You may have heard the story that Henry VIII knighted a cut of beef he was about to enjoy by naming it "Sir Loin," but that's not true. Good story though.

The Loin

Now everyone take a step to the right to meet the loin primal. The loin consists of the tenderloin, the sirloin, the New York strip, and the tri-tip. Loin cuts start at the point where the round ends and continue to the right on the diagram, ending where it meets the rib primal. A steer has 13 ribs. If you count them starting from the head, ribs 1 through 12 are considered part of the forequarter primal, but the 13th rib is part of the loin primal (*don't ask why*).

Parts of the Loin

The tenderloin (or filet mignon) is a long, roughly cone-shaped muscle that is situated inside the loin and runs horizontally on each side of the backbone. The tenderloin runs nearly the full length of the loin. If you want the entire tenderloin, it must be removed *before* separating the loin into the sirloin and the short loin. Otherwise, you end up with a sirloin that contains one portion of the tenderloin meat, and a short loin with the other portion (this is what's shown on the CD).

So this is your first decision—do you want the whole tenderloin or not? Many people do, since it's the most tender cut in the animal. That's because it's a support muscle for the spine, so it doesn't get the same kind of vigorous exercise that leg and shoulder muscles do. In my opinion, it's not the most flavorful cut, but it's definitely the most tender.

The sirloin, top butt, or hip (all names used for this subprimal) is what I call the Beginning of the High-End Steaks (*putting on my tux here*). This is one of those cuts you can subdivide in more than one way. You can make steaks by slicing the sirloin bone-in with a saw, or you can debone the sirloin first and then slice it. These steaks are fabulous on the grill "as is" (no need to marinate), and deliver unique flavor all their own. Or instead of steaks, you can use this cut (boneless) to make a nice oven roast, known around Vermont as a spoon roast.

Another option you have is to remove the cap, which is hiding on top of the sirloin. To find it, turn the sirloin fat-side up and look at its larger face. On top, there's a piece of meat just held on by connective tissue and some fat. This is the cap.

Removal of the cap is shown in detail on the CD. Once the cap is removed, the solid piece left is the heart of the sirloin.

The heart of the sirloin is used to create restaurant-style sirloin steak—sliced about 2 inches thick and cut into portions about 3½ inches square. The cap meat itself can be cut into cubes and used for kebab meat, or sliced ⅛ to ¼ inch thick and threaded on skewers for quick grilling in your favorite marinade or grilling sauce.

There's one more bit: When you remove the tail end of the sirloin in one piece, it's known as the triangle tip or tri-tip roast. Awesome on a grill.

The short loin is also what I call an "either/or" cut, depending on whether or not you remove the portion of the tenderloin that is contained in the short loin. If you do not remove the tenderloin, you can use a saw to slice the short loin into porterhouse and T-bone steaks, each including a chunk of tenderloin. The porterhouse steaks will contain more of the tenderloin than the T-bone. Why? Because the tenderloin tapers, kinda like a very long ice cream cone, as it gets closer to the T-bone end of the loin. So when steaks are cut from a short loin that still contains the tenderloin, on average you'll get approximately seven porterhouse and seven T-bone steaks about 1 to 1¼ inches thick. This may vary from animal to animal.

But if the tenderloin *was* removed, then you'd have tenderloin steaks or roast. Removing the muscle meat from the opposite side of the short loin would then give you boneless New York strip steaks. Or you can leave this meat attached to the bone and slice it bone-in as bone-in New York strip steaks.

Don't freak out. All will become clear on the attached CD, where I show in detail how to remove a portion of the tenderloin from the short loin and how to cut bone-in New York strip steaks.

So—to recap and, hopefully, unconfuse you— from a single short loin you can have *either* porterhouse and T-bone steaks *or* New York strip steak and tenderloin, but not both. Unless you opt to have the short loin done half and half—that is, slice

Cole's Notes

What does *tri-tip* mean to you? To me, it means California. That's where I learned about it, and that's where it's really popular. Here in Vermont, when someone asks for a triangle tip roast, I usually say, "What part of California are you from?"

Tri-tip is actually the tail of the sirloin. In California, they take the tail off the sirloin *before* they slice it so it comes off in one piece and it's shaped like a triangle. Tada! Tri-tip roast. A fabulous cut of meat.

Lots of people ask me to explain the difference between steaks. I get it—it's confusing. Some of that's because names change over time, and many cuts of meat have several different names (tied to different geographic areas). Typical high-end steaks are the New York strip, the filet mignon, and the rib eye. Then you drop down to the restaurant-style sirloin.

New York strip and filet mignon come from the loin of the animal . . . in fact, from the same piece of meat. The difference is that the New York strip is cut from the *top* of the loin and the filet mignon from *inside* the loin. Since the inside muscle is particularly inactive, the filet mignon will be the most tender. In fact, it's the tenderest cut on the animal. But the New York strip delivers better flavor. It's a matter of personal preference.

Rule of thumb when ordering: Anything from the animal's loin and ribs will be a tender piece of meat because these muscles work the least.

The term *face* always refers to the exposed portion of a cut of meat. For butchers, *face* is also a verb. When I face the exposed surface of a piece of meat, I trim it with my knife so that it's smooth and even. This is also called giving it a nice face.

off seven T-bone *or* seven porterhouse steaks— then remove what's left of the tenderloin and the New York strip.

Roast Advice

- **Is my roast too rare?** Probably not, provided that the internal temperature of the meat is held at 145°F (63°C) for about 35 minutes to kill harmful bacteria. At 150°F (66°C), bacteria will be killed in one minute; at 165°F (74°C), in about 10 seconds. Rare meat is safe as long as it's cooked correctly.
- **Bone in or bone out?** This may surprise you, but whether the bone is attached will not affect the taste. So again—personal taste and convenience. No bone = easier to carve. But ah, the joy of gnawing on those bones. That caveman thrill. That feel of grease on the face. The looks of horror from the other diners. Isn't it all worth it?

You'll Love Me for This: Flat Iron Roast

1. Take four to five well-trimmed flat iron steaks.
2. Lay one on top of the other till you have a nice stack-o'-steak.
3. Tie it all together *tightly* with the strings no more than 1 inch apart.
4. Trim each end just a teeny bit so everything's nice and even (no bits hanging out).
5. Season with kosher salt and black pepper. (I use kosher salt because it is light and flaky. It sits on the meat surface and draws the juices to the surface, but it doesn't absorb them. Regular salt gets into the nooks and crannies of the meat and absorbs moisture, which can make the meat dry.)
6. Cook uncovered in a 400°F (204°C) oven for about 40 minutes.
7. Remove and let it rest for 5 to 10 minutes.

This is better than prime rib.

The Rib

Another step to the right. We've moved a little farther toward the head of the animal. If you look at a beef cow standing in a pasture, it's obvious that the rib section of the animal doesn't work as hard as parts like the legs, neck, and buttocks. Ah, tenderness.

This is one reason why rib is one of the most popular cuts—think rib-eye or Delmonico steak, or standing rib, or boneless prime rib roast—all best suited for uncovered dry cooking (grilling, broiling, or oven roasting).

The Chuck

Take another step, and just to the right of the rib is the chuck primal. The chuck primal contains the shoulder, the chuck, the brisket, and the foreshank. The chuck is the portion closer to the spine of the animal; the shoulder is the lower portion. Below is your brisket and your foreshank. Remove the brisket and foreshank from this primal and you are left with the square-cut chuck. The square-cut chuck contains the shoulder and the chuck (yes, I know—chuck and chuck). On our diagram, the vertical line between the rib and the chuck marks the face of both pieces (the face of the chuck is to the left; the face of the rib to the right).

The chuck is very versatile—let me count some of the ways. The center part of a muscle is called the eye; thus, the first 4 or 5 inches of center meat extending into the muscle from the face of the chuck is called the chuck eye. This is the very center of the face of the chuck. You can separate the chuck eye from the rest of the chuck by first cutting a 4- to 5-inch-thick slice off the face of the chuck. Then you can separate out the eye from that slice by releasing the connective tissue that holds it in place (this is shown on the CD), a process called seaming the eye. Slice the eye in half and you'll have two very nice chuck-eye steaks, also known as chuck-eye Delmonico or poor man's rib eye. I like poor man's rib eye because it defines the cut very well. It's much less expensive than a rib eye, but since it's the very next cut to the rib eye, it shares some of the rib eye's qualities. Thus, it's quite tender and extremely flavorful.

The rest of the chuck may be sold bone-in (as I like it), although most supermarkets sell it boneless. (Supermarkets buy cases of vacuum-packed boneless chuck rolls, as they're called.)

The chuck makes a terrific pot roast—slow cooked with moisture—or the very best, tenderest juicy stewing beef. Even a chuck steak can be fairly tender if you marinate it.

When the chuck is done boneless, you can also get my third favorite steak—the flat iron. This comes from the inside of the shoulder blade, so its movement is limited. This lack of movement makes flat iron the third most tender piece of meat on the animal (next to the tenderloin and the teres major)—but with much more flavor. Flat iron is an awesome grilling steak, but is also incredible broiled or pan-seared. Note that the flat iron is removed while still in square-cut chuck form, as it comes from the shoulder blade.

Finally, underneath the flat side of the shoulder blade lies another cut called the teres major, or petite tender—about which, until recently, I knew very little. In fact, in 46 years as a butcher, I had never eaten a teres major. Couldn't even pronounce it. But recently, a chef friend of mine cooked one up and I was blown away. This was one of the most incredible pieces of meat I'd ever eaten—more tender than a flat iron, with a different texture, flavor, and mouth-feel. So I am going to consider the teres major the second most tender piece of meat in a beef animal, and the flat iron the third most tender. Teres major has moved up to Position 2 as my second favorite steak.

The Shoulder

Just below the chuck primal cut, where the front upper leg of the animal begins, lies the shoulder.

The shoulder makes my very favorite pot roast when sliced with the bone left in. When I first

Making Sure Your Burger Won't Kill You

Most bacteria live on the surface of meat. When meat is ground, the surface bacteria are ground up and mixed throughout the product. To prevent issues (yeah, it's those issues that will get you), be sure to cook ground meat to an internal temperature of at least 165°F (74°C) (this includes meat loaf).

started cutting, it was called the arm pot roast. It's generally sold in supermarkets, boneless, as a shoulder pot roast, or sliced for shoulder London broil or shoulder steaks. The shoulder also gives you the leanest and meatiest short ribs for braising, as well as very nice stewing beef.

The cuts I've described above are the most commonly known—the ones the average consumer is probably most familiar with. But there are other cuts, too, that just aren't offered for sale anymore, such as an arm pot roast or a neck roast.

A final note: Both the chuck and the shoulder produce the best grade of ground beef, providing the fat content doesn't exceed 15 to 25 percent.

Okay—let's back up a little. I want you to move all the way left—back to the ass-end of the animal where we started, but closer to the ground (you may need to crouch a little). We're going to start at the feet and move right to the belly, then the front.

The Hindshank (or the Bottom of the Ass-End of the Beef)

Back to the left end of the steer. The part of the leg below the round is known as the shank; in this case, it's the hindshank. The shank should never be ground into burger. Do not do it! Why? It's full of cartilage and tendons that turn hard as bone when the meat is cooked quickly (such as burgers). Instead, the shank should be sliced on a power saw approximately 2 inches thick, then slow moisture cooked. There are many excellent recipes out there for beef shanks: osso bucco, peposo notturno—or soups. Note that when cooked properly, the cartilage and tendons transform into gelatin, which itself can be very tasty.

The Flank

Now let's move up the leg where it curves toward the belly of the animal. This is the flank. The popular flank steak is buried in this section; it can be grilled as it is or marinated. It's very flavorful and pretty tender, providing it's sliced properly across the grain, fairly thinly, and at a bit of an angle.

The flank also contains flap meat. This meat used to be used primarily for ground meat, but now is often marketed in supermarkets as sirloin tips. (See how nomenclature can fool you?) Of course, it's not sirloin. It's quite flavorful, but it should be marinated for better tenderness. Flap meat is usually cut into strips about 2 inches wide and 1½ to 2 inches thick.

The Plate

Now we move to the belly of the animal and closer to the head. From the belly up to the rib is called the plate. The plate is typically used for plate short ribs. Approximately three strips 3 to 4 inches wide are sliced on a power saw, then each short rib is separated by knife and trimmed of excess exterior fat. These short ribs are fattier than those from the shoulder, but they're very flavorful and, in my opinion, just as good.

The Brisket

Moving forward a bit more to the head . . . below the shoulder and just before the front or foreshank (front leg), we have the brisket, sometimes called the breast. Those who know it, love it. Brisket can be brined for corned beef, smoked (ever had Montréal smoked meat?), or cooked as fresh brisket. This cut is recommended for slow, long moisture cooking, often coated with marinade and wrapped in aluminum foil. Ask any Jewish mother for her brisket recipe . . . it'll knock your socks off. I once had a brisket that had been slow-cooked in a low oven for 20 hours. Wow.

As for which cut of the brisket is best, it's a matter of taste. You can ask for the tip (also called point cut), which will be slightly fattier. Or you can ask for the flat cut, which is leaner. Your cooking method will often determine which cut is best for the final dish, as well as length of time in the oven. Leaner cut = more cooking time.

The Foreshank

The foreshank should be treated the same as the hindshank. Braised.

Offal

Let's not forget the **offal**. This term comes from "off-fall"—items that fall off the animal carcass when it's butchered. Offal are parts of the animals that are not skeletal muscle, and generally include things like the heart, lungs, liver, abdominal organs, and extremities like head, tail, feet, brains, and tongue. Other terms for offal you may see are *organ meats* and *variety meats*. (In poultry, offal is called giblets.) Offal also includes sweetbreads, which in butchery does not refer to baked goods! There are two types of sweetbread: stomach or belly (the animal's pancreas) and neck or throat (the animal's thymus gland).

These are best when used as fresh as possible. And don't tell me that offal is awful, because it isn't. Properly cooked, tongue is delicious—I have a friend who remembers loving it as a child. She used to beg her mother to give her the very tip.

'Nuff said. My point is that if we're going to eat an animal, we ought to honor it by consuming all of it.

Brisket à la Karen

Okay, Passover. For those seders with a zillion people talking at once and the women arguing about whose gefilte fish is whiter and whose matzo balls are lighter, here's a nice brisket that should quiet things down, at least for 30 seconds.

You want the fattier part of the brisket, which should be as large as possible (all those relatives, plus leftover cold brisket sandwiches, right?).

Place the brisket onto a doubled heavy-duty length of aluminum foil. Rub with powdered mustard, and then coat with (1) Lipton's onion soup mix (several packages, depending on the amount of meat); (2) lots of ketchup; and (3) as desired, a little Worcestershire sauce plus chopped onions (scant). Some garlic salt couldn't hurt.

Wrap everything up *tight* and place into a low oven (250 to 275°F, or 121–135°C) for hours. Hours. Cannot overcook this.

When ready to serve, release the steam and let it sit, wrapped, for about 15 to 20 minutes. Then open the package of delight and slice against the grain.

Mazel tov.

Rocky Mountain Oysters— for Intrepid Meat-Eaters

Rocky Mountain oysters aren't seafood, they're the testicles from bull calves. Other terms for this delicacy are *prairie oysters*, *Montana tendergroins*, *cowboy caviar*, and *swinging beef*. Assuming you can find a source for calf testicles (not bull testicles!), here's how to cook them: Peel off the skin, marinate in beer for about 2 hours, bread with a cornmeal/flour/egg mixture, then deep-fry till done (they'll rise to the surface). You'll either love 'em or hate 'em; I've never tasted them.

To get you thinking creatively, here are a few easy recipes . . .

Kidney

I know, you're thinking, *Like, I'm ever going to eat this?* Trust me. Place whole beef kidneys into a stockpot half full of cold water. Bring to a boil. When you see a layer of scummy foam on top, drain the kidneys, throw out the water, refill, and repeat the process until there's no foam left (always starting with cold water). Then remove the kidneys, cut them into ¼-inch slices, and sauté with butter and onions. You can add a few strips of bacon for added flavor. Salt and pepper to taste. This was often lunch at the IGA store I apprenticed at.

Heart

Beef heart is also a tasty delight, as is veal heart (which is a bit more tender). Just slice it ⅛ to ¼ inch thick and sauté in butter and onions. This was also a common delight I remember from my childhood. Delicious, and inexpensive.

Brains

A caveat: Brains are often scrambled with eggs or battered and fried. I personally won't eat them, despite the fact that they are a very common dish in Europe. I want to see more research on the possible health effects of eating calf brains. Ever since the increase of mad cow disease (bovine spongiform encephalopathy), I prefer not to recommend brains or any nervous tissue until research proves there's absolutely no health risk.

BEEF BREEDS FOR BEGINNERS

Quick! Name 10 beef breeds! Holstein? (No, it's a dairy breed.) Angus? (Yes!) Jersey? (Nope, another dairy breed.) Texas Longhorn? (Well done!) Bet you couldn't do it.

There are over 70 beef breeds—or maybe even more—out there. So while you may pride yourself on knowing about Angus or Hereford, it's too early to pat yourself on the back.

The variety of beef breeds and how they came to be is a fascinating story in itself, one I didn't want you to miss. So sit back (there won't be a test) and read about how cattle came to be.

Capturing the Aurochs

About 10,000 years ago, somewhere in what's now Iran, a few enterprising (and fearless) souls captured a wild ox, kept it alive, and bred it, thus producing . . . another one. This was no small feat. That wild ox was technically an **aurochs**—the ancestor of today's cattle, and not a beast to trifle with. Standing up to 6 feet at the shoulder (that's shoulder, not head), weighing more than a ton, and armed with horns 30 inches long, it was definitely not tame. Aurochs killed people quite regularly.

And people killed them. Their meat was delicious and a concentrated source of protein, providing much more nutrition pound for pound than the wild grains, nuts, and fruits our ancestors gathered. Aurochs were prized game animals and had been revered for thousands of years. This was certainly the great mythic bull of early religions.

We'll never know what was in the mind of those first aurochs catchers, but we do know that somebody actually did it. And we know a surprising amount about where the event happened and how long it took to turn fearsome aurochs into what we now call cows . . . smaller, calmer, easier to manage. Domesticated animals.

The domestication of the cow—one of humanity's greatest achievements—happened in two small Neolithic villages in what are now Iran and Turkey, respectively. These two villages less than 160 miles apart were among the very first places where people began to give up the

hunter-gatherer lifestyle and settle down. And there is now absolutely no doubt that the ancestors of *all* present-day domesticated cows originated in these two tiny villages.

How do we know? Because of the work of a team of British, French, and German scientists who analyzed DNA from cow bones from these two sites, dating to more than 10,500 years ago. What they discovered will blow your mind.

The differences between the 10,500-year-old DNA from the two villages and those of today's cattle were so tiny that the only way to explain them was if the original cattle population had been extremely small. How small? *About 80 animals.* How's them apples? From 80 original animals, painstakingly bred over 1,000 years, to more than 1.3 billion cattle today.

Obviously, this happened gradually. But imagine the delight as news of this new, easy-to-manage mini aurochs began to circulate. A source of power (oxen), a source of milk, a source of meat—all right at hand in your backyard. This profoundly changed history.

From the original aurochs, or *Bos primigenius*, to use its scientific name, we eventually got to the cattle we know today, *B. taurus* or the European type (the cattle we're used to here in America) and *B. indicus* or the Asian/African-type, like the zebu. The rest of this chapter will deal with *B. taurus*. Known to you and me as cows, cattle, steers, heifers, et cetera. Also fondly known as prime rib.

Actually, let's be more specific. What you call a *Bos taurus* defines it. For example, an intact (uncastrated) adult male is a bull. A calf is a baby cattle animal. An adult female that has had a calf is a cow; if she hasn't calved and is less than three years old, she's a heifer. In the United States, a castrated male is called a steer; if it's going to be used as a draft (work) animal, it's an ox. A polled animal is one with no horns. Cattle raised for meat are called beef cattle, and those bred for milk production are dairy cattle. And a dogie is an orphaned calf (if you're a cowboy).

BRINGING BACK THE DISTANT PAST

Wild aurochs have been with us through most of human history—in fact until 1627, when the last aurochs died in Poland. Caesar described aurochs in Germany's Black Forest: "They are but a little less than elephants in size, and are of the species, color, and form of a bull. Their strength is very great, and also their speed. They spare neither man nor beast that they see. The people, who take them in pitfalls, assiduously destroy them; and young men harden themselves in this labor, and exercise themselves in this kind of chase; and those who have killed a great number—the horns being publicly exhibited in evidence of the fact—obtain great honor." (No wonder they became extinct.)

Aurochs were the "flagship species" of Europe. In the Greek myth about the founding of Europe, Zeus—in the form of an aurochs—seduces the virgin Europa. And the rest is . . . prehistory.

I'd loved to have seen one. And maybe, soon, I will. Because an initiative called the TaurOs Programme is bringing them back to repopulate wild Europe. The project began in 2008, and already a herd of 100 first- and second-generation crossbreeds is living in Holland. The ultimate aim is to see free-roaming herds across the continent . . . to bring the aurochs back.

How are they doing it? By using primitive cattle breeds from isolated regions in Spain, Italy, and the Balkans that retain characteristics of the aurochs, and through DNA analysis, and because they know what the aurochs looked like. It was huge, and its body was different from modern cattle. Its legs were much longer, and its neck and shoulders more muscular. And bulls had a white stripe down their backs.

All this brings us to the big question—why? We called the co-founder of the program, Ronald Goderie, and his answer was unexpected. All across Europe, vast regions are being abandoned as people move away. By 2020, four out of five Europeans will live in cities. Leaving the countryside to revert to . . . what? Enter the aurochs.

Apparently, an abundance of big grazing animals is a condition for restoring wilderness. Grazers facilitate biodiversity. The TaurOs foundation already had a herd of other grazers helping to manage "rewilded" land. These are primarily semi-wild ponies and Highland cattle. But Highland cattle aren't fully self-sufficient. So the foundation decided to create the best wild bovine grazer to fill the ecological niche left when the last aurochs died.

The breeding is based on the best science: genetics, ecology, archaeozoology, cave paintings, and more. The end goal is to see thousands of Tauros across wild Europe. Project partners include California University–Santa Cruz, Rewilding Europe Initiative, Wageningen University, the European Cattle Diversity Consortium, and others.

The new cattle won't be exactly aurochs (that's why they're calling them Tauros). But pretty close. Neat, huh?

And here is the result: "Manolo Uno." This magnificent animal is named after Manolo Pelotas—one of the last living Spanish cowboys—whom the TaurOs founders met in Spain while seeking the ancient breed-line Pajuna cattle. It was the *vaquero* culture brought to the New World by men like Pelotas that created the mythic cattle-drive tradition of the West.

Manolo Uno. ©*Ronald Goderie, TaurOs Programme.*

When Is a Herd Not a Herd?

A little-known fact about Highland breeders is that they don't call their herd a herd. It is called a *fold* of Highland cattle, because in winter in the olden days the animals were brought together at night in open shelters made of stone called folds to protect them from the weather and wolves.

From Buef to Beef

Why is cow's meat called "beef"? The word *cow* comes from the Old English *cū*. The Latin word for cow is *bōs*, which became the Old French word *buef*.

When the Normans conquered England in 1066, they brought their language with them. Soon, all fashionable people were speaking French, not that "primitive" English language of the Anglo-Saxon natives. When the meat of the native *cū* was served to the French nobles, they called it "beuf."

And the result is history. We eat beef from cows.

Full-Blood or Purebred? Or Landrace?

If the animal is a full-blood, it is a 100 percent purebred whose parentage can be traced through both sire and dam sides in the breed herdbook (which records every ancestor) back many generations—often beginning as early as the 18th century. The breed herdbook might originate in the UK, Europe, or elsewhere.

If the animal is purebred, then it still qualifies as a pure strain of the breed, except that its parentage is less than 100 percent true back through the generations. It will contain genetics from some other breed, and will have been created by crossing a full-blood with another animal.

Okay, so what's a landrace? A landrace is a local variety of a domesticated animal (or plant, but let's not go there) that developed through natural adaptation to its environment. This is quite different from a formal breed, where human selection has played a role. However, don't go thinking that a landrace is the same as an ancestral species; it's not. Ancestral species are unmodified by human breeding activities, while landraces are often formed through the reversion of a domesticated animal escaping into the wild and becoming a feral species that evolves into a new—landrace—breed. (Oh, and to further confuse you, there's a very popular pig breed called the Landrace.)

Some (but Not All) Beef Breeds

The role of breed as a component of beef production plays a role, but not as large a role as boosters of a particular breed might argue. Beef breeds were developed to suit different regions and climates, as well as different consumer preferences. Simply stated, there are good cattle and poor cattle of any breed, whether raised on grass or grain. Beef farming is a huge business, worth literally billions of dollars, so it's competitive. There is everything to (potentially) gain by asserting that your particular cattle breed is better than your competitor's. Breed associations and marketing organizations spend millions promoting their breed and commissioning studies from animal scientists to prove that their meat is more nutritious, lower in cholesterol, and so on than others. But the fact remains that there are as many differences in quality within one breed as there are differences between breeds; it's a matter of individual animals.

Photo by Cassie Dorran. Courtesy Canadian Angus Association.

Aberdeen Angus

The Aberdeen Angus (or Angus as it's usually called) evolved in Scotland, and is one of the most famous beef breeds. The earliest animals can be traced to the 1700s in the counties of Aberdeenshire and Angus. Angus cattle are naturally polled and come in red (Red Angus) and black (Black Angus). It is the most popular beef breed in the United States. Angus are noted for good maternal qualities and a high carcass quality. Angus cattle are also recognized for their ability to forage under rugged conditions.

Courtesy HC Sims Farm, Pennsylvania.

Belted Galloway

Often called "Belties," Belted Galloway cattle can be black, red, or dun . . . but always sport that white middle. It's believed they originated in the 12th century in the Galloway district of Scotland. Generally, Belted Galloway bulls weigh from 1,800 to 2,000 pounds, and cows about 1,100 to 1,300. Belted Galloways are one of a subset of Galloway cattle: Galloways, Belted Galloways, and White Galloways. All Galloways are polled and thrive under harsh conditions. Beef from this breed is said to be lower in calories and fat and higher in protein than many other breeds.

Courtesy American Blonde d'Aquitaine Association.

Blonde d'Aquitaine

Blonde d'Aquitaine cattle developed in the southwest of France and arrived in North America in the 1970s. They're "blonde"—cream- or fawn-colored. The breed is recognized for its beef characteristics and high-yielding carcasses. They are long-bodied and heavily muscled with very lean beef. Mature bulls weigh from 1,800 to 2,200 pounds; females range from 1,100 to 1,400 pounds and carry 60 to 70 percent of all their fleshing in the hindquarters. Blonde and Blonde-cross animals dress out 60 to 70 percent of their live weight and have impressive yield grades, with the majority of carcasses grading yield grade 1 and 2.

Courtesy V8 Ranch, TX.

Brahman

The Brahman (also known as Brahma) originated from the original *Bos indicus* cattle from India, the "sacred cattle of India." All *B. indicus* cattle, including the Brahman, have a hump above the shoulder and neck. This animal is the famous bucking bull of rodeos. The Brahman is naturally adapted to hot climates, so it's a popular "contributor" breed for crossbred warm-climate cattle. Brahmans vary in color from gray to red to almost black.

Courtesy Canadian Charolais Association.

Charolais

Charolais cattle are named for the region of Charolles in France, where the breed became established in the 12th century. The first Charolais came into the United States from Mexico in 1934, and attracted attention for its color—pure white—as well as its size and muscling (bulls can weigh up to 2,400 pounds). Charolais are popular in meat cattle breeding programs, producing heavier calves than many other breeds.

Courtesy Pam Malcuit, Morning Star Ranch, Texas.

Dexter

The small black Dexter originated in Ireland, where they were kept by small landholders and called the "poorman's cow." A 1845 report claimed that the breed originated with a land agent named Mr. Dexter, who created the breed in the late 1700s by selecting the best of the hardy mountain cattle of the area. In the early 20th century, Dexters became the show cattle of the English gentry, but by the 1970s they were endangered. Since then, they've regained popularity. The reason is their meat—of excellent quality and flavor, with good marbling. Dexter cattle are more economical than larger cattle: Their small size means that more cattle can be grazed on less acreage. Of more interest to modern farmers, they are particularly suitable for grass raising without supplementary feeding.

Photo by Jeannette Beranger, The Livestock Conservancy.

Florida Cracker

Descended from animals brought to Florida by the Spanish in the 1500s, Florida Cracker cattle are one of the oldest breeds in the United States. Gradually, the animals developed into the heat- and humidity-tolerant breed of today, capable of living on poor forage in Florida's challenging environment. Florida Crackers are small animals whose colors and horn shapes vary widely. These cattle are considered endangered by the American Livestock Breeds Conservancy.

Courtesy American Gelbvieh Association.

Gelbvieh

Gelbvieh evolved in the 19th century in northern Bavaria as a milk/beef and draft animal. They are large and heavy-muscled, with high fertility and milking ability and a very good, well-muscled carcass.

Courtesy American Hereford Association.

Hereford

The Hereford (sometimes called the Whiteface Hereford) is one of the oldest cattle breeds, possibly founded on the draft ox descended from the small red cattle of Roman Britain. Herefords get their name from the county of Herefordshire, the agricultural region where the breed evolved in the early 1600s. These early Herefords were much larger than today's animal. Cotmore, a winning show bull and famous sire shown in 1839, weighed 3,900 pounds. Herefords are red-dish brown in color with a white face, horned or hornless, and popular; there are more than five million of them in over 50 countries.

Highland

Highland cattle originated in the northern high country and western islands of Scotland, areas known for their climate—when it isn't raining, it's raining. The animals are supremely adapted for bad weather (maybe that's why they're quite popular in my home state of Vermont). Highlands have a double coat—a downy undercoat and a long outer coat that can reach 13 inches, and is well oiled to shed rain and snow. This breed is exceptionally hardy, with a natural ability to convert poor grazing efficiently. They're also known for living a long time—in fact, there are Highland cows over 18 years old that have continued to produce calves. Because they do so well in rough outdoor conditions, Highlands are often chosen by northern farmers as grass-fed and -finished beef animals.

Courtesy Canadian Limousin Association.

Limousin

The golden- to red-gold-colored Limousin comes to us from south-central France, where it's known for the quality of its meat. The Limousin has always been selected for its meat qualities, yielding high quality, lean, and large carcasses.

Courtesy American Lowline Registry.

Lowline Angus

Australian Lowline cattle were developed from an Aberdeen Angus herd established at the Trangie Research Centre in 1929 to provide breeding stock for the New South Wales cattle industry. The Lowline is a black, naturally polled animal. Mature bulls measure about 45 inches at the hip; mature females, 40 inches. They produce very high quality meat and are exceptionally docile (read about Paul List's Lowline grass-raised herd in chapter 4).

Courtesy American Maine-Anjou Association.

Maine-Anjou

The Maine-Anjou is one of the most popular breeds in France. This is a large, horned animal, which often has a white underline and small white patches on the body. Cattle producers like Maine-Anjous because they grow fast and yield lean carcasses with high cutability (lots of usable meat).

Courtesy Patrea L. Pabst, Beaver Creek Farm.

Piedmontese

This breed originated in Piedmont in northwest Italy, developing through natural selection. In the late 19th century, the breed's characteristic "double-muscling" appeared. These cattle carry one or two copies of the *inactive* myostatin gene (a growth regulator). This results in a higher lean-to-fat ratio and less marbling, as well as less connective tissue in the muscles. So . . . less fat but higher tenderness. Piedmontese bulls are 1,800 to 2,000 pounds at maturity; cows, from 1,000 to 1,100 pounds. Since they do not fatten on corn, they're ideal for grass farming.

Photo by Jeannette Beranger, The Livestock Conservancy.

Pineywoods

Pineywoods cattle descended from original stock scattered along the Atlantic and Gulf Coasts by Spanish explorers in the early 1500s. The breed is heat-tolerant, long-lived, resistant to parasites and diseases, and able to forage on rough vegetation that commercial cattle won't touch. Pineywoods are also "dry land" cattle and have evolved to avoid predators by spending only a minimum of time at their water hole. This makes them very low-impact cattle, as they do not contribute to bank erosion and fouling of streams like most domestic stock. The Pineywoods is critically rare, with only 1,500 to 2,000 individuals alive today.

Courtesy Joe Henderson, Chapel Hill Farm.

Randall Lineback

Sounds like a football position, but it's a beef breed—actually, the rarest breed in America. It was developed about 200 years ago by US farmers whose aim was to create an animal suited to subsistence farming in the Northeast. Later, a few animals were acquired by the Randall family of Vermont, who created a herd that stayed pure through several generations of the family. The descendants of these animals are now called Randall Lineback cattle. This is the colonial American breed that hauled cannons to George Washington to reclaim Boston from British invasion in the Revolutionary War of 1776. The Randall Lineback is an all-purpose breed, developed before the split into dairy versus beef breeds.

Red Devon

The first Devon came to the States with the Pilgrims in 1623. But the breed is far older, with records describing them in Devon, England, in 200 b.c.e. The Devon was known in England as the all-round cow. An 18th-century rhyme describing the Devon goes,

> Broad in her rigs and long in her rump,
> Straight flat back with never a hump,
> Fine in her bone and silky of skin,
> She's a grazier without and a butcher within.

Today's Devons are often chosen for grass operations. They do well in all climates, and produce high-quality beef. An interesting point: Devons have the thickest hides of all cattle, giving them high resistance to external parasites and temperature extremes.

Courtesy Jeremy Engh, Lakota Ranch.

Photo by Jeannette Beranger, The Livestock Conservancy.

Red Poll

The modern Red Poll breed was developed by combining two British landrace strains: the Suffolk and the Norfolk. Red Polls are quiet, medium-sized cattle that are able to finish to choice grade on grass alone. Their beef is well marbled, and the carcasses are high yielding.

Courtesy American Shorthorn Association.

Shorthorn

This breed originated in England and is often called the "foundation breed" because it's been used in the development of more than 30 cattle breeds around the world. It's a red, white, or red-and-white animal that may be horned or hornless. Shorthorns are hardy and mainly used for the production of beef.

Courtesy Paulette Cochenour, American Simmental Association.

Simmental

Simmental cattle were developed in the Middle Ages in Switzerland (the name refers to the Simme Valley there). These cattle are popular because of their adaptability to varied environments, as well as their high beef yields with a minimum of waste fat. Simmental cattle are the second largest breed, after the Brahman. They may be horned or hornless. They're heavily muscled and grow fast.

Courtesy Dickinson Cattle Co.

Texas Longhorn

We all know these. Including their horns decorating Cadillacs. This is a true American animal. But it wasn't "bred"; instead, it's the result of natural selection, generation to generation, deriving from ancestors brought to America 500 years ago. Texas Longhorn cattle can be pretty much any color; owners claim that no two are alike. Because they will eat a wide variety of natural forages and weeds, farmers can put them on pastures that need less fertilization and weed killer. You can ride them, too. Really.

Courtesy American Wagyu Association Inc.

Wagyu

Wa is the Japanese word for "Japanese," and *gyu* means "cattle." Thus, *Wagyu* = "Japanese cattle" or "Japanese beef." To be more precise, the word refers to specific types of cattle raised for meat in Japan. This includes the famous Kobe beef, which is Wagyu raised in the province of Kobe, Japan. In the United States and several other countries, producers have created "Wagyu-style" or "Kobe-style" beef. Wagyu (and Kobe) are known for their unbelievable marbling, yielding a cut of meat that's soft and buttery, with so much internal fat flecking that the meat is often more white than red. The legendary richness and mouthfeel have made Wagyu pricy—to put it mildly. And you've also heard about the way the animals are raised: drinking beer in the summer to give them more appetite (doesn't work for me), and massaging them to relieve stress (does work for me).

Watusi

Get a load of those horns! They can reach up to 8 feet tip-to-tip. This breed is native to Africa, where it eats leaves and grass in water-poor savannas and grasslands. The breed traces its ancestry back more than 6,000 years, but here in North America, a breed registry was only formed in the early 1980s. Watusi cattle are medium-sized and very heat-tolerant—their horns actually act as radiators, moving heat away from their bodies. Of more interest to meat eaters, recent studies have discovered that their meat is very low in fat and may contain less cholesterol than any other cattle breed.

Courtesy Dickinson Cattle Co.

Photo by Jeannette Beranger, The Livestock Conservancy.

White Park

These white cattle were wild in early Britain, and were actively hunted. In the 10th and 13th centuries, British kings awarded huge land grants to the church and the nobility, who then "emparked" their land—building walls to keep out poachers, and confining the white cattle along with the other game animals. In 1919 the first British White Park Registry Association was formed; in 1940, one bull and five cows were sent to the United States as seedstock in case of a Nazi invasion of England. In 1999 the American British White Park Association was formed.

Composite Breeds

In recent years, there's been a growing trend toward "composite" breeds. A composite is really just an animal made up of two or more component breeds. For example:

Angus + Simmental = Simangus
Brahma + Angus = Brangus

Why? To retain **heterosis** (also called hybrid vigor)—the tendency of a crossbred individual to show qualities superior to those of both parents.

Composite breeds that have been around for a while include Shaver Beefblend, Hays Converter, and Brangus.

Photo by Jessica McBride. Courtesy Merrill Cattle Company/Beefalo Meats.

Beefalo

Beefalo cattle are a species cross between buffalo and domestic cattle. The breed standard is three-eighths buffalo (bison) and five-eighths bovine (cattle). Beefalo breeders say that the breed mixes the hardiness of bison with the quality of beef—a winning combination of lean, tasty meat that is low in bad cholesterol, low in fat, and high in protein. The breed is considered cheaper to raise and maintain than many others, because its buffalo genetics deliver excellent hardiness and the ability to forage on its own with the ease of handling of domestic cattle.

Courtesy International Brangus Breeders Association.

Brangus

The Brangus combines the traits of Angus and Brahman cattle, with genetics stabilized at three-eighths Brahman and five-eighths Angus. Recent tests conducted by Texas A&M University confirmed the ability of Brangus to produce exceptionally high-quality carcasses.

Courtesy Dickinson Cattle Co.

BueLingo

The BueLingo is a composite breed developed on the Bueling Ranch by Russell Bueling and R. B. Danielson of the Animal Science Department of North Dakota State University, and can now be found across 38 states and three Canadian provinces. The aim was to produce hardy cattle that are economical to maintain with high fertility and outstanding carcass merit. BueLingos can be either Black or Red Belted. The belt must circle the entire body. A BueLingo cow weighs from 1,100 to 1,200 pounds and will wean a 600- to 700-pound calf that will continue to grow fast—weighing 1,000 to 1,100 pounds at one year.

Courtesy Hays Ranches, Calgary, Alberta.

Hays Converter

This suggests an animal that's great at the conversion of hay, right? Slightly wrong.

The Hays Converter is actually named after Canadian senator Harry Hays, and is the first beef breed recognized by the Canada Livestock Pedigree Act and developed by a Canadian livestock producer. The senator was looking for a leaner cow that would thrive in western Canada, and that would gain weight as efficiently as possible—getting to market weight in one year. The name of the breed refers to the efficiency with which it converts feed to meat.

Santa Gertrudis Breeders International, Kingsville, Texas.

Santa Gertrudis

The Santa Gertrudis is named for the Spanish land grant on which Captain Richard King established his famous King Ranch. The breed was developed specifically for the tough climate and native grasses of South Texas. It's the result of crosses between Shorthorn and Hereford cattle with heat-tolerant Brahman cows. The Santa Gertrudis was recognized by the USDA in 1940, making it the first official American beef cattle breed.

Courtesy Brian and Sonja Harper, Circle H Farms, Brandon, Manitoba.

Shaver Beef Blend

It took nine different breeds to produce the Shaver Beef-blend breed (named for its creator, Dr. McQ. Shaver). The nine breeds are Galloway, Highland, Red Devon, South Devon, Lincoln Red, Gelbvieh, Saler, Blonde d'Aquitane, and Maine-Anjou. This breed is called the "ultimate" composite: designed for high yield and early maturing, with an excellent temperament for easy handling.

Chapter Eight

A Side of Pork

If pork competed in a high school "most popular" contest, it would win. Of all the meats out there, it's consumed by more people than any other; in fact, about 40 percent of the world's meat consumption is pork. Who eats the most? You've probably guessed the Chinese, and you'd be right, but Americans aren't far behind. Why is this meat so popular? Read on . . .

This lovely sow is being raised by Sterling College students.

WHY EAT PIGS?

(1) Pigs are delicious. (2) They're a great source of lean and healthy protein. (3) Their meat is inexpensive compared to other meats. (4) Because they're there. Consider the other common North American meat animals and mentally list the products they're known for (I'm not talking by-products like glue here). Cattle give us milk and its many related products such as yogurt and cheese . . . as well as meat. Sheep produce milk, yogurt, cheese, fiber . . . as well as meat. Pigs give us pork.

Pork is good for you. It's a source of protein, vitamins, and minerals. Pork is low in sodium and contains no sugar. It's an excellent source of

A pig omnivore illustrated by Leonard Leslie Brooke, from *The Story of the Three Little Pigs* circa 1904. *Image from Library of Congress (LC 84181093).*

selenium, niacin, vitamin B$_6$, thiamin, phosphorus, and zinc and a good source of riboflavin and potassium. And it delivers 5 percent of your daily total of iron, a mineral often lacking in infants, teenagers, and women.

In our distant history, pork was a preferred meat because pigs will eat anything, including "waste" materials (peels, rinds, old vegetables, et cetera). Because of their adaptability, pigs became a staple in many cultures. The Chinese Meishan pig is a perfect example. Meishan pigs are very quiet—almost as shy as I am—so they prefer a solitary existence and don't roam widely to forage. As adults, heavy wrinkles of skin develop on their heads, essentially rendering them blind. They don't seem to mind and are perfectly content lying in a comfy spot all day as long as they have access to food and water, which they easily locate with their extraordinary sense of smell. They are also remarkably fertile, often giving birth to 20 or more piglets! And curiously, unlike most pigs they also have the ability to digest fibrous foods. With the exception of fertility, these are not traits that typically fit well into conventional production systems, so most farmers find little value in the breed.

Why are these pigs unique? They were bred to be. In ancient China, every family kept a pig to deal with leftover food and other waste. Since pigs were also a primary source of meat, flavor was critical; it was also important that sows be prolific mothers. In China, the Meishan is considered a national treasure and is an extraordinary example of the ancient "art" of animal breeding and husbandry.

In 1900s America, farmers used to call hogs "mortgage lifters" because they were such efficient converters of feed that they were almost always profitable. Nearly every farmer in the 1930s, '40s, and '50s raised pigs for meat and lard. In those days, most people cooked with lard, so fat pigs were preferred. Today's pigs are much leaner and more heavily muscled than their predecessors.

Raising a pig is still a cheap, efficient way to produce a lot of protein. So from the perspective of a meat farmer, the pig is a glorious creature. It bears litters frequently, producing from 8 to even 15 piglets or more per litter. Those piglets grow fast. In six months, a tiny 3-pound piglet can increase its weight to 260 to 270 pounds, yielding over 100 pounds of high-quality, nutrient-dense protein.

Pigs are **omnivores** (they'll eat anything—even meat), but they prefer and thrive on roots, nuts, and grain. So farmers aren't limited to one or two feeds that may rise in price. Pigs can eat surplus crops, which helps other farmers. They can't digest most fibrous plant matter, though; cows and the other ruminants rule here.

What Is Pork?

Pork (as sold in stores) is meat of the domesticated pig. If the meat comes from a nondomesticated (wild) pig, it's usually referred to as wild boar.

The term *pork* includes fresh as well as cured, smoked, or otherwise processed pig meat. Generally, *pork* is used for fresh meat, as opposed to processed meats such as ham or bacon (although there is such as thing as fresh ham, which I'll get into later).

And although pork is touted as "the other white meat," the USDA considers and treats pork as red meat, because pork is red when it's raw—although not as red as beef—and only turns pale when it's cooked. The term *white meat* generally refers to veal, rabbit, and the breast of chicken and turkey.

Get Fresh with Me

Here's something you might not know. When you're buying pork, *fresh* has a particular meaning when applied to cuts like bacon, shoulder, or ham. It means that the meat is not smoked. Thus, if that piece of pork is called "ham," it's smoked. If it's called "fresh ham," it's not. Other examples are pork steak versus fresh pork steak, picnic shoulder versus fresh picnic shoulder, bacon versus fresh bacon. You get my gist. And keep in mind that if the pork is smoked, it may not be fully cooked. Read the label or ask to find out whether it's "partially cooked" or "fully cooked."

Scorched Gravy

Here's a recipe from my cousin Judy Butler.

After frying your side pork, add flour to the grease left in the frying pan and brown it (scorch) until it turns a nut-like brown. Add water to make gravy. Season with kosher salt and pepper. Johnnycake is corn bread. And, to quote Judy, "After eating this my father always used to say, *Side pork and johnnycake makes a Frenchman's belly ache.*" (He was from Québec.)

Lots of folks ask me about fresh bacon—as in, isn't it just bacon with an adjective? Nope. Fresh bacon is not smoked (it's often called side pork in supermarkets). If you get it from a butcher, it'll be called fresh bacon or sliced pork belly. And it's delicious. You bacon lovers—as in all of humanity—ought to try it sometime.

Fresh bacon couldn't be simpler to cook: Season the slices with salt and pepper. Place into a *hot* cast-iron skillet—*no grease needed*, for obvious reasons. Fry till the edges are crispy. Fight over leftovers.

I like to serve this with corn bread and boiled potatoes mashed with milk gravy or scorched gravy.

Using the example of fresh pork tenderloin, we see that a 3-ounce serving provides about 102 calories. This is good. That one serving also provides about 23 grams of protein. Even better.

What about fat? Pork itself contains no natural trans or hydrogenated fats. The only meats that contain natural trans fat are from ruminant animals like beef, lamb, or goat. And pork fat is higher in "good" unsaturated fatty acids than beef, veal, or lamb fat.

That same serving of tenderloin contains about 3 grams of fat (of which about 1 gram is saturated). That's only 5 percent of your daily requirement. Also of note is that trimmed pork has less total fat and fewer calories than many other meats. Pork's downside for some people is cholesterol—about 21 percent of your daily total is lurking in that 3 ounces of lean tenderloin.

But of course, that's tenderloin. What about—dare I say it—something like braised pork belly or barbecue ribs? They will still deliver the nutrients described above, plus lots more fat—and thus more calories. Even so, I'm having a pulled pork sandwich for lunch.

One important note about lean pork: Modern pigs have been selected to meet consumer demands for lean meat with little fat. This has reduced the flavor, tenderness, and eating quality of the meat. To respond to this challenge, the pork industry has focused on genetics to improve meat quality. It has also developed a process known as enhancement.

Enhanced pork—also known as "seasoned" or "pumped"—results from a process in which whole muscle groups such as the leg or loin are injected with a mixture of water, salt, and sodium phosphate during processing. This mixture binds moisture in the meat, improves tenderness, increases shelf life, and adds a degree of salty flavor.

Enhancement is nothing new. It's been used in the poultry industry for decades and also appears in some beef products. Enhanced pork and other meat products are considered safe, but must be clearly labeled in the meat case as WATER ADDED or SALT ADDED. Note that they may contain as much as 10 percent added weight due to the enhancement process.

So when you are buying an enhanced product, you are getting 10 percent added water and a product that has already been salted. If you're planning to brine your pork before barbecuing it, keep this in mind, because the final result may taste too salty.

The majority of grocery outlets in the United States and Canada now sell enhanced pork products, many of them exclusively. It is becoming quite difficult to find non-enhanced pork (another reason for learning how to butcher your own pigs).

Range Pork and Garbage Pork

Traditionally, pigs have been raised by two methods: free range or confined. In the free-range method, pigs are released into forests or rough pastureland to find their own food like mushrooms, roots, worms and grubs, all kinds of nuts, wild fruit, and even meat like birds and mice. This method of raising pigs is sometimes called pasturing.

Since pigs are omnivorous and enterprising, this worked well historically, and still does today in the relatively few pastured pork operations out there. In colonial times, pigs roamed free, squeezing through the crude post-and-rail fences of farmers and wandering through neighboring farms and forests. They soon became such a nuisance that laws were passed to identify which pig belonged to which farmer. The cliché image of the ring in the pig's nose dates from these early days; rings prevented the pigs from rooting up farmers' crops.

The term for pasturing in a forest—practiced for centuries in Europe—is **pannage**. In the Middle Ages, the owners of forests made more money selling mast rights than wood. **Mast** refers to the nuts or fruit of forest trees like beech, oak, and so on, which are eaten by wild animals—and pigs—as they forage. Mast feeding survives in Spain, where cured pork derived from free-range pigs is prized—and very expensive.

The second method of swine raising involves bringing the food to the pigs. One of the earliest—and still extant—uses of the pig was as a garbage eater. Our garbage. In early China and Korea there was a "privy pig" that ate—oh, yuck—human excrement. And then was eaten by the humans. I'll leave you to ponder that for a moment, and move on.

The garbage pig has been kept by many civilizations. Its role was to eat a family's leftover scraps, as well as waste from mills and garbage from hospitals and large institutions. Pretty much anything we didn't want. In the days before municipal sewage systems, it was the peripatetic pig that took care of things (as Charles Dickens observed about 1842 New York).

Was it this association with garbage that led to the avoidance of pork by certain cultures? No other animal is saddled with so many strictures. About a fifth of the world's population won't eat pork. That includes Muslims, Jews, Ethiopian Orthodox Christians, and many Buddhists and Hindus.

TROTTER TERMINOLOGY

Hogs, pigs, and swine are all general terms for the animals that give us pork. A **sow** is a female pig that has given birth or **farrowed**. Her offspring are **piglets** or **shoats** (the term *shoat* has been largely replaced by *nursery pig* or *grow-finish pig*). A **weaner** is a piglet that's been separated from its mother and is eating only solid food (generally from three to six weeks old). After weaning, a young female pig is a **gilt**. Male pigs grow up into more nomenclature: A male pig castrated before puberty is a **barrow**. A male adult pig that has *not* been castrated is a **boar** . . . he's ready for breeding. And a male pig castrated later in life is a **stag**. An average wild pig can live up to 20 or so years; a domestic pig, 8 to 10 years (if it isn't eaten much earlier). Try to learn some of these terms used by swine raisers so you won't look too much like what Vermonters call a "flatlander" when you're negotiating with a farmer to buy a pig.

Pork is the meat of pigs. In pork production terminology, a *suckling pig* refers to a piglet slaughtered before weaning. A ***feeder pig*** is a weaned gilt or barrow that weighs from about 15 to 20 pounds at weaning to 80 pounds or so at 10 to 12 weeks of age. A *grower pig* is a pig from 80 pounds to 180 or 200 pounds. And a *finisher pig* is a pig over about 200 pounds that is in the process of being finished for slaughter. Market weight for pigs (the age at which they'll be slaughtered) is generally around 250 to 280 pounds, although suckling pigs will be slaughtered (obviously) at a much, much lower weight.

Pork Inspection and Grading

Like all meat sold to consumers, pork is either USDA- or state-inspected. The process is the same as the beef inspection process described in chapter 4. The PASSED AND INSPECTED BY USDA seal ensures the

Painting of Saint Anthony by Piero di Cosimo, c. 1480.

Born in 251 C.E. in Egypt, Saint Anthony the Abbot is the patron saint of pigs and gravediggers. Why? Anthony worked with those stricken with skin diseases and in his time skin diseases were treated with pork fat, so after his canonization, he was often depicted with a pig. Late in life Anthony became a close friend of Saint Paul the Hermit. When Paul died, Anthony buried him, thus becoming the patron of gravediggers. In medieval Paris, the monks of St. Anthony enjoyed special rights to keep pigs within the city walls.

pork is wholesome and free from disease. That seal will display an establishment number identifying the exact slaughter facility in which the animal was processed. You can look up establishment numbers numerically on the FSIS website; for details, see the resources section (USDA FSIS Meat, Poultry, and Egg Product Inspection Directory). This list covers processing facilities that slaughter all meat animals; not just pigs.

Cole's Notes

Just so you know . . . it's pig or **swine**; a litter or farrow of piglets; a drift, mob, or herd of pigs; a **sounder** of wild boar; a parcel of hogs; and a drove of pigs (but only if they're moving).

The USDA grades pork carcasses on only two levels of quality: acceptable and unacceptable. Pork sold as acceptable quality is the only grade of fresh pork sold in stores, and it should have a high proportion of lean meat to fat and bone. After being deemed acceptable, pork is graded for yield, meaning the ratio of lean meat to fat and bone. Unacceptable quality pork has meat that is soft and watery and is graded US utility. Pork graded as utility is mainly used in further processed products and is not available in stores for consumers to purchase.

In Canada, a pig carcass either passes inspection and is certified, or it fails, and will be segregated for a decision about use or non-use.

However, when we get to quality grading, it's quite a different story. Although there are certainly variations among pork carcasses, there is no quality grading system in place similar to the "visible" beef grading system (with its prominently labeled grades of prime, choice, and so on). Savvy raisers understand that the presence of some marbling in pork enhances flavor, but may make the product less attractive in a market that focuses almost exclusively on leanness. A fat pig used to be a desirable pig, but today's consumers demand lean animals (often at the expense of flavor and tenderness).

Or do they? The increasing popularity of naturally marbled, fatter heritage breeds may herald a new attitude toward this wonderful meat. The marbling that develops in pigs is a result of genetics and usually indicates a mature animal that has had the opportunity to spend more time on feed . . . thus developing some intramuscular (marbling) and also extramuscular (trim) fat.

Many heritage-breed pigs aren't as attractive to pork raisers because they take two to three times longer than modern breeds to grow to a usable size, and they must consume two to three times the amount of feed to gain that weight (most of which will be fat). Unless enough customers are willing to pay a premium and accept more fat, these breeds are usually economically unsustainable—hence their heritage status. And often the increased fatness of heritage breeds doesn't translate into better flavor, juiciness, and tenderness. Like any pork product, the animals must be fed, housed, handled, processed, and prepared with care to ensure wholesomeness and quality, and must also have the genetics to produce a superior product.

So is pork graded or not? The answer is no in terms of comparison with beef grading. However, although there's *no* USDA quality grading system for pork, there are certain characteristics within pork meat that relate to quality. Most processors will voluntarily inspect pork for color, marbling, and ultimate pH, the three most important factors in determining lean firmness and texture. This is unlike other meats and is an important point to absorb.

Pork carcass quality is determined almost exclusively by the ultimate pH of the meat after 24 hours of chilling post-slaughter. It's complex, but I'll make a stab at explaining. Basically, the quality of pork is directly affected by two factors: the activity in the animal's muscles in the 48 hours *before* slaughter; and what happens in the muscles for 24 hours *after* slaughter.

The process is driven primarily by glycolysis, the conversion of muscle sugar (glycogen) to lactic acid. When blood stops flowing (that is, the animal is dead), lactic acid builds up in the muscle and affects pH. Lowering of pH is one of the most significant changes that occurs in muscle during its conversion to meat. When heat is present, falling pH will denature proteins. Meat is about 70 percent protein. The rate and extent at which pH declines is highly variable, and it's these variations that create the differences in meat quality. Low pH alone will not affect meat quality as much as the combination of heat and falling pH will together. Getting the heat out of the carcass within the first 30 to 60 minutes is critical to helping prevent a precipitous drop in pH that will denature meat proteins.

The ultimate pH level of the meat affects two of its visible qualities: color and water-holding capacity. The higher the pH, the darker the meat and the higher its water-holding capacity will be. The lower the pH, the lighter the meat and the less water it will retain.

These qualities drive pork quality determination, divided into four levels.

Denaturation

Denaturation is a phenomenon that occurs when the molecular bonds that maintain the three-dimensional structure of a protein are broken. Denatured proteins have a looser and more random structure. Heating can cause denaturation, as can vigorous shaking, or exposure to detergents, acids, or alkalis.

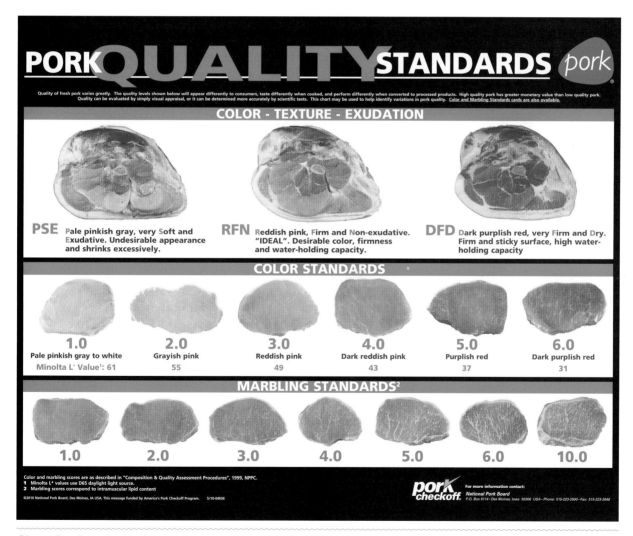

Chart of pork quality levels. *Courtesy National Pork Board.*

- **PSE (pale, soft, exudative):** The meat is pale and soft, exudes moisture and cannot hold introduced moisture, and when cooked is tough and dry.
- **RSE (reddish pink, soft, exudative):** The meat is a more desirable color of pork—reddish pink—but remains soft and exudative. When cooked, it will shrink excessively. Note that this category does not appear in the Pork Quality Standards chart on page 187.
- **RFN (reddish pink, firm, non-exudative):** The meat is reddish pink and firm and doesn't exude moisture. When cooked, it's tender and juicy. This is ideal pork.
- **DFD (dark reddish purple, firm, dry):** The meat is a dark reddish purple and is very firm. It's dry and may feel tacky. It spoils very quickly and is mushy to eat. The DFD condition is similar in characteristics to a beef dark cutter—but in pork, the condition is most often caused by the animal being off feed for an extended period of time.

By keeping animals well fed, comfortable, and stress-free, then processing the animal after a brief fast of 8 to 12 hours and getting the meat quickly into the chiller, the rate and extent to which pH declines within the meat can be controlled, resulting in much better quality.

And what about marbling? There are 10 degrees of marbling grades used to measure intramuscular fat in pork production. They are designated as 1 through 10 and closely correlate to the percentage of intramuscular fat or marbling within the lean. And you know—having read the beef chapter of this book—that increased marbling improves the eating experience.

Is Pork Aged?

Almost all freshly slaughtered pork is rapidly chilled for 24 to 48 hours, then cut and vacuum-sealed to prevent contact with oxygen, and sold within 7 to 10 days. This is what you'll find at your supermarket or meat store.

But some pork is dry-aged for 10 to 14 days. Two things occur during this process. First, the enzymatic decay (what I often call "controlled rot") of the meat causes it to become more tender as the connective tissues begin to break down. Second, the outside fat dries. This creates flavors that are said to be similar to those in salt- and air-cured products such as prosciutto. Unlike beef, dry-aging pork for more than 14 days doesn't produce any noticeable benefit.

Buying and Using Pork

Pork is generally thought of as being from a young swine that is less than a year old, and in fact most swine are slaughtered at six to seven months of age. This results in lean, tender meat. When buying pork, you should look for meat that is pink to light red in color and that has a modest amount of marbling along with white fat.

Market hogs don't vary in live weight as much as beef cattle and can be subjectively estimated with more accuracy. The normal industry weight at harvest is 260 to 270 pounds with an average of 268. US hogs tend to grade heavier and fatter than Canadian hogs, where the average weight is about 213 pounds. However, niche pork programs or those using heritage breeds often harvest their pigs at much lighter weights—down to 190 to 200 pounds.

One notable exception to the above slaughter weights is suckling pig, which is a very young pig slaughtered while still on its mother sow's milk—at generally between two to six weeks old. Suckling pigs are traditionally cooked whole, as a special dish for special occasions. Hence the proverbial image of a piglet with an apple in its mouth.

Is There Such a Thing as "Natural Pork"?

The USDA definition of *natural pork* relates to the way the pork was processed, *not* how it was raised. It must contain no added artificial ingredients and be *minimally processed* (a term you may

Most prefer pork that's not too dark.

have heard). This means that the meat must not be fundamentally altered; specifically, it must not have been pumped or "enhanced" with liquids (as described earlier in the chapter).

Naturally raised or similar terms are are often seen on labels, but they are not regulated and vary widely. In fact, the specifics of the term *natural* are the responsibility of the individual meat processor. Some labels may state that the pigs were raised without hormones or antibiotics or were fed all-vegetable diets. Others may state that the pigs were raised in outdoor facilities or in barns that provided bedding, and so on. A number of individual pork marketing organizations have established their own guidelines for methods of pork production that they term "natural." Do your homework on brands of interest if this is important to you.

Note: No pigs are ever fed or injected with hormones. Some pigs are fed products that contain a

Cole's Notes

Trichinosis is a disease caused by eating undercooked meat from animals infected with the microscopic parasite *Trichinella*. It's often associated with pork, and I get a lot of questions about it . . . specifically, what is the correct internal temperature to cook pork to? I'm vigilant about buying only the best meat from a properly raised animal (which is why I like to know about the farmer behind the product), so an internal temperature of 145°F (63°C) is what I look for. This gives me a tender, delicious result. Note that the Trichinosis organism dies at a temperature of 138°F (59°C). However, there still are people who prefer to cook pork to a higher internal temperature of 155°F (68°C) . . . it provides peace of mind.

feed additive drug called ractopamine for the last few weeks before slaughter. Ractopamine stimulates muscle growth and reduces fat, and has been designated by the USDA and Canadian government as generally safe for pigs and people, but it has been banned in many other countries.

You'll be interested to learn that almost all fresh pork (and poultry) sold in stores is "enhanced"; water and other ingredients have been added to it to improve its flavor, lengthen shelf life, or add weight. This is ubiquitous; in fact, it's almost impossible to buy in-store fresh pork or poultry that has *not* been enhanced.

If this concerns you, I urge you to read the very fine print on the next package of fresh pork you pick up at the supermarket. If you're not satisfied with the ingredients, then seek out a farmer who raises heritage pigs (see the breed list at the end of this chapter).

HOW ALMOST ALL PIGS ARE RAISED

Every week, more than two million pigs—the progeny of about eight million sows (mostly) and boars—are slaughtered to feed Americans. Do the math—that's over 104 million pigs per year killed to feed our appetite for pork. To satisfy this enormous demand, a system has developed that I'm quite comfortable terming "industrial."

The system is tuned for efficiency, so sows are artificially impregnated. The semen comes from breeding stock that is continually being tweaked to deliver desired traits in offspring.

This is what's termed "conventionally raised" pork, and almost all pork offered by supermarkets and other meat outlets is produced in this system. In conventional pork production, the goal is efficiency and sustainability and the best return on

investment, just as it is in most industrial pursuits. What does that mean for the pigs?

The first thing to know is that science plays an enormous role in commercial pork production. It starts with genetics—specifically, the production and sale of semen from swine genetically selected for specific traits that will be of use to the pork raiser. These traits include such things as the quantity of milk a sow produces (swine litters are large), lean growth, mothering ability, meat quality, litter size, the soundness of feet and legs, and behavior. As for behavior, it's important to select for less aggressive animals. Pigs can be mean. Very.

Pork produced in conventional facilities comes from commercial crossbred strains of swine, whose genetics are developed by a *very* small number of highly specialized pig genetic companies across the United States and Canada. In North America the big players are Danbred, Hypor, Genesis, Genetiporc, and the largest—PIC (Pig Improvement Company). Tabatha Jeter of PIC explained why the word *genetics* is so important to modern pork production:

When you go to a meat market or supermarket anywhere in North America, there's a 40 percent chance that the pork product you select will carry the influence of PIC genetics. But the word genetics *doesn't mean what you think. No pig genetics company out there is involved with genetically modifying swine. It's quite the reverse: The animals we breed are* non–*genetically modified. The word* genetics *comes in because of what we do. We map the genome of the breeds we work*

That's a Big Pig

The biggest pig on record was Big Bill, a hog from Tennessee in 1933. Big Bill weighed 2,552 lbs and was 9 feet long.

with to identify traits that will confer an advantage to the offspring. Once those genetic traits have been identified, we mate the best animals to each other in order to produce the best crossbreeds. It's just like traditional farming, except harnessing the science of genetics to ensure high success. It's basically natural selection . . . we look at the DNA but don't mess with it.

The genetics firms crossbreed to select for desired traits and sell the semen to commercial pork producers, who then use it to inseminate their sows.

Modern pig facilities vary in size. A small operation might have about 600 sows, each of which will produce from 10 to 16 piglets two or sometimes more times a year. A large facility might consist of several barns, each holding as many as 5,000 to 10,000 sows.

The largest farms have a series of barns organized in very efficient pods with 5,000 to 10,000 sows in each. In the United States, the trend is toward larger farms, which tend to maximize productivity and efficiency and limit environmental impact.

Modern Stall-Based Housing

You've probably heard horror stories about the confined quarters of conventional pork facilities. Here is how Denise Beaulieu, research scientist at the Prairie Swine Centre in Saskatoon, Canada, described modern stall-based housing, which is the general commercial method of raising conventional pork. In this method (which is actually only about 30 years old), sows are impregnated and placed into **gestation** stalls. Each stall is just big enough for the

These are gestation stalls. Why do this? According to conventional swine raisers, gestating sows are very aggressive. Additionally, they want to eat too much when they're pregnant (sound familiar?). So to feed them individually and properly and to ensure their health and welfare they are kept in stalls during their pregnancy. *Courtesy National Pork Board.*

And this is a farrowing stall. *Courtesy National Pork Board.*

sow to stand up and lie down—about 6½ feet long by 2 feet wide. Floors are usually smooth concrete and have narrow slots to allow waste to drop through. Sows are placed into the stalls when they're bred, and since the average gestation time is 114 days and sows are pregnant most of their lives . . . well, you get the picture.

When the sow is about to give birth, she's moved to a **farrowing** stall. This is larger than the gestation stall, but the sow still cannot turn around. Why is confinement still needed? Because sows often weigh 500 pounds or more and can easily roll or step on their 3-pound piglets, killing or injuring them without even knowing it is happening. So much for the maternal instinct. (Even in the wild, 30 to 40 percent of piglet deaths can be due to crushing by Mama.) The farrowing crate is designed to encourage the sow to get up and down more

slowly and to limit her movement to avoid crushing her babies. It also provides a warm area away from Mama where the piglets are safe and warm (100°F, or 38°C) while keeping the sow much cooler at her desired temperature of 65 to 70°F (18–21°C).

The sow and piglets remain in the farrowing stall for about a month. Then the piglets are weaned and moved away from the sow to a nursery group of about 20 to 30 individuals. Lots of room, at least when they're small; much less as they grow.

However, you mustn't assume that all traditional pork-raising facilities are as restrictive as described above. I spoke with one out west where farrowing stalls are used, but the sows can enter and exit as they like. And very often the younger sows prefer the stall. Why? Because they're intimidated by the older, dominant sows. In swine social order, pigs remain in family groups that do not associate with others. In each group, those on top are bullies, and fights can get nasty. So by their second or third pregnancy, sows often come and go (from the farrowing stall) as they please.

Modern Group Housing

Increasingly, pork producers are moving to a production system called group housing. In this method, sows are kept in stalls for the first three or four weeks of their pregnancy, then moved into group accommodation in a barn. Sounds easy, but it's a bit thorny, because of the nature of pigs.

A dominant sow—and there's always at least one "sow boss"—will grab most of the food, pushing aside weaker or lower-status sows. So scientists are investigating different methods of ensuring that each sow gets her share. One way is electronic sow feeding, where each sow wears a transponder that "reports" back to a central computer about her feeding. In this method, the sow walks into a feeding stall, her tag is read, food pops out, and she eats what she wants and leaves. The reason she's monitored is to keep her in good health and condition—neither overfat nor skinny. Ah, if someone could just monitor my feeding . . . I'd be perfect.

Another method that is actually growing more quickly in the industry is the "free access" stall (the method I referred to briefly above). In these farms, sows can enter or leave a private stall as they please. A door closes behind them, preventing other sows from entering. They are fed individually in the stall, which also prevents a dominant sow from eating more than her share and depriving others. Ironically, most sows spend the majority of their time each day in a stall. They like the peace and quiet it affords them.

It's likely that the way swine are raised will shift from the conventional gestation/farrowing stall method to the newer small-group model because of customer demand from individuals, animal welfare organizations, and even big buyers like McDonald's pushing for legislation to make group housing mandatory. People are uncomfortable about the idea of a sentient animal spending its life in close confinement and are expressing their opinions more vocally. Group housing is already mandatory in some states, and as of January 1, 2013, a European Union mandate decreed that all European pigs must be group housed—no more stalls.

So keep your eye out for packaging that says "our pork comes from a non-stall facility" or something like that. And if you're uncomfortable with stall-based practices, become an advocate for group housing.

No matter which is used, stall and open-pen systems are designed to provide sows living environments with comfortable temperatures and clean water and feed, while protecting the baby pigs in the litter. Sows in well-managed conventional farms do not suffer from intimidation from dominant penmates, heat or cold, parasites or pests, predation, or sunburn (yes, pigs get sunburned). They're monitored daily and injuries and illness are treated immediately, which is often more difficult—or even unfeasible—in outdoor and alternative housing systems.

Many people would prefer that their pork come from pigs raised outdoors in facilities that allow the pigs to express "normal behaviors" along with access to bedding, lovely mud wallows, and so on. Thing is . . . simply moving a pig outdoors doesn't ensure its welfare. Outdoor production systems require excellent animal husbandry skills—a fact that's often not well understood by farmers who are new to the specific health and welfare requirements of a pig.

In fact, it's worth examining the common belief that "outside pigs" are happier and healthier than "inside pigs." This is a generalization that assumes ideal environmental conditions and a caring farmer, which is not always the case. Pigs are very sensitive to cold and heat—just as we are—so keeping swine outside in extreme conditions may not be conducive to either their health or their piggy happiness. I've seen pigs outside in winter suffering from the cold in polluted dirty conditions. My point, of

Sweat like a pig. Go ahead, try it. Well, you can't. Because pigs have no sweat glands. That's why they wallow in nice cool dirt. So maybe you're sweating like . . . a human.

course, is that to really be certain the animal you buy from a farmer is worth the investment, it might be a good idea to examine its living conditions for yourself, and not just on a nice, warm sunny day.

But one thing I *do* know. Many farmers, including "heritage" or niche pork farmers, raise pigs very successfully outdoors and do an excellent job. Just as in any animal-raising system, there are far more good producers than bad ones. And a happy outdoor pig is a very happy pig. (To learn more about how pigs are raised on pasture, see the section on Walter Jeffries and Sugar Mountain Farm in chapter 4.)

Can Pork Be Organic?

Yes. Both the United States and Canada have standards in place for the production of organic pork.

They're similar in most aspects, especially in the requirement for formal certification.

Generally, organic pork standards include housing that meets this social animal's living and feeding needs: enough space to move around in, access to the outdoors, and so on. Organic operations may not house pigs on slatted floors or in stalls and must provide dry bedding materials such as straw or wood shavings—from organic sources, if available.

Pigs for organic pork production have to be born from parents raised organically throughout their lives. Vaccinations are allowed only to prevent communicable diseases that cannot be controlled by other means. When pigs become sick or injured despite preventive measures, they must be treated, isolated, and removed from the organic program.

Because the pigs have access to the outdoors, organic pork operations must have a plan to minimize parasite problems: preventive measures like fecal monitoring and emergency measures in the event of a parasite outbreak. Castration and ear tags are allowed, but at the youngest possible age to avoid any undue suffering.

All feed must be organic, and pig diets may not include feed medications, growth promoters, lactation promoters, synthetic appetite enhancers, animal by-products, preservation agents, coloring agents, and genetically engineered or modified organisms (GMOs) or their products. And so on. You get the picture; it's pretty strict. However, there's no requirement in the organic standards that pigs must be on pasture; conversely, not all pastured pigs are raised organically.

In the United States, pork must carry the USDA seal for organic products to be sold as certified organic pork; a third party must certify that feeding and production requirements listed in the US National Organic Standards were followed.

In Canada, certified organic products will carry the official Canadian organic seal as well as the name of the body that has certified the product as organic.

So What Do You Think?

I know what I think. I would source local heritage or high-quality pastured pork from a farmer I know well, or opt for organic pork. Why? I want to know where my pork comes from and who raised it. I also want to know that the animal was well cared for and "lived the good life" as long as possible. Pigs are intelligent—many say smarter than dogs. They are social animals capable of forming close bonds with other creatures.

I believe that the creatures who share this world with us deserve respect. If we are going to eat animals, then we have an ethical, moral, and rational obligation to consider our actions, and I want to know for sure that the farmer who raises my pork takes very good care of the pigs. So—whether raised inside or out—as long as the method is humane, the decision is yours.

This Little Piggy Says

- Oink, oink in America, Britain, Spain, and Italy.
- Roncar in Portugal.
- Groin, groin in France.
- Grunz in Germany.
- Knor, knor in the Netherlands.
- Øf-øf in Denmark.
- Hrgu, hrgu in Russia.
- Nöff, nöff in Finland and Sweden.
- Snork in South Africa (my personal favorite).

A Heroic Pig

In 1915, two British vessels trapped the German naval ship SMS *Dresden* off the coast of Chile. The German crew escaped to shore, sinking the *Dresden* behind them, but leaving on board the ship's pig. An hour later, the pig was spotted swimming toward one of the British boats. Two sailors dove in and rescued the pig, which was named Tirpitz (after the German admiral Alfred von Tirpitz). Tirpitz, who became the ship's mascot, was eventually awarded the Iron Cross by the ship's company for sticking to his ship after everyone had deserted it. Tirpitz's head is on display at the Imperial War Museum in London.

Look into these eyes and tell me this animal is dumb. *Photo by Heidi C. Normand. Courtesy Mosefund Farm, Branchville, NJ.*

Bringing Home the Bacon

The phrase *bring home the bacon* is thought to originate with the tradition of the Dunmow Flitch (a flitch is a side of bacon). The story goes that in 1104, a local couple impressed the prior of Little Dunmow with their mutual devotion so much that he gave them a flitch of bacon. This became a tradition: The town would give a side of bacon to every married man who swore before the congregation that he had not quarreled with his wife for a year and a day. Any husband who succeeded was said to "bring home the bacon." The tradition continues in Dunmow today. So, how many of you could claim that bacon?

TOURING A SIDE OF PORK

Time to get specific. The following section will make you happy, because a pig carcass is nowhere near as complicated as a beef carcass. In fact, when I teach beginners, I always start with pork. If you don't eat pork, I'd suggest skipping to lamb. Both are much simpler to deal with (but hey, don't hesitate to plunge into beef if you want to). The attached CD butchering course is as comprehensive as I can make it and will guide you through each step of the butchering process for each animal. It'll be just as though I'm standing right beside you, spilling coffee.

Before you get cutting, it will help to familiarize yourself with the animal and its various components. Take a look at the cuts illustration, then read the next section before you begin the butchering process. You will have a much better concept of where all the parts are on your carcass.

An entire pig carcass will vary in weight, depending on the breed and finishing time. You

should expect a typical hanging weight (carcass) of about 200 pounds, which should yield about 100 pounds of usable meat, or about 50 percent of the hanging weight. Don't forget to use all the bits—lard, tail, feet, head, bones. Think head cheese, soups, stews.

Follow along with the illustration, the pig's head to your right, tail to your left.

Feet

Although there are many recipes for pigs' feet, I haven't touched on this aspect of a pork carcass, nor demonstrated (in the attached CD) how to trim them—because my carcass came with its feet removed. This is common in Vermont slaughterhouses. Why? Because the carcass wasn't scalded after killing. Pigs delivered "skin-on" are always scalded to remove hair and bristles, and—more importantly—to clean the toe pads and toenails.

Hocks

Let's start just above the back feet. From the foot, we move up to the first joint where the leg bends. From this joint to just below where the leg meets the ham or butt is the hock. Hocks can be smoked to be used as a flavoring for soups like pea soup or lentil soups and a variety of other dishes. Braising is the best way to cook fresh (unsmoked) hocks. Try braising them in sauerkraut.

Ham

Next, we look at the ass of the pig, which is called the ham. This cut presents you with choices, just as beef cuts do, and as I explained in "Beef Cuts from Tail to Nose" in chapter 7. You can smoke a whole ham. Or you can slice a fresh ham into pork steaks. You can remove bones and seam out various cuts just as you would a round of beef, and you'll end

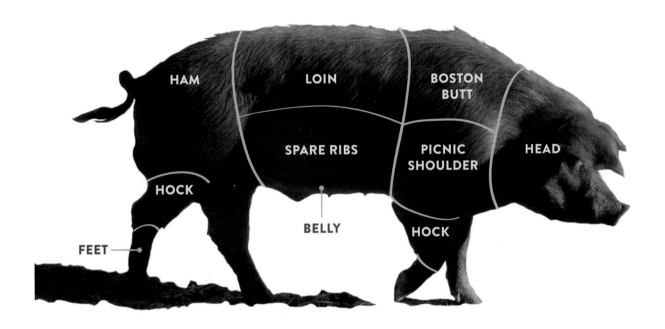

This visual guide shows which part of a pig is the territory of origin for bacon, ham, and other cuts of pork. *Courtesy Walter Jeffries, Sugar Mountain Farm.*

up with a top round, bottom round, eye of round, and sirloin tip. This process is demonstrated step by step on the CD. All of these cuts can be sliced into steaks or thin cutlets, used as boneless roasts, or smoked individually for small mini hams.

Loin

Moving right from the ham toward the head we come to the loin. The loin typically consists of a sirloin end (the part closest to the ham), a center cut (which is in the middle), and the rib end (which is the portion closest to the head). The pork tenderloin lies inside the backbone part of the loin closest to the sirloin end of the loin, running horizontally along the spine on each side. On the CD, I demonstrate how to remove a pork tenderloin from the loin.

The sirloin makes a nice oven pork roast, or it can be boned out and sliced into steaks (this is

shown on the CD), cutlets, or a boneless roast. The center loin is usually cut into chops. However, it can also be used for roast, boneless chops, or thin cutlets. The rib section of the center loin can be used for a crown pork roast, providing at least two are used. The rib end also makes a nice flavorful roast or chops, or it can be used for country-style spare ribs.

Cole's Notes

Head cheese is not cheese. It's the curious name of a cold cut made from the meat from an animal's head, usually jellied or in aspic. No eyes or brain—but things like jowls and, sometimes, feet. Sure doesn't look like cheese to me, but tastes yummy. Try it.

Pork Tenderloin

Here's a terrific recipe for pork tenderloin that's actually easy to do (no fancy opening it up, or coring and stuffing).

1. Cut one pork tenderloin into 1¼ to 1½ inch slices. Brown the slices on all sides in olive oil, then place them in a baking pan (nice tough oven pan).
2. Now slice "pie" apples (somewhat tart and hard-fleshed) and place them all around the tenderloin slices. Sprinkle everything lightly with nutmeg (not much—a little nutmeg goes a long way) and black pepper.
3. Cover with about ½ inch of sour cream. Bake in a 350°F (177°C) oven, uncovered, for 30 minutes.

Caul Fat

If you watch *Iron Chef*, you may have seen the chefs wrapping things in a kind of shredded white substance. That's caul fat. It's the lacy membrane that surrounds the viscera—the central internal organs of the animal. In a recipe, it's normally used to hold together a rolled or loose piece of meat. It eliminates the need for string, butcher's twine, or netting. During cooking, the caul fat partially melts away.

If you're buying a pig directly from a farmer, you can ask the farmer or slaughterhouse to save the caul fat for you. Or you might be able to find caul fat for sale at a decent meat market.

Shoulder

The shoulder is the area between the end of the rib end loin and the ear. The shoulder consists of two different cuts. The upper half near the backbone is known as the Boston butt; the lower half closest to the leg is known as the picnic shoulder.

The Boston butt can be sliced into shoulder chops—very flavorful—or boned out and tied to make my favorite boneless pork roast. The lower portion, the picnic shoulder, can be boned out for sausage (to learn more about this, refer to chapter 11) or slow-cooked with the bone for pulled pork. There are also other great recipes for this cut.

Head

The head has a lot of useful meat. The cheeks are called hog jowls, and they can be smoked and used for bacon or dry-cured for Roman bacon; there are also many slow-cooked recipes. Pig snouts will weigh about a pound and have more meat on them than you might think. They can be used in soups, or smoked, or roasted, and they're quite tasty. Really they are—just don't look at them. Pigs' ears are often used for dog treats—but check with your vet first. There are some folks who don't recommend them. They can also be pickled, fried, or deep-fried. Mmmm—pork fried in pork.

The remainder of the head can be cooked and used for recipes like good old Vermont head cheese, which we always used as a sandwich filler.

Belly

Now let's move a bit lower on the pig to the belly, from the ham to just behind the front leg. This is the pork belly or fresh bacon; it also contains the spare ribs. Once the spare ribs are removed, we are left with fresh pork belly or bacon, which of course is most often smoked for smoked cured bacon. Often the end nearest the ham for approximately 6 to 8 inches is used for salt pork—fat belly meat that is brined or cured and used for flavoring baked beans and other dishes. This concludes our tour of the pig.

SOOEY, SUID, SUS

Suidae is the scientific name for the family of animals that includes wild and domestic pigs, which make up the genus *Sus*. This genus includes warty pigs, the wild boar, and the domestic pig or *S. scrofa*.

Sus scrofa is found pretty much everywhere, and is generally considered the only Suid species that has been fully domesticated. In fact, recent genetic studies prove that almost every domesticated pig in the world comes from wild *Sus scrofa* ancestors.

The Pigs Who Came to Dinner

Interesting concept, domestication. As we generally think about it, the term implies conscious thought and effort on the part of human actors. This was certainly so in the case of the domestic cow. But pigs are a somewhat different story.

The traditional view of domestication is based on the principle of captivity. You have to have control over an animal in order to change its genetic makeup. Works for cows. But considering the pig, scientists question whether the term *domestication* can be applied at all. Because all through history, pigs have coexisted with us both in and out of captivity, back and forth. A kind of mutually beneficial relationship—at least (considering it from the pig's perspective) till the industrial pig farm.

Scientists theorize that there was an initial stage of domestication, during which tamer and tamer wild boar began foraging closer to human settlements (delicious garbage). Gradually, wild boar began to change, as they became more dependent on scavenging from people. From an early villager's point of view, this was great. Not-quite-so-wild boar near at hand. No need to trudge through the forest, spears in hand, terrified of being trampled by a dangerous animal. Dinner waiting right there at the garbage dump.

A **genus** is a group of very similar species. The term is used in the science of taxonomy, which identifies, describes, classifies, and names living beings.

Ah, the Domestic Life

Domestication is a process wherein people select certain wild animals and gradually acclimatize them to breeding and surviving in captivity. The aim is generally to create an animal suited to close association with humans. Aspects of the animals' behavior—particularly mating—are controlled to select preferred traits that will be passed on to future generations. Eventually, over generations, the appearance and behavior of the animal changes to conform to what its human "designers" want. Think about dog breeds.

Of course there was more to it than that. But you get the point. Eventually, those not-quite-so-wild boar became part of the village, happily snarfing down scraps in pens. And so the true domestic pig was born.

The debate about pig domestication isn't about how it happened, but rather where it happened. Unlike sheep, whose wild ancestor lived only in specific regions, pig ancestors were everywhere. There are nine possible geographic centers—some in Europe, some in Asia—where pig domestication may have occurred. One theory is that pig domestication took place during the beginning of the early Holocene period in several isolated locations in the Near East and China. Another indicates that domestication took place independently in multiple regions in addition to the Near East and China, including Europe and Japan.

So which is it? The very latest evidence—DNA samples from modern European and Asian wild boar and domesticated pigs—shows that at least

AN ITALIAN TREASURE:
ANTICA MACELLERIA FALORNI

Greve in Chianti, Italy. *Courtesy Macelleria Falorni.*

Courtesy Macelleria Falorni.

The vineyards, olive groves, and ancient hill towns of Tuscany, along with its Renaissance cities of Florence, Siena, Lucca, and Pisa, define the classic image of Italy.

In the center of Tuscany is the region of Chianti, where the great Chianti Classico wine is produced. And in the middle of Chianti lies the small village of Greve—home of the Antica Macelleria Falorni.

Founded in 1729 by Gio Botta, the shop—now in the hands of the eighth generation of the Falorni family—is exactly where it has always been, in the central piazza of the town. Caterina Bencistà Falorni handles the family's marketing (this is a big butcher shop, with more than 50 employees). Caterina says:

> In the beginning, the shop was simply a place to purchase meat. The turning point came in 1806 when Lorenzo di Angelo became the first "true butcher" of the family. My father, Stefano, and his brother Lorenzo have taken the business international, transforming the shop into a company that exports all around the world.
>
> Our most important principle is to work only with meat whose quality is superb, and whose provenance we know. This means using local suppliers and focusing on certain specific meat breeds. Our beef is from the famous white Chianina cattle; our

> pork from the rare Cinta Senese swine; and our wild boar is sourced from a nearby farmer who raises local Tuscan boar in the little village of Colle Val d'Elsa.
>
> Our second principle is to bring the countryside into the products we make. Nature has provided Chianti with a wonderful range of scents and flavors: fresh fennel, juniper, garlic, parsley, sage, rosemary, to name a few. This has led to the creation of typical salumi, characterized by a unique smell and taste. Sometimes we add Chianti Classico wine to create special products like Greve salami and sausages.

Two Special Falorni Recipes: From Our Family to You

"When we were young and we smelled that 'special' scent in the kitchen, we knew it was a special occasion," Caterina says. "Grandmother Beppa was preparing her roasted pork with milk!"

Roasted Pork with Milk

Take a good piece of deboned and trimmed pork, and tie it with string in 1-inch intervals so that it keeps

Salumi Is Not Salami

Salumi are Italian cured meat products, usually pork, but also beef (in which case they're called *bresaola*). Salami is a specific kind of salumi.

Cinta Senese Swine

The Cinta Senese pig gets its name from its broad white band behind the head (*cinta* means "belt" in Italian). Native to Tuscany, the breed dates to the Middle Ages, but almost disappeared in the 1990s. It was rescued from the brink by a group of local families. The breed's meat was recently awarded a special European Community classification as a product of "high quality tightly tied to tradition." To be certified, the meat must be from pure Cinta Senese pigs, raised only in Tuscany, free range, and fed with a combination of natural forage and GMO-free pelleted cereals. Every piglet is tagged at birth, and meats must carry certification that allows buyers to trace the product back to the original animal.

its shape. Insert rosemary and garlic cloves under the string. Season with salt and pepper. Cook it with olive oil, diced onions, carrots, and celery until it reaches a beautiful hazelnut color. Once the vegetables reduce, add enough milk to cover the meat. Simmer with the lid on until the milk reduces into a thick cream. Slice the meat and place on a serving platter. Strain the sauce through a fine sieve and pour over the meat.

Stinco di maiale al forno (Pork Hocks)

Pork hocks are a perfect dish for the winter months, served at a table by the fire with homemade bread and the happiness brought by a good Chianti wine.

Brush hocks with salt dampened slightly so that it adheres to the meat; rub in minced garlic, rosemary, salt, and pepper. Pour red wine (it's better if it's fairly young) into the bottom of a double boiler fitted for the oven. Place pork hocks on the upper grill of the double boiler and put the dish into the oven preheated to 425°F (250°C). When the skin becomes crispy, lower the temperature to 275°F (150°C). Serve steaming hot accompanied by an appetizing side dish of sautéed turnips, lentils, or beans.

Lorenzo and Stefano Falorni. *Courtesy Macelleria Falorni.*

The family (that's Caterina on the bottom left). *Courtesy Macelleria Falorni.*

All pigs like to wallow, like this wild boar snoozing. *istockphoto.*

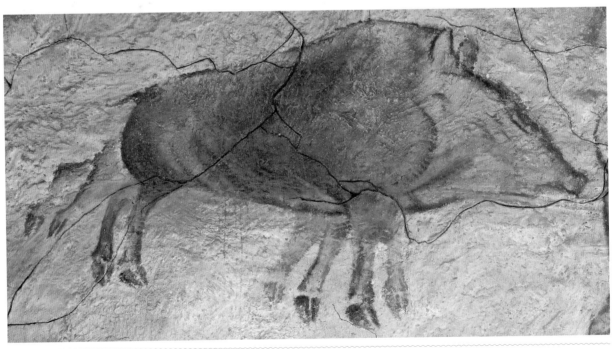

This ancient painting of a boar is from the Altamira cave in Spain. It dates to the Upper Paleolithic period, about 18,000 years ago. *Photo by HTO, Wikimedia Commons.*

two populations of wild boar in western and eastern Eurasia were independently domesticated, and that Asian domestic pigs were later bred with European domestic pigs. Dr. Greg Larson, a reader (assistant professor) in the Department of Archaeology at Durham University in the UK, is one of the scientists who discovered all of this. According to Larson, "Based on both genetic and archaeological evidence, we now know that pigs were domesticated at least twice: once in the Near East and once in East Asia about 10,000 years ago."

Butchers Yes, Farmers No

At one site of early domestication of our friend the pig, the ancient pre-agricultural settlement of Hallan Cemi in southeast Turkey, pig molars found during excavations told an interesting story: At least 10 percent of the pigs were under six months old when they were killed and eaten; 29 percent were less than a year, and only 31 percent lived to three years. This pattern of consumption is consistent with the practice of raising pigs as domesticated food animals.

What surprised the archaeologists most was what they *didn't* find at this early human settlement of small stone houses. They didn't find grains of wild wheat or barley. This flies in the face of everything we learned in school: namely, that people in the ancient Middle East were first attracted to the abundant grains that grew wild in the area, then began gathering the grain seeds and planting them. Resulting first, in settlement, then agriculture, then the domestication of animals.

Except in Hallan Cemi, it didn't happen this way. The earliest inhabitants weren't farmers; instead they depended on gathering nuts and seeds, hunting wild sheep and deer and . . . raising pigs. Scientists know the pigs were domesticated by the smaller size of their molars and by the fact that most of the slaughtered pigs were under a year old and male. This is consistent with domestication, since you'd need the females for breeding.

The Holocene Period

The Holocene period is the geological time span that began at the end of the last Ice Age. The name comes from Greek and means "entirely recent." Pretty much all of human civilization as we know it developed during the Holocene (writing, technology, pizza.)

A Very Weird Pig

This lovely creature is a babirusa. Yes, it's a pig. *istockphoto.*

The babirusa is a forest-dwelling wild pig with curly tusks found only on the islands of Sulawesi, Togian, Sula, and Buru of the Indonesian archipelago.

What's really weird are its tusks. There are four. The animal's 10 inch long upper canines grow *through* its snout and curl over its forehead, and its lower canines protrude from both sides of its jaw.

A babirusa lives on whatever it can find in its hot, humid climate: fruit, nuts, leaves, and insects. It's a good runner and swimmer, loves—like all pigs—to wallow, and chatters its teeth when it gets excited.

Cured Ham Olé!

Jamón ibérico is a cured ham produced mostly in Spain from pigs that must be at least 75 percent Iberian. Piglets are fattened on barley and corn, then released into pasture and oak groves. Just before slaughter, they're fed only acorns or olives. Hams are salted and dried for two weeks, then rinsed and dried another four to six weeks. Step 1 complete. Step 2 is the curing process, which can take more than two years.

Hams are labeled according to the pigs' diet, with an acorn diet being most desirable.

- The finest is *jamón ibérico de bellota*: hams cured for 36 months, from free-range pigs that eat only acorns.
- Next is *jamón ibérico de recebo*: from pastured pigs fed acorns and grain.
- Finally, there's *jamón ibérico de cebo*, or *jamón ibérico*. This is from grain-fed pigs and is cured for 24 months.

Is it good? You betcha. And good for you. That is, if you can afford it; it can cost almost $100 a pound! Most of its fats are oleic acid, which lowers bad LDL cholesterol and raises good HDL cholesterol.

Pig Ancestors Aplenty

Unlike us humans, pigs are still sharing the planet with their wild ancestors, because the wild boar is very much alive and eating, kicking, charging, and goring. The domestic pig's wild ancestor can be found pretty much everwhere . . . across Europe, Asia, the Mediterranean, North Africa, Australia, and the Americas.

Many of these populations are a traditional part of the environment. In other places, wild boar have been introduced for sport hunting. This has had its ups and downs. Good sport, but rapidly increasing boar populations (they can have two litters a year). To give you an idea of how fast one wild boar can become many, a few years ago several wild boar escaped from game farms in western Canada. Now there are thousands out there, and the government is offering bounties for pairs of hog ears.

Wild boar easily interbreed with domestic pigs, which happens quite often, resulting in feral populations of pigs creating environmental chaos: damaging trees, eating farmers' crops, and spreading disease. Perhaps that's why in many places you don't need a license to kill wild pigs.

So what can you do with a wild boar? Well, eat it, for one. In Europe, you can find boar meat for sale in the best butcher shops. In France, it's called *sanglier* and is very popular.

A Swineology of Breeds

The word *pig* comes from the Anglo-Saxon word *pecga* and the early German word *bigge*. The Medieval Dutch referred to the animal as *bigge*, *big*, and finally *pigge*, which became the Middle English *pigge*. Today, we use *pig* to refer to any domestic swine.

In 1930, there were 15 pig breeds listed in the *USDA Agricultural Yearbook*. Over half of those breeds have since disappeared as the drive to more "efficient" breeds has come to dominate pork production.

Today, pigs are generally grouped into two categories: commercial and heritage breeds. Industrial processors prefer the former; smaller farmers may opt for heritage, or a mix of both. Every producer has his or her preferences, and will select breed hogs with a view to his or her ultimate goal: volume meat processing, "gourmet" cuts, bacon, method of farming, and so on.

Here's where the genetics of certain European breeds comes in, since many Old Country breeds are known for producing huge litters.

Commercial Breeds

Today, most pork for sale in stores comes from "designed" animals, which you learned (earlier in this chapter) are developed by specialized swine genetics companies and sold to pork raisers in the form of boar semen or live animals. That boar semen comes from what are called **terminal sires**: animals of superior quality used in crossbreeding programs for market pigs. Among the breeds listed in the next section are the Duroc and the Berkshire—the number one and two terminal sires in commercial swine production.

Courtesy National Swine Registry, IN.

American Landrace

About 90 percent of hybrid gilts (female pigs) produced in Europe and the United States contain Landrace bloodlines. Landrace pigs are white, with drooping ears that slant forward over the pig's snout. The Landrace has a long back (perfect for bacon) and large hams—so you can see why its genes are sought after. The sow is an excellent mother, rearing large strong litters.

Courtesy National Swine Registry, IN.

American Yorkshire

American Yorkshire is the US name for the English Large White. So what color is it? You guessed it. Today, it's the most recorded breed in the United States, found in nearly every state, and has become the basis of many commercial hybrids. It's probably the principal meat-producing pig out there. The American Yorkshire has erect ears, a dished face, a long back that produces excellent bacon, big hams, and a high proportion of lean meat. It makes an excellent terminal sire, much sought after for crossbreeding.

Courtesy Sally and Michael Knight, Strattons Farm, Stirling, ON.

Berkshire

It's believed that Oliver Cromwell's army discovered this pig in the shire of Berk, England—hence the name. The Berkshire arrived in America in the early 1800s. This is a medium-sized black pig with white spots, four white-stockinged feet, ears pointing up, and a short snout. The breed is noted for its hardiness, as well as performing well as a terminal sire. Because the breed contains a high amount of intramuscular fat, its meat is highly marbled, producing juicy pork suited to high-temperature cooking.

Courtesy Ken Bauer of Woodbine, Maryland (swine breeder).

Chester White

The Chester White originated in Chester County, Pennsylvania, in the early 19th century, when large, white pigs from the northeastern United States were bred with a white boar from England. This medium-sized pig is—you guessed it—white. It has a long body, straight back, and floppy ears. The Chester White is a popular terminal sire for commercial crossbreeding operations because of its longevity, but it's not as popular as other breeds.

Courtesy National Swine Registry, IN.

Duroc

The all-American Duroc made its public debut at the 1893 Chicago World's Fair. Named after a famous 1820s racehorse, this large, red pig with floppy ears and a calm disposition converts feed to meat faster than just about any other breed. It's an ideal outdoor pig, extensively used in the production of outdoor hybrids. As a terminal sire the boar produces heavily muscled finishing pigs, the succulent meat displaying good marbling. That meat is extremely tasty, producing superb shoulders and spare ribs. It's among the top three breeds in the United States.

Courtesy National Swine Registry, IN.

Hampshire

One of the oldest American breeds, the Hampshire sports a white belt around its body, erect ears, and a nice curly tail. It's a large animal known for its ability to forage, its hardiness, and its high-quality meat. It's one of America's most popular pigs, heavily muscled and lean, with less back fat and large loin-eye areas. As a result, Hampshire boars make good terminal sires. Their carcass quality and thin skin make processing easier.

Courtesy Canada Agriculture and Food Museum, Ottawa, ON.

Lacombe

The Lacombe is a Canadian pig, eh? It's white, medium-sized, and long-bodied; it has large droopy ears and is noted for its high-quality meat. The Lacombe was developed at the Canadian Agriculture Research Station at Lacombe, Alberta (aha!), about 50 years ago by crossing Landrace pigs from Denmark, Berkshires from Britain, and Chester Whites from the United States.

Courtesy Ligon's Polands, Old Hickory, TN.

Poland China

The Poland China is one of America's oldest—and most common—breeds. These were the "walking pigs" that were driven hundreds of miles to market in the 1800s. They are big-framed and long-bodied, black with white faces and feet, and with a white tip on the tail. These lean hogs lead the industry in pounds of hogs produced per year per sow. The beautiful animal in this photo comes from the oldest Poland China hog herd in the United States, founded in 1916 by James Duncan Ligon and still in business.

Heritage Breeds

The following is a sampling of heritage breeds. Typically, heritage bloodlines go back to the time when swine were raised in open pastures. Because of their environments, they became known for characteristics like the rich and hearty taste of their meat, distinct marbling, bacon flavors, and creamy fat. Because they're not well suited to commercial farming, they have been dying out. But many farmers are rediscovering the delicious quality of these breeds, and they are increasingly in demand by discerning chefs and diners.

Photo by Jeannette Beranger, The Livestock Conservancy.

Choctaw

This is one of the descendants of Hernando de Soto's original 13 pigs. The Choctaw is small (about 100 pounds) and black with occasional white markings. Its has wattles on each side of its neck; its toes are fused into a single, mule-like hoof. Curious. Their long legs make them fast and athletic. Choctaws are self-sufficient, foraging on their own for roots, nuts, and berries. They're almost all found in the region of southeast Oklahoma that was formerly the Choctaw Nation. There aren't many Choctaw pigs left.

Photo by Jeannette Beranger, The Livestock Conservancy.

Gloucestershire Old Spot

This medium-sized pig is blond with black spots and huge floppy ears. Gloucestershire Old Spots hail from England, where they're so well known for their foraging abilities that they were called the "orchard pig." They began to arrive in the United States in the 1930s, and have made a genetic contribution to the American Spot and the Chester White. They're sought after for the higher fat ratio and taste of their excellent meat, but today they've almost disappeared in America. Not in England, though.

Photo by Jeannette Beranger, The Livestock Conservancy.

Guinea Hog

The Guinea Hog is a small black pig that originated from West African pigs brought to America on slave ships in the early 1800s. These superb foragers are also known as Pineywoods Guineas, Guinea Forest Hogs, Acorn Eaters, and Yard Pigs, and were once the most common pig on southeastern small farms. Guinea Hogs produce excellent hams, bacon, and lard. They're also gentle and easy to care for. The Guinea has a small carcass, but is well worth the effort. A typical Guinea pork chop is roughly the same size as a lamb chop. The flavor is complex, tender, and delicious. Guineas produce a ham with a substantial fat rind that makes the ham ideally suited to long curing or slow roasting.

Photo by Colleen Primmer, Primmer Pasture Pork, Brashear, MO.

Hereford

The Hereford pig—like the Hereford cow—has a reddish brown body and white face. It's a medium-sized pig unique to the United States. Developed in Iowa and Nebraska during the 1920s from Duroc, Chester White, and Poland China bloodlines, they're valued for their use as pasture animals, as well as for their good disposition. The meat's good, too.

Public domain.

Iberian

A black, almost hairless pig, the Iberian is found in the Iberian Peninsula, where they've adapted to pastoral settings, feeding on acorns from four different types of oak. Their capacity to accumulate fat under the skin and between muscle fibers, along with their diet from the nutrient-rich land, produces the typical white streaks that make Iberian hams prized for making *jamón serrano*.

Photo by Jeannette Beranger, The Livestock Conservancy.

Large Black Hog

Large and black, this pig is a forager that is happy to graze on grass and so is perfect for a pastured pork sustainable farm; it's also docile and easy to raise. It's famed for its bacon, described by eaters as "wonderful." Unfortunately, it's very rare; at the time of this writing, there were not even 1,000 Large Black Hogs in the United States. But a breed association exists and is trying to increase that quantity.

Photo by Heidi C. Normand. Courtesy Mosefund Farm, Branchville, NJ.

Mangalitsa "Wooly Pig"

The Mangalitsa or Wooly Pig is a critically rare pig from Hungary. There are three distinct types: the Blonde, the Red, and the Swallow Bellied (black with a white belly). Mangalitsa meat is well marbled and juicy. Their high-quality fat is claimed to contain a healthier balance of omega-3 to omega-6 fatty acids than seed oils. (And you thought pork was bad for you.)

Photo by Jeannette Beranger, The Livestock Conservancy.

Mulefoot

Mulefoot pigs are named for their feet; they have hooves like a mule (like the Choctaw pig). This is a medium (200- to 600-pound) solid black animal with pricked-forward ears, noted for its ease of fattening and high-quality meat, lard, and hams. A few years ago, Mulefoot pork won a blind taste test comparing the meat of eight heritage breeds, so maybe it's time to bring it back. Fast. There may be only a few hundred purebred Mulefoots out there.

Photo by Jeannette Beranger, The Livestock Conservancy.

Red Wattle

This is a large reddish pig with a fleshy wattle on both sides of its neck (you can just see them in this photo). They can weigh up to 1,200 pounds and measure 4 feet high by 8 (yes, 8) feet long. Luckily, they're gentle. These are hardy, good foragers that produce lean and tender meat. They have a mild temperament. Red Wattles adapt to a wide range of climates. The consensus is that the breed originated from large, red, wattled hogs in eastern Texas.

Courtesy James Harley, Harley Farms.

Tamworth

This increasingly popular breed originated around 1810 on Sir Robert Peel's estate in Tamworth, England, when his pigs interbred with Irish pigs. Tamworths are considered one of the closest breeds to the original European forest swine. They are reddish, deep-sided pigs with a long neck, long legs, and a long nose. Tamworth hogs typically walk and stand with an arch in their back; they have medium-sized erect ears. If you like bacon, try some Tamworth.

Tamworths are rugged animals extremely well suited for forest grazing. They also graze compatibly with cattle.

Chapter Nine

A Side of Lamb

The peak period for American sheep production was in the 1940s and '50s, when the nation raised over 55 million head a year. This number subsequently dropped, but the sector is making a comeback as more and more people discover just how delicious this meat is. Today, there are over 80,000 sheep operations in the United States, most of them small family farms. Sheep and lamb have been staples in diets of people around the world for generations and are the "center of the plate" featured entrée in many countries like Australia.

Lamb is a versatile meat that's exceptional when it's prepared properly. Thing is, mutton—meat from very mature sheep—was a World War II staple fed to many GIs, who remember its chewiness and gamey flavor with horror. This is one reason why the taste for prime lamb was lost for a whole generation. But it's coming back!

Grazing sheep—doesn't this make you feel peaceful?

WHY EAT LAMB?

One 3-ounce serving of lamb is only 175 calories, and falls into the USDA's definition of *lean meat* (meat containing less than 10 grams of fat, 4.5 grams of saturated fat, and 95 milligrams of cholesterol per 100 grams, or 3.5 ounces). Lamb is also an excellent source of protein, vitamin B$_{12}$, niacin, zinc, and selenium, and a good source of iron and riboflavin.

Why don't we eat more of it? I've heard people complain that it has a gamey taste they don't like, particularly in the trim fat. But this has more to do with how it's raised and butchered; good lamb is delicate and flavorful, with no gamey taste.

What's My Grade?

Just like beef, lamb/mutton is USDA quality- and yield-graded. The grades of lamb and yearling mutton are prime, choice, good, and utility. Mutton carcasses are graded choice, good, utility, and cull.

Quality Grade

To determine quality—how good your lamb chops will taste—a carcass is first placed into the proper age class, then evaluated for certain factors. There are three age classes.

- **Lamb:** 2 to 11 months of age
- **Young or yearling mutton:** 12 to 24 months of age
- **Mutton:** Over 24 months of age

The quality score is based on maturity, lean quality, and carcass conformation. Maturity is determined by evaluating lean muscle, rib bones, and the break joints. Lamb carcasses have break joints; flat rib bones; and pink to light red, finely textured lean muscle. Yearling mutton carcasses will have at least one spool joint; moderately wide, flat rib bones; and darker red, slightly coarse lean muscle. Mutton carcasses have only spool joints, wide flat ribs, and muscles ranging from dark red to very dark red.

Lean quality is evaluated by looking at the amount of fat streaking on top of and within the

You'll Be Gland to Know This

Every leg of lamb comes with a little something special—a gland that sits in the shank end of the leg between the bottom round and the eye of the round. This gland is embedded in a little chunk of suet (fat) inside the leg. Unless you've taken a butchering course, it's unlikely that you'll find it.

This gland should be removed before you cook the leg because it's what gives the meat a gamey flavor. Removing the gland changes the taste of the lamb, making it mild and delicious.

If you're buying a leg of lamb rather than cutting it yourself, keep in mind that most meat departments sell leg of lamb with the gland in. So it's up to you, the educated consumer, to request "oven-ready" leg of lamb—with the gland removed. Your options? Either ask your supermarket meat department to remove the gland (don't assume that it's not there!) or find a good butcher you trust.

Let Me Deconfuse You

- **Cannon bone:** A leg bone in hoofed animals between the hock joint (a joint in the hind leg corresponding to our ankle) and the fetlock (the projection just above the animal's hoof).
- **Break joint:** A cartilaginous (flexible) area of the cannon bone that is not yet turned to bone (ossified). This joint ossifies with age to become what is called a **spool joint**.

inside flank muscles. Prime-graded carcasses must have a minimum firmness score of "moderately firm," and choice-graded carcasses at least a "slightly firm" score.

It's not uncommon for ewes to give birth to triplets.

Not So Dumb

Experiments at the University of Cambridge in the UK show that sheep are about as smart as monkeys. They recognize our facial expressions, respond to their names, and navigate by memorizing their surroundings. Farmers in West Yorkshire have also reported flocks of sheep that learned how to cross cattle grids by rolling across on their backs. I'm really not making this stuff up.

Carcass conformation is evaluated by looking at aspects of the body related to its ability to yield a higher proportion of edible meat. Thus, wider, thicker carcasses are graded higher than thinly muscled carcasses.

Yield Grade

The yield grade—amount of edible cuts—of a carcass is determined primarily on the basis of fat thickness, measured between the 10th and 13th ribs over both rib eyes, then adjusted.

There are five yield grades ranging from 1 to 5, with yield grade 1 providing the greatest amount of edible meat and 5 the least amount (with the most fat) from a carcass.

The vast majority (nearly all) of lamb carcasses are graded prime or choice. And consequently most lamb purchased off the farm is also going to be prime or choice as long as it is young and healthy, and has been fed well.

BASIC SHEEPERY

Typically, an ewe (female sheep) weighs from 100 to 220 pounds; a ram (male sheep), 100 to 350 pounds. The average life expectancy of a domestic sheep is about 10 to 12 years, although they can live much longer.

In a lamb-raising setting, ewes are generally culled when they reach age seven or older, usually because their highest productivity is between three and six years old. Often they're culled because they lose or wear out their teeth (termed **broken mouth**). You can tell how old a sheep is by its teeth. Lambs come into the world with eight baby teeth arranged in four pairs on their lower jaw—no upper teeth at all. At about a year old, the central two baby teeth are replaced by two permanent incisors. At about two years old, the second pair of baby teeth is replaced, and so on. By four years old, the sheep sports a full set of permanent teeth, which immediately start to wear down as the sheep chews its way through its pasture. Gradually teeth are broken or lost, and the animal has more difficulty eating. So sheep kept in tough environments are often culled at an earlier age.

Sheep farmers have names for the stages of teethdom:

- **Lamb:** Eight milk teeth, no permanent teeth (lambs are generally up to about one year old, but can occasionally be older)
- **Two tooth hogget or yearling:** Six milk teeth, two central incisors (one to two years old)
- **Four tooth:** Four milk teeth, two central incisors, two middle incisors (24 to 36 months old)
- **Six tooth:** Two central incisors, two middle incisors, two lateral incisors (36 to 48 months old)
- **Full or solid mouth:** Two central incisors, two middle incisors, two lateral incisors, and two corner incisors (more than four years old)
- **Broken mouth:** An ewe that has lost some of her teeth
- **Gummer:** An ewe that has lost all her teeth

A female sheep is an **ewe**; a male is a **ram**. A **lamb** is a sheep that is less than a year old. A castrated male sheep is a **wether**. When an ewe has babies, she is lambing. A finished lamb is one that has completed meat development to a certain weight, at which point it creates a cover of fat, improving both the tenderness and the flavor of the meat. (See "Feeding and Finishing: Fact and Fiction" in chapter 7, "A Side of Beef.")

In North America, most domestic sheep are either meat animals or dual-purpose (meat and wool). Meat producers sell either slaughter lambs or feeder lambs. **Slaughter lambs** range in weight from less than 100 pounds (popular with many ethnic communities) up to about 135 pounds. A **feeder lamb** is one that's sold to someone who intends to feed it to a heavier weight before slaughter; naturally, these weigh less than slaughter lambs—from 50 to 100 pounds.

Since a meat farmer's profitability depends on a steady supply of high-quality carcasses, important factors in meat sheep farming include the productivity of the ewe (how often she lambs and the number of lambs per birth), lamb growth rates, and the cost or availability of feed for the sheep, although many breeds will only breed and lamb during a single period of the year.

SHEEP PRODUCTION METHODS

Sheep may be raised in several ways, depending on what the farmer's aim is, the size of the flock, and the local environment. Large commercial sheep operations are either range band flocks or farm flocks. Very large flocks are often called mobs.

Range band flocks are generally groups of over 1,000 sheep grazing in large fenced or open-range pastures. Since the areas are so large, the sheep eat only what they can find in the natural environment. Range band management is the main type of sheep operation in the United States and places like Australia and South America. In these large bands and in rugged terrain, sheep are difficult for the shepherd to count and keep track of. To assist with this chore, a black sheep is often added according to a ratio of 1 to 100 or 1 to 50. That way the shepherd only has to count the black sheep daily to know that the flock is theoretically intact (is that cool or what?). Since high-quality nutrition isn't necessary for the production of good wool, wool sheep are often kept in poorer climates or arid regions.

Farm flocks can also be large (although not nearly as large as range band flocks), but differ in that they're kept in much smaller fenced pastures. Often farm flock sheep are fed supplements like grain, but they also thrive on well-managed and intensively grazed pastures.

There are also other kinds of flocks: for example, flocks comprising only purebred sheep, hobby

Illustration by Sir John Tenniel for *Through the Looking-Glass and What Alice Found There.*

Opposite her was an old Sheep, sitting in an armchair knitting, and every now and then leaving off to look at her through a great pair of spectacles. "What is it you want to buy?" the Sheep said, looking up a moment from her knitting. "I don't quite know yet," Alice said very gently.

Farm flocks are typically found in places where large tracts of open land are less common.

Huge range band flocks are managed by shepherds and sheepdogs, who often live with the flock as it moves. *Photo by Stephen Ausmus. Courtesy USDA.*

flocks, fiber flocks (sweaters!), domestic pets (sheep instead of lawn mowers), and so on.

Just Eatin' Some Forbs Here

Sheep are herbivores. They're designed to eat plants, grazing on grass and the soft parts of other plants. Like cows, they're **ruminants**, with a complicated four-part digestive system that lets them break down plant cellulose into carbohydrates. Like the cow's, the main parts are the rumen, reticulum, omasum, and abomasum. (For a complete explanation of ruminant digestion, see "How Cattle Work" in chapter 7.)

Sheep do best on mixed low pasture, where they crop close to the ground (one reason sheep can overgraze an area faster than cows). They can also manage on hay, but do best as foragers.

A sheep will normally graze about seven hours a day in two sessions: early morning and late afternoon. Given their druthers, they'll seek out broad-leafed plants in the pasture first (the forbs), then eat grass. Sheep will eat a greater variety of plants than cattle and actually complement a cattle grazing program quite well. When fresh forage is not available, farmers typically feed sheep stored hay and other types of pre-harvested food like silage. Sheep may also be fed grain (in moderation), as well as protein sources like soybean meal.

Lambs may be finished before slaughter on pasture alone (grass-fed), or with grain supplements. Meat from pasture-finished lamb is leaner with a stronger flavor. It may also be a more healthy option for some people. Remember that what a lamb eats will affect how it tastes.

Natural? Organic? Grass-Fed? The Options

Just like beef and pork, there are options in selecting your lamb. Let's take a look.

In the United States—unlike many other countries—the average lamb carcass is larger, about 135 pounds. This is because lambs are slaughtered at a slightly older age in order to provide a higher meat-to-bone ratio (larger cuts). If you're buying a lamb directly from a farmer, ask about slaughter weight; if you want a smaller animal, say so.

Most American lamb is finished on grain, which gives the meat a milder flavor, and the majority of lamb sold in stores will grade USDA choice. If you

In many places, sheep are left to wander as they will. This sheep was encountered on a very steep hill in Scotland, happily surrounded by nice fresh grass and many miles from its home.

want more marbling and the highest quality, seek out prime, which may be hard to come by.

Can Lamb Be Organic?

Certainly, if it conforms to the national USDA organic requirement. Not to beat this drum too hard, but USDA organic standards are rigorous. All products (including eggs and dairy products as well as meat) must come from animals that are not given antibiotics or growth hormones, or fed any GMO (genetically modified organism) product. Nor can most conventional pesticides be used. A government-approved certifier must inspect and pass the farm before the organic label is granted. Note that organic lamb may be given grain during its life.

Natural Lamb

You guessed it (well, if you read the previous chapters) . . . there's no legal definition of *natural meat*. Often, "natural" products are similar to organic products; they just don't have any third-party certification. Just as often, they're not, and may be conventionally produced with variations such as no supplemental hormones, or no growth promotants, or . . . whatever. Generally speaking, natural lamb is pasture-raised, eating mostly grass. There's just no way to know. Ask.

Grass-Fed Lamb

Yes, there is such a creature. Grass (defined as a natural forage mixture) is the natural diet of sheep. The USDA defines *grass-fed* as "animals that receive the majority of their nutrients from grass throughout their life, while organic animals' pasture diet may be supplemented with grain." But a grass-fed animal can be given antibiotics or growth hormones, so if this is an issue for you, look for a label that reads GRASS-FED ORGANIC.

Whatever your lamb has been fed, you can be (almost) sure that it will not include meat or bonemeal. Feeding any ruminant (sheep, goat, cattle) bone- or meat meal is against the law in the United States.

A Touch of the Sea

Ten miles off the north coast of Scotland lies Orkney—70 small islands, only 20 of which are inhabited. One of these is North Ronaldsay, where lamb is raised in a special way. The entire island is surrounded by a 6-foot-tall stone wall—not to keep the sheep *in* the pastures, but to keep them *out*, forcing them to live entirely off the seaweed that lies everywhere on the coast. Aficionados claim that the meat of this lamb reflects the taste of the sea. North Ronaldsay lamb is a true Scottish treat.

North Ronaldsay sheep on shore rocks. *Photo by Darren Cassie.*

TOURING A LAMB CARCASS

There are four major primal cuts on a lamb carcass. They include the leg, loin, rack, and shoulder. These cuts can be processed further. In front of you is a lamb—head to the left, ass to the right . . . We'll start at the hind foot and move upward to the ankle.

From the ankle to the knee joint is the hindshank. This is great for braising (slow moisture cooking).

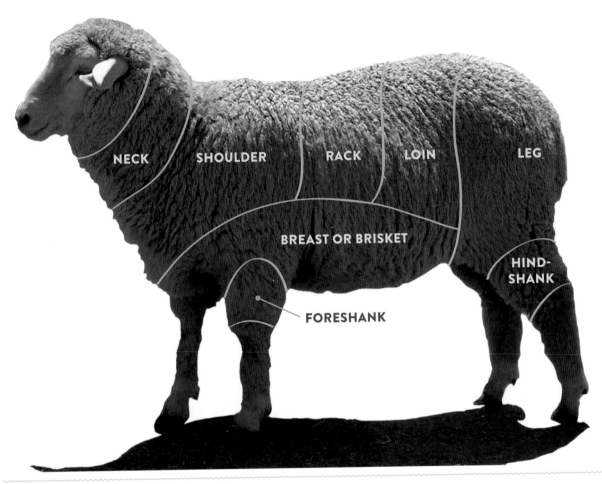

A slaughter-age lamb. *Courtesy American Sheep Industry Association.*

Leg

Roughly from the knee up to the tail is the leg. Anyone who eats lamb loves leg of lamb. Lamb leg is best suited to dry uncovered cooking: oven-roasted, butterflied and grilled, et cetera. The leg can also be seamed out like a round of beef or a fresh ham on a pig to give you a top round, bottom round, eye of round, and sirloin tip. The individual cuts will be much, much smaller then they'd be on beef or pork. However, it certainly increases your recipe options. On the CD, I show you two different ways to prepare a leg of lamb: as a semiboneless leg of lamb and as a boneless leg of lamb.

Loin

Moving left from the end of the leg to the first rib is the loin. This is where loin lamb chops come from—and who doesn't love loin lamb chops? By the way, there are three types of lamb chops: rib chops (which are usually the highest-priced at the store), loin chops (midpriced), and shoulder chops (much less expensive than rib chops). It's just like the range of quality available with steaks. All of these chops can be very tasty as long as they're prepared properly. I like to cook lamb chops on the grill: 3½ to 4 minutes per side. Under the broiler, the timing would be about the same, but it will depend on your particular oven—broiler temperatures and responsiveness vary.

Rack

From the first rib to the beginning of the shoulder blade (which is approximately a hand's width from the neck) is the incredible rack of lamb or rib, which yields wonderful rib lamb chops. The prime rib of lamb, so to speak. I generally cut these into double chops, with two ribs per chop, and again I recommend a dry cook. Another fun thing to do is cut the rack into individual chops (just one rib per chop), sprinkle with sumac or another Middle Eastern spice, and just mark them on the grill, quickly. It'll just take a minute or two per side. And then serve them as an appetizer, like little lollypops, with yogurt sauce for dipping. One bite each.

Shoulder

From the shoulder blade to the head is the lamb shoulder or chuck, which is a very versatile cut. The top part closest to the spine is the blade bone, which gives us blade bone shoulder lamb chops. The lower section is the round bone shoulder, which gives us round bone shoulder chops. The whole section can be made boneless and roasted like a leg of lamb—it's often called "poor man's leg of lamb." If you try roasting this, you may find that it has a different texture than a true leg of lamb, and there may be more waste, but it will be very tasty. Cuts from the shoulder are also good for braising.

Now let's drop a little lower to where the very top part of the front leg starts from the body of the lamb. Voilà! This is the foreshank, to be treated the same as the hindshank.

Breast

Next is the belly of the lamb. Above the foreshank and running all the way back to the leg is the breast. Lamb breast can have a pocket cut into it for stuffed lamb breast, or it can be sliced between each rib for lamb riblets. It can be used for ground lamb combined with any other lamb trimmings, or it can become part of a lamb stew.

Karen's Crusted Roast Lamb

1 trimmed* leg of lamb, approximately 5 pounds (make sure the butcher has removed the gland within the leg; if you're doing it yourself, consult the "Leg of Lamb Number Two" segment of the CD)
1–2 garlic cloves, crushed
1 teaspoon salt
2 teaspoons coarsely ground black pepper
½ teaspoon powdered ginger
1 bay leaf, crumbled into bits or crushed in a small mortar and pestle
Equal quantities of dried thyme, sage, and marjoram (amount depends on size of leg; use at least ½ teaspoon of each)
1–2 tablespoons soy sauce (you can use light soy sauce if you're trying to cut down on salt)
1–2 tablespoons olive oil

1. Preheat the oven to 300°F (149°C).
2. With a pointed knife end, stab small slits all over the lamb. Combine all the remaining ingredients into a paste and rub it over the meat, massaging as you go to make sure it gets into the slits.
3. Place the leg on a rack in a roasting pan and roast, uncovered, 18 minutes a pound for well done (170°F, or 77°C), 15 minutes a pound for medium (155°F, or 68°C), or 12 minutes a pound for rare (140°F, or 60°C). *I strongly suggest not roasting till well done! Medium rare to medium is best.*
4. Transfer the lamb to a warm plate and let it stand for 15 minutes before carving. Serve with pan gravy.

* By *trimmed*, I mean trim away as much of the outer fat layer as you can, as well as the silverskin membrane.

SEATTLE: DON AND JOE'S MEATS

This Seattle shop started in 1906 when Polish immigrant Dan Zido set up a butcher shop with a partner, eventually moving to the present location in Pike Place Market—a local favorite filled with farmers, shops, and restaurants. In 1969, Don Kuzaro, Sr., and his brother-in-law Joe Darby purchased the meat market and changed the name to Don and Joe's Meats. The current owner is the grandson of Don Kuzaro. His name? Don, Jr.

If people buy it, we carry it. Lamb tongues, kidney, head, hearts, liver, fresh-ground lamb, beef sweetbreads, tripe, feet, Rocky Mountain oysters, suet, hanger and skirt steaks, pork liver, chitterlings, caul fat, pork fat, neck bones, veal sweetbreads, hindshanks, and ground veal.

We usually have fresh ducks and rabbits, duck confit, capons, and baking hens on hand. With one day's notice, other specialties are available most Tuesdays and Fridays. Things like pheasant, quail, wild boar, and venison.

Our choice and prime beef comes from Washington Beef in eastern Washington. Choice American lamb is from Superior Farms on the West Coast and the Rosen Company out of Colorado. Washington-grown chicken and some free-range chicken are from Draper Valley. We're currently featuring Painted Hills Natural Beef in a few selected items. We look for the best-quality product out there.

Unfortunately, we no longer sell hanging beef by the half or quarter for "locker meat." Believe it or not, beef processors won't sell it to us that way

Don Kuzaro Jr. *Courtesy Don Kuzaro Jr.*

anymore. It's kind of sad. But you can still buy a whole lamb or suckling pig, or a larger BBQ or luau pig (with a few days' notice). We're proud of the quality of our ground meat. While others grind and regrind, we use only fresh meat prepared each morning to guarantee quality.

Most of our customers live or work downtown, but the market is such a draw that many people drive in to Seattle to shop. I'm sure you can find meat cheaper at other stores, but I believe we're worth it. I own this place and work here full-time, so I make sure we put out a good, fresh product and treat you right! Need help with the menu? What cut works best? Need it boned or frenched? You'll get it.

You think all meat tastes the same? I've tried a beef tenderloin from a warehouse store—sorry, it tasted like cardboard compared to our filet! We dry-age our rib and T-bone steaks; make all our own fresh sausages on-site; and have fresh veal,

American lamb, and lots of hard-to-find products such as sweetbreads, hanger, and skirt steaks. I think that large selection and customer service keep people coming back.

The debate over grass-fed beef will be settled by the public. For me, grass-finished beef is just not as tender or flavorful as grain-finished, but may be better for you. There is some debate if wet-style aging is effective. We believe it helps some, but is not near as good as dry-aging. Most of our steak meat has 21 days wet, and a few cuts have 6 more days of dry-age. We don't do more days of dry-age as there is loss of weight due to shrinkage and the outside of the meat must be also trimmed off. Also, once cut into steaks an aged piece of meat will turn dark faster, which gives you much less case life. Some high-end steak houses can dry-age the full 21 days, but it is hard to do retail unless you can get the extra money for your weight loss and sell the product fast enough.

Don Kuzaro Jr. (*left*); Ellery Heer (right); Carey Heer (*background*). *Courtesy Don Kuzaro Jr.*

Rack of Lamb, Cole Style

Add a little oil to a really hot pan. Sear the meat presentation-side (fatty-side) down first. Then sear the rest, and don't forget to sear the meaty part, too.

Season with kosher salt and pepper, sprinkle on a little rosemary, and place into a 450°F (232°C) oven. Yes, that hot! With a piece of meat as choice as this, hot and quick's the way to go. You can go up to 500°F (260°C), but do not go lower than 425°F (218°C).

Roast till nearly done. Then, for the last 3 minutes, drizzle the lamb with honey and sea salt. Should take about 8 to 10 minutes tops.

Cole's Lamb Shanks

Best to cook six at a time—this is a company meal!

Front lamb shanks usually have more meat on them than hindshanks. But they both taste great.

Soak the shanks in chilled red wine for at least 6 hours, with bay leaves, fresh thyme, onions, carrots, and celery. Drain and pat dry.

Heat a large pan over medium-high heat, add grapeseed or canola oil, and sear the lamb on all sides, so you get a nice crunchy caramelization. Then add braising liquid (any kind of stock) and braise the shanks at a low temperature in your oven (325°F, or 163°C) until they're just falling off the bone.

You can serve them whole on a plate, or you can remove some of the meat and present it artistically on top of the shank.

Oops, one last thing. Trot with me back to the front of the lamb, because we almost forgot a very tasty bit—the neck. This is incredible for bone-in lamb stew—possibly my favorite stew.

THE WILD AND WOOLLY HISTORY OF SHEEP

Wild sheep were some of the most successful early mammals and were distributed widely across Europe and Asia, as well as North and South America. There are probably six wild sheep species: the Argali and Urial of Central Asia, the Mouflon of the Middle East and Europe, the Bighorn and Dall of North America, and the Snow Sheep of Siberia.

But it's *Ovis aries*—the domestic sheep—that we're interested in. All several hundred (some say more) breeds of domestic sheep are thought to have descended from the wild Mouflon. Today, there are easily over a billion domestic sheep out there.

Sheep were the second type of animal domesticated by humans (dogs came first). This happened about 11,000 years ago in the area called the Fertile Crescent—the region of western Iran and Turkey, Syria, and Iraq where the first villages evolved.

How did it happen? More easily than the domestication of cattle. Scientists present an interesting theory that might explain how sheep were domesticated. It's highly likely that an early hunter found an adorable orphaned wild lamb and brought it back to his village. Who could resist? And since lambs are generally docile—even wild lambs—they could easily have been kept by people long enough to reach breeding age. The advantages are obvious: Sheep provide not just meat, but also milk, leather, and wool.

They vary widely in size, shape, and color. This is, of course, because they've been selectively bred for particular traits such as hardiness, docility, meat quality, milk, or wool. The original ancestors of today's sheep had long, coarse hair; over time, this became wool, and the long hair disappeared.

Or did it? Many domestic sheep breeds don't really produce much wool. Called hair sheep, they

Sheep were domesticated before this Neolithic stone circle was erected on the isle of Mull in Scotland.

generally come from the tropics, where wool would be a little too toasty. Hair sheep are becoming increasingly popular as meat animals . . . efficient and easy-care . . . no shearing required.

Hair? Wool? Fact is, all sheep produce both. The difference lies in the percentage of hair to wool, or vice versa. More hair fibers = hair sheep. Some newer breeds are crosses that grow more wool than traditional hair sheep, but make up for it by shedding their coats annually. These are called—no surprise here—shedding sheep.

The first domestic sheep weren't very woolly; they were kept for meat, milk, and skins. Little by little, people began to seek more wooliness in their sheep; by about 6000 b.c.e., archaeological sites in Iran reveal evidence of woolier sheep. Three thousand years later, wool sheep were widespread, and wool became the first commodity valuable enough to be internationally traded.

Sheep traveled with Columbus on his second voyage to the New World and were left purposely behind in Cuba and Santo Domingo to serve as easy-to-catch "food banks." In 1519, when Hernán Cortés began his explorations of Mexico and the western United States, he brought along the descendants of Columbus's sheep. Today's Navajo Churro and Gulf Coast breeds are thought to be the distant descendants of sheep brought to the new country by these early explorers.

In the early American colonies, the primary function of sheep was to produce wool. This angered England, whose internal economy was closely tied to the wool industry, and England began trying to stop wool production in its increasingly troublesome American colonies.

Didn't work. Soon colonists were smuggling sheep in under English noses. By 1664, there were more than 10,000 sheep in the colonies. Spinning and weaving were considered patriotic, and Massachusetts passed legislation that required every young person to learn how to spin and weave.

Very quickly, America was an exporter of wool goods . . . competing directly with England, which responded by outlawing the American wool trade; if a colonist was caught, his right hand was cut off! We all know what happened next, but what many don't know is that the restriction on sheep production was just as much an irritant to the colonists as the Stamp Act, and was a primary factor in the Revolutionary War.

Modern Sheep Breeds

In the United States, breeds of sheep are classified into six categories: meat, fine wool, long wool, dual-purpose (meat and wool), hair, and double-coated/minor breeds. Since we're talking butchery here, let's take a look at the main meat breeds (which include both wool and hair sheep).

A Selection of Primary Meat Breeds

Many meat breeds are named after regions of the United Kingdom where the breed originated. So if you've ever vacationed in England or Scotland, you'll recognize some of these names . . .

Courtesy Rod and Bernadette Nikkel, Barrhead, AB.

Canadian Arcott

This is one of three breeds developed by the Canadian government and released to farms in the late 1980s. The breed is the result of a crossbreeding program including the Ile de France and Suffolk, resulting in a new breed with strong meat characteristics. It is a medium-sized sheep that produces an excellent carcass with good meat-to-bone ratio. The ewes are easy lambers and require low to medium maintenance. They adapt well to either pasture or confinement management. The rams make excellent terminal sires to improve meat characteristics on many other breeds.

Courtesy Cheviot Sheep Society of New Zealand.

Cheviot

Cheviots developed in the bleak Cheviot hills of the UK, where they still can be found grazing at high altitudes. A small white-faced sheep with bare head and legs, its primary use is to produce meat. Cheviots are adaptable, do well on poor pasture, and also produce nice wool.

Courtesy Rich Fitz, Ugly Dog's Farm, Davison, MI.

Clun Forest

This breed originated in the Clun Forest region of the UK. The breed is hardy and long-lived; ewes lamb easily. This is a multipurpose sheep producing meat, wool, and milk. Its lamb and mutton are very high quality, and the breed does well on grass alone.

Photo by Jeannette Beranger, The Livestock Conservancy.

Dorset Polled and Dorset Horn

Two types: with or without horns. The hornless (polled) are very popular with farmers. These are medium-sized animals that are prolific lambers, often producing two crops of lambs a year . . . a boon to the small meat farmer.

Courtesy Richard J. Davis, Hampshire Down Sheep Breeders' Association, UK.

Hampshire

Widely used for terminal breeding sires, Hampshires are large sheep with black faces and medium wool fleeces. They're good milkers and produce excellent carcasses. The Hampshire is named for the English county of Hampshire, where it evolved from a variety of breeds. Hampshires are known for their fast growth and lean, heavy-muscled carcasses; they're also efficient converters of forage into meat. They adapt well to small farm flocks but do not do well in big range flocks.

Courtesy Elizabeth Cavey, Cavey Family Montadales, Carroll County, MD.

Montadale

An American sheep popular with farmers, developed in the 1930s to marry the best qualities of farmed mutton sheep with western range sheep. It's medium-sized, covered with white wool, and has a bare head, bare legs, and black hooves. Montadales are prolific, producing lambs with lean carcasses.

Courtesy A. J. Hambley, Oxford Down Sheep Breeders' Association.

Oxford

Oxfords are medium to large, with dark brown faces and wool on their legs. They're used primarily as terminal sires in farm flocks. The breed is fairly prolific, with good mothering ability, producing large, meaty carcasses.

Courtesy Montana Jones.

Shropshire

Shropshires are medium to large with a dark face, and are popular for farm flocks focusing on meat production. They are prolific, possess good milking and mothering abilities, and are used as terminal sires in market lamb production. The lambs are hardy, fast growing, and produce lean, well-muscled carcasses.

Courtesy Gary Jennings, American Southdown Breeders' Association.

Southdown

One of the oldest purebred sheep breeds, the Southdown is a small to medium-sized sheep with a light brown face. The breed is popular for smaller farms, because they need less space than other sheep and are economical to maintain, producing excellent meat on less forage. Southdown sheep are known for the tenderness and flavor of their meat, and have often won carcass competitions in livestock shows.

Courtesy Tracy Hagedorn.

Suffolk

Black-faced Suffolk sheep are the most popular breed in the United States and are raised primarily for meat. This is the largest-sized breed in the US, with rams ranging from 250 to 350 pounds and ewes from 180 to 250. This, along with its fast rate of growth, makes the Suffolk a good sire breed for crossbred slaughter lambs. Suffolk lambs grow rapidly and produce carcasses that yield a high percentage of usable meat. They are adaptable to farm flocks but do not thrive well in large range flocks.

Courtesy Eugene and Niki Fisher, Fisher Texels, Indian Valley, ID.

Texel

Texel sheep originated on the island of—you guessed it— Texel, on north coast of the Netherlands. The 19th-century Dutch breeders of this sheep were seeking an animal that would produce well-muscled lambs with superb meat. They succeeded. Today, the Texel is known for its superb muscle development and lean fat content; it's a top choice for excellent lamb. The Texel sheep is white-faced with a black nose and black hooves. It's medium-sized, hardy, and adaptable to natural forage conditions.

Photo by Jeannette Beranger, The Livestock Conservancy.

Tunis

Surprise! Not white! The reddish Tunis is the result of crossing sheep imported from Tunisia in the 18th century with local American sheep . . . making this one of the oldest American breeds. Tunis sheep are very productive and produce vigorous lambs over an extended breeding period. The carcass yields a high percentage of meat that is exceptionally fine-flavored—not only the lamb, but also the mutton.

Combination Breeds (Meat and Wool)

This little group is also called dual-purpose. So . . . great meat and great wool.

Courtesy Half Diamond Farm, AB, www.canadiancharollaissheep.com.

Charollais

This sheep breed originated in France, where they were raised alongside Charolais cattle. Charollais (note the extra "l") are medium-sized, heavy sheep with a clean head and fine to medium, dense wool. As terminal sires crossed with other ewe breeds, they produce lean high-quality lambs, weighing from 8 to 11 pounds (depending on whether they're single lambs, or twins/triplets). Carcasses are lean and heavily muscled, particularly in the loin and hindquarters, with an above-average yield.

Courtesy Bryan and Gina Vinihg, Federer Corriedales, Cheyenne, WY.

Corriedale

Corriedales are medium-sized, white-faced sheep. Developed on the Corriedale Estate on the South Island of New Zealand in the 1860s, they are the oldest of all the crossbred sheep. Corriedales were developed to be true dual-purpose sheep, producing excellent wool as well as a high-quality carcass. They are prolific, often have multiple births, and adapt easily to a variety of climatic conditions. The quality of both their fleece and their meat offers farmers the opportunity to maximize flock profits.

Courtesy Elizabeth H. Kinne Gossner, Stillmeadow Finnsheep.

Finnsheep

Finnsheep are a true multipurpose breed. In Finland, where they originated, they're raised for meat, wool, and pelts. In the United States, their primary use is in crossbreeding programs aimed at increasing the lambing percentage of commercial flocks. This is because the breed naturally has a high incidence of multiple births—three, four, five, or even more lambs at a time. Finnsheep are sought after to produce the extremely lean meat preferred by many ethnic communities.

Photo by Jeannette Beranger, The Livestock Conservancy.

Gulf Coast

One of America's oldest breeds, Gulf Coast sheep originate from flocks brought to the New World by Spanish explorers in the 1500s. The breed evolved across the Gulf Coast, and are small to medium-sized white to tan animals, with an occasional darker sheep. They're used both for their non-scratchy wool and their lean and succulent carcasses.

Courtesy US Targhee Sheep Association, Mardy Rutledge.

Targhee

Targhee sheep are medium- to large-sized animals that thrive in most climate and forage conditions. They were developed by the USDA for use in western range flocks in the United States, and so are very well suited to American western range country. They produce good-quality market lambs.

Hair Meat Breeds

These are sheep that don't produce wool, so they don't need shearing. Hair sheep are used primarily for meat and hides. They generally produce lean, excellent meat with no "muttony" taste. There are two categories of hair sheep: true hair sheep and shedding sheep. True hair sheep reflect their original ancestry. Like the wild Mouflon, domesticated about 10,000 years ago, their coat is mostly hair. True hair sheep dominate in hot or humid regions like the Caribbean and West Africa. An example of a true hair sheep is the Barbados Blackbelly. Advantages of hair sheep include excellent vigor, better heat tolerance, and parasite resistance as well as year-round breeding.

Hair sheep are preferred for halal (Muslim) slaughter, because halal buyers look for uncastrated ram lambs, and hair sheep are often left whole. They are also liked by organic and grass-fed producers because of their greater parasite resistance (which means less use of veterinary medicines or antiparasite chemicals).

Photo by Colleen Kozlowski. Courtesy Barking Rock Farm.

Barbados Blackbelly

This distinctive sheep can be seen everywhere in Barbados—generally, in the middle of the road freaking out drivers (happened to me). It has a black belly, and its red-brown hair coat doesn't need shearing. The Barbados Blackbelly is a medium-sized, hardy, year-round breeder, with exceptionally lean, mild-tasting meat. Grass-fed farmers like them because they can be easily raised on lower-quality forage.

Photo by Linda Bennett. Courtesy Ebbers Red Sheep.

California Red

A cross between the Tunis and Barbados breeds, the distinctive California Red begins life a darkish red-brown, gradually turning light tan except for their gold to cinnamon-colored legs and head. Rams often sport red manes that contrast with their beige bodies. Rams typically weigh 225 to 250 pounds; ewes, from 130 to 150. Their meat is particularly tender and contains less fat than the meat of many other breeds.

istockphoto.

Dorper

The Dorper is solid white or white with a black head. Highly fertile, Dorpers are hardy and adapt to all kinds of environments (they were originally bred for harsh South African conditions). Dorpers are not quite hair sheep. They have a mix of hair and wool, but don't need shearing; instead, they shed their wool in late spring. Their lambs grow quickly to maturity, yielding excellent carcasses.

Public domain.

Katahdin

Katahdin sheep originated in Maine. Their woolless, easy-care hair coat comes in many colors (surprise!), and the sheep are extremely tolerant to both hot and cold conditions, as well as to parasite infestation. These are the largest of the hair breeds, producing a lean, well-muscled, very tasty carcass. They can be any solid color or color combination. The purpose of the breed is to efficiently produce meat, which is tender and mild-flavored.

Courtesy Josef Slejtr, Czech Republic.

Romanov

A pure breed, the Romanov sheep originated in the Upper Volga region of Russia, eventually making its way from Europe to Canada, and then the United States. Romanovs are born black and gradually turn gray as they age. They are increasingly popular as a meat breed, since ewes will often produce quadruplets—or more—at one lambing (the breed record is nine). This, plus the fact that they become sexually mature at three months old and will breed throughout the year, makes them the ideal sheep for meat farmers. Unfortunately they produce very poor carcasses and are usually used in crossbreeding programs to produce marketable lambs with acceptable lean muscle content.

Courtesy Reed Farms, TX.

Royal White® Sheep

Notice the symbol? The Royal White® is currently the only sheep breed with its own registered trademark; it's actually registered as a living animal with the US Trademarks Office in Washington, DC. It's a genuine pure hair sheep (no wool in its past), and has been trademarked to protect its qualities from dilution through crossbreeding. The Royal White® is a large sheep whose lambs grow extremely fast, yielding a carcass of very tender, high-quality, low-cholesterol meat. Its secondary use is for the premium leather garment trade.

Courtesy River Bend Ranch, Day's Creek, OR.

St. Croix

The St. Croix was developed in the Virgin Islands, where it's known as the Virgin Island White. It's completely wool-free, so no shearing. The breed lambs unassisted, and the carcasses yield mild, tender meat.

Courtesy James Harley, Harley Farms.

Wiltshire Horn

The ancient Wiltshire Horn breed is often classified as a hair sheep, but is in fact a wool breed that sheds its wool. The hair coat is thick enough to handle northern winters. Both sexes are horned. The breed has good meat conformation and a good growth rate. They are productive (twins are common) and noted for their ease of lambing. Generally, lambs yield about a 6 percent higher meat-to-bone ratio than do other sheep breeds.

Chapter Ten

Cluck–I'm Your Domestic Chicken

I know what some of you are thinking. Everybody knows how to cut up a chicken!

Well . . . nope. (And a note here from coauthor Karen: "I thought I knew how until I watched Cole's method. Apparently, I had no idea!")

Chickens are one of the most adaptable and palatable meats out there, with literally thousands of recipes for roasting whole chickens, braising quarters or halves, frying up legs, poaching breasts, stewing chicken livers, and so on. Plus, of course, there's spicy buffalo wings.

But—as with other meat animals—there's more to chicken than meets the eye. By now, you know that I'm a believer in knowing as much as possible about the meat we eat so often. We've taken a closer look at beef, pork, and lamb. Time for chicken . . .

A homebody chicken wanders a seldom-traveled road in northern Scotland.

WHY EAT CHICKEN?

Like all meats, chicken is high in protein. One serving (defined as 3 ounces) of roasted chicken thigh delivers about 21 grams of protein; a roasted breast, about 25 grams. Neither contains carbohydrates. Fat content varies, too, between thighs and breasts. The same 3-ounce serving of roasted breast has about 7 grams of total fat; 2 grams of this is saturated. The thigh contains 13 grams of total fat and about 3.5 grams of saturated fat. If you're trying to keep your fat intake down, simply remove the skin. Both thighs and breasts are quite low in calories: 170 in roasted breasts, 210 in roasted thighs.

You'll notice that I specified *roasted*. That's because how you cook your chicken will affect its nutritional values, which I'm sure you know. So if you're coating it in sour cream and deep-frying it in lard, or marinating it and cooking in a sauce that includes maple syrup or honey, all the numbers will go up.

HOW THE COMMERCIAL POULTRY SECTOR WORKS

In the United States and many other countries, the poultry industry is vertically integrated. This means that all the processes involved in producing meat chickens are handled by different member firms owned by one company. This consolidation of many small companies means that most of the world's commercial broiler chickens are produced by three large companies (Cobb-Vantress, Aviagen, and Hubbard). Each company sells a variety of different branded crosses in the form of chicks, which are distributed worldwide via their own specialized

Cole's Notes

If you want to check how fresh a chicken is, smell the *bone* parts first. If it's a whole chicken, try to smell inside the cavity. They'll love you at the supermarket.

fleet of environmentally controlled chick trucks or by commercial airline carriers.

Each company owns breeder farms that produce hatching eggs, a hatchery where chicks are born, a broiler grow-out facility where they are raised, a feed mill to produce their food, a processing plant where they're slaughtered and processed, a rendering plant, a further processing plant for by-products, and a retail distribution system that delivers products to store warehouses or individual stores.

With this degree of vertical control, you'd guess right in assuming that the elements of a commercial meat chicken don't just happen, but are carefully designed. Each company produces millions of pedigreed chickens a year. As the birds are grown, they are assessed for "best traits"; only the top 3 percent are kept for use as pedigree breeders.

What attributes are they looking for? To find out, I talked with Professor Peter Ferket of North Carolina State's Prestage Department of Poultry Science.

Poultry breeding companies work with very sophisticated programs that develop genetic lines or families with specific traits aimed at commercial markets. Male and female lines are created. The male line is typically selected for growth rate, feed efficiency, and meat yield, and the female line for reproductive performance, fitness, and compatibility with the male line to produce high meat-yield traits.

Great-grandparent stock are crossed to produce the male and female line parent stock. Then these parent stock lines are crossed to produce the commercial broilers that are grown for meat. Each line and cross is developed for specific commercial objectives. For example, a Ross 308 strain [Ross is one of Aviagen's brands] would be suitable for small broilers sold as whole carcasses or KFC fryers and grown to about 36 to 42 days of age. The Ross 708 strain is a slower-growing high-yield white meat bird.

In general, poultry breeders select for growth rate, feed efficiency, health and disease resistance, feather development, carcass confirmation, and meat characteristics. Modern poultry breeders work with molecular biological (genetic) markers associated with economic and consumer-important traits. But—like swine breeding companies [as you read in chapter 8], no genetic splicing, modification, or cloning takes place. In other words, there's no such thing as a GMO chicken. All strains are developed through genetic selection and crossbreeding.

Since none of the broiler breeding companies use artificial growth promoters, growth hormones, or GMO or cloning technologies, they must stringently monitor the health of their birds. Attention to detail is intense. According to John Hardiman of Cobb-Vantress, broiler breeding companies focus on improving the growth rate and meat yield of the bird while reducing carcass fat and reducing skeletal, skin, and feathering defects. Starting in the 1990s, additional technologies have come online, including X-ray examination of the skeleton for improved joint health, measurement of the birds' walking ability, blood oxygen measurement to assure adequate cardiovascular fitness, meat quality testing, and ultrasound estimation of live bird yields.

How the Birds Live

In conventional meat chicken production, birds are raised in what are called free-run systems. They live in barns that vary in size according to the size of the flock. The chickens are not de-beaked (a practice that relates to egg production systems); nor are they caged. Instead, they roam freely around the facility on a floor covered with wood shavings or straw (termed a deep-litter system). Water and feed are always available, there's good airflow, and the temperature is carefully controlled, 'cause chickens don't like being too cold or too hot. In fact, in free-range operations where chickens may go in- or outdoors as they please, when it's cold or hot outside you'll usually find them in the barn.

Space per bird is determined by the size of the bird being grown and ranges from about half a square foot per bird for small birds to 1 square foot or more for large birds. The chickens are free to roam throughout the litter floor house, take a dust bath, and find a comfortable spot. Nest boxes or perches are not needed for broilers.

Their lives aren't long. Birds are slaughtered when they reach the appropriate size for the product being produced. This ranges from about 28 days for "Cornish" chickens (2.2 pounds) to about 62 days for the largest broilers used for deboning (8 pounds). Broilers between these weight ranges are used for eight-piece cuts, portion-controlled operations, or for sale as roasters.

As soon as chickens are slaughtered, they're chilled to take the carcass from its live temperature to the proper "meat temperature" of about 32°F (0°C). Chilling is done either via a cold-water bath or an air-chilling procedure. Most chickens in the United States are water-chilled—placed into a vat of ice water. In air chilling, carcasses are blown upon with cold air as they travel slowly along a track, a process that takes longer than water chilling. Many people prefer air-chilled chicken, claiming that the taste is better or that the chicken loses less water during the cooking process. You'll have to judge for yourself.

Generally, chicken is shipped from the slaughtering facility to its various destinations (restaurants, supermarket distribution centers, et cetera) on the same day it's slaughtered. As for the recommended time to consumption, this, too, like everything in meat and life, varies.

Assuming the chicken is frozen and remains so, it will be good for up to a year or more. If it has been packaged fresh using the MAP (modified atmosphere packaging) method described in chapter 5, then it will remain fresh in its unopened package for up to three weeks and remain perfectly edible for another week after the package is opened. If the chicken is packaged fresh in a regular container (not MAP), then it should be good for a week to 10 days.

Is There Another Option?

Of course there is. If you're buying chicken from an individual farmer, it is very unlikely to be a commercially developed and distributed bird. Instead, the chicken will be one of several meat breeds commonly used for small operations. Or you might decide to raise your own birds by purchasing eggs or chicks from a hatchery that deals directly with individuals and small farmers, like Murray McMurray Hatchery in Iowa. We spoke with owner Bud Wood about the birds they raise there.

We sell many breeds of chicken here, but only three of those are meat birds (the others are collectors' birds or egg producers). The best-known meat chicken is the Cornish Rock cross. This is virtually the same bird as commercial chickens and comes from the same commercial lines. Bred for large breasts and super-fast growth, Cornish Rocks are usually butchered at about eight weeks.

We also sell the Red Ranger™ (yes, this breed is trademarked). These are a heritage cross without the big breasts of the Cornish Rock and the meat is a little darker. Many people think the flavor is better—more "chickeny"—and I agree. Since the Red Ranger™ doesn't grow as quickly as a Cornish Rock, it takes a little more time to get them to slaughter; I recommend butchering them at from 10 to 12 weeks. They're extremely popular with our customers.

Our third meat chicken is the Dark Cornish. These are the old-fashioned Cornish birds—the

Cock of the Walk

A cock of the walk is someone who struts around in a pompous way. The expression derives from the rooster's strut around the farmyard—master of his domain.

Bird Brain?

Do chickens think? Is there anything in those tiny little heads? You may be surprised to learn that they're actually quite smart.

Chickens recognize other flock members from their features (my, what a pretty beak you have), and have more than 20 distinct vocalizations. They've even been taught to complete agility courses. Now, that's something I'd pay to see.

Courtesy Tim Daniels, poultrykeeper.com.

ones originally crossed with White Rocks to develop the first commercial broilers. They still retain the big Cornish breast and deliver more chicken flavor. People buy them because they don't want a commercial broiler.

Just a Little Terminology

Chickens are social. They prefer to live in groups of hens dominated by one adult rooster. Juvenile roosters are tolerated by the head rooster until they're mature and start competing for hens.

- A **bantam** is a miniature chicken. Bantams are usually about a quarter the size of a regular chicken.

- A broiler is a young chicken bred for its meat, and generally killed at about 9 or 10 weeks old. Broilers typically weigh between 2½ and 3½ pounds.
- A female chicken over one year old is called a **hen**.
- A female chicken under one year of age is a **pullet**.
- A Cornish hen is simply a young ordinary chicken butchered very young (about four weeks old).
- A broody hen is a hen that wants to sit on eggs to hatch them—kind of a motherly chicken.
- A male chicken over one year old is a cock or **rooster**.
- A **capon** is a castrated rooster. They're delicious, but almost impossible to find nowadays.
- A male chicken less than a year old is called a **cockerel**.
- A baby chicken is a chick.

A comparison of a Red Ranger™ (left) and a Cornish Rock (right) at the same age of eight weeks. *Courtesy Murray McMurray Hatchery.*

Feed and Labels

I know, I know—you're fed up with logos and ads about "free-range," "organic," "pasture-raised," "natural," "grain-fed," and on and on. We discussed many of these labels in earlier chapters, but there's a real reason for knowing more about labeling practices in poultry raising because, frankly, the way chickens are raised, fed, and slaughtered has just as much effect on the final palatability and flavor of the meat as do raising methods for cattle, pork, and lamb.

Let's Talk Chicken Feed

All chicken feed is grain-based (corn, wheat, some barley, oats, sorghum, soybean meal, canola meal, vegetable oil, and various grain by-products). Note that some formulas may contain animal protein by-product meals.

Feed formulations also contain nutritional additives that supply calcium, phosphorus, vitamins, and trace minerals, as well as substances that inhibit mold and toxicity in the feed.

Free Range

This term may give you a mental image of flocks of chickens scratching their way to the horizon, but it actually means that the birds have "access to the outdoors" for some part of the day. Could be five minutes, could be all day. There's no official definition. By the way, chicken labeled ORGANIC must be free-range, but not all free-range chickens are organic. And for you northern folks, here's an interesting note from the Chicken Farmers of Canada: "Be wary of 'fresh' free range chicken in stores when it's -30 degrees outside, it may have been frozen product defrosted for sale and should not be re-frozen."

Free Run

Free run—as you learned earlier in this chapter—is the conventional way of raising meat chickens. It's different from free range in that chickens don't have to have access to the outside but must be able to move around freely within the barn. Though there is no legal definition of this, all chickens raised for meat are considered free run.

A free-range outdoor chicken wanders through the front yard of a small inn.

Farm-Raised

You may see this on labels. It's a nice way to get you feeling all warm and fuzzy, but really (and technically) all chickens are farm-raised—even those raised in what activists term factory farms.

Natural

Under USDA regulations, a "natural" product must contain no artificial ingredients, coloring, or chemical preservatives, and must be minimally processed.

Organic

As noted in previous chapters, the USDA only allows use of the word *organic* for foods produced according to certain stringent specifications, and organic foods must be certified (as explained in chapter 7). Your call whether it's worth the extra price or not; many people opt for grain-fed chicken that's not organic. Of course, if you can find a farmer who raises chickens ethically and in a sanitary fashion, that's by far the best route. Go visit and check for yourself.

No Hormones Added

The Food and Drug Administration of the United States forbids the use of hormones in poultry production. So does Canada. So if you see a label stating proudly that NO HORMONES WERE USED in the raising of the chicken, take a closer look at the package. There should be another label somewhere that explains that no hormones are ever used in poultry production. Ah, the bragging that goes on in this business . . .

Raised Without Antibiotics, but What About Other Stuff?

RAISED WITHOUT ANTIBIOTICS means that no antibiotics were administered during the raising of the bird. Generally speaking, the use of antibiotics in feed for poultry is decreasing significantly because of consumer demand. However, it's normal practice to raise young birds on medicated feeds.

So while the birds may not have been treated with any antibiotics, they may have received

Medicated Feed

Medicated feed contains a pharmaceutical drug that is controlled by the FDA. There are two categories of medicated feed additives: Category I are drugs that are considered safe and don't require a withdrawal period before slaughter; Category II drugs do. Feed additive medications can either be antibacterial (antibiotics) or antiparasitic (for treatment of coccidia and worms).

Coccidiosis

Coccidiosis is an intestinal infection of birds and domestic animals such as dogs, caused by a microscopic parasite that attaches itself to the lining of the gut. Small poultry raisers may use preventive acidifying preparations like apple cider and garlic tonic administered in the chickens' water to combat the disease.

medication via their feed to prevent or control common diseases like coccidiosis or to increase their rate of growth.

This is an interesting topic, because until very recently, the poultry feed additive of choice to combat coccidiosis as well as to improve weight gain and feed efficiency was roxarsone (3-nitro-4-hydroxyphenylarsonic acid), under the brand name 3-Nitro. This drug is a source of organic arsenic, similar to the form of organic arsenic that is present in almost everything in nature, including a glass of drinking water, and until very recently thought to be safe for humans.

In 2012, an FDA study found trace amounts of inorganic arsenic in chicken livers of birds that had been fed roxarsone. Although the FDA claimed the meat was safe to eat and poultry raisers endorsed the use of roxarsone, in 2013 Maryland lawmakers passed a bill banning the use of all arsenic-based drugs in chicken feed, citing links between inorganic arsenic and human illnesses such as neurological problems in children and possible links to cancer, heart disease, and diabetes.

What is interesting is that Pfizer, the distributor of roxarsone, voluntarily suspended its sale, but because the recall was voluntary, no legal requirements were ever set in place regarding the use of the drug. Thousands of chicken farmers simply stockpiled the feed additive and continue to use it.

However, the supply is quickly running out and the drug is no longer commercially available in the United States. In addition, many poultry-raising companies are using a vaccine in place of coccidiostats. To me, this story illustrates how substances we eagerly assume are safe may later prove not to be. Yet another cautionary story.

Grain-Fed

This claim needs questioning, since chicken feed—as you now know—is mostly grain. The term is probably used to convey the idea that the chicken never consumed any animal by-products. In this case, the term is misleading, since many grain feeds do include by-products. What you want to look for is a label that is much more specific, such as ALL-VEGETABLE DIET.

All-Vegetable Diet or Vegetarian Grain-Fed

Commercial poultry feed often contains processed protein, fats, and oils from meat and poultry by-products (chickens eating chickens!). If you want to be sure that what the chicken ate never included meat or poultry by-products, look for labels like ALL VEGETABLE or VEGETARIAN GRAIN-FED. Although chickens are omnivores, they can be raised as vegetarians, as long as there's enough protein in their feed.

TOURING A CHICKEN

I'm going to take you on a tour of a chicken carcass, and this tour is different from the others we've been on—I'm going to illustrate it with a whole carcass instead of a live bird, because for most of you, the de-feathered version will be easier to deal with.

Voilà! The whole chicken.

First, the wings. Wings are wonderful for (as you all know) buffalo wings, or other recipes like deep-fried wings, braised wings, and more. Generally, the small wing tips are discarded, but don't—instead, reserve them for chicken stock.

Onward to the chicken legs, which can be divided in half into thighs and drumsticks. You know this already, right?

Chicken breasts deliver wonderful lean meat that usually works best when lightly cooked in braising

Jewish Chopped Liver

This is also fondly known as "heart attack on a plate."

    Chopped onion
    Oil or rendered chicken fat (which I know you won't
        use, but you might as well know how this used to
        be made)
    Small pile of *fresh* (!) chicken livers
    2–4 hard-boiled eggs
    Salt and pepper to taste

Sauté the onion in the oil till soft. Add the livers and cook, turning, till done (about 8 minutes). Then chop or grind (do not puree) with the eggs. Season.

As Karen's mother used to say, "Try it, you'll like it. Eat, eat." *What? You think it looks icky?* You should try it . . . it'll put hair on your chest. Anyway, if you don't use the giblets in a chicken, why pay for them?

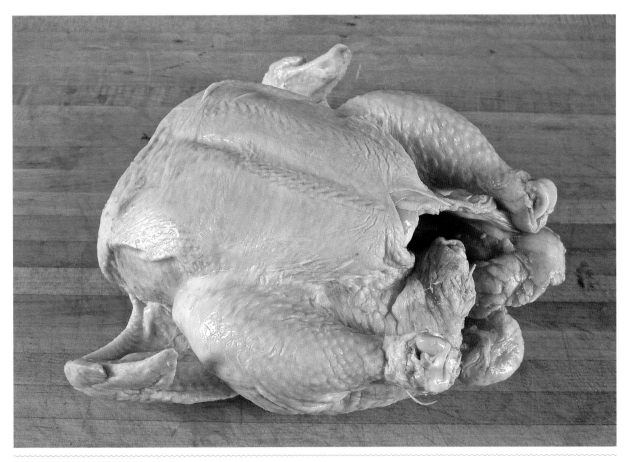

The whole enchicklada.

liquid, then dressed with a tasty sauce. Cook them gently, because overcooked breasts are dry and can be a little stringy.

You'll be left with the bones, the back, the wing tips, and the neck. These are the starter ingredients for wonderful stock, which can be made today and frozen for soup stock. Simply place the bones, neck, and back into a large pot; add one onion, a few cut-up carrots, herbs to taste, and water; and boil down into chicken stock. Remove the vegetables and cool the stock, then skim off the layer of fat from the cooled broth. Use now or freeze.

There's more, of course, because you have insisted that the farmer give you the chicken giblets: the liver and gizzard. Gizzard is very tasty, and can be sliced or chopped and cooked in chicken broth as part of the soup-making process.

It's also good on its own (when I was a kid, this was one of my favorites).

As for chicken liver, there are many recipes for this, either whole, sliced, or ground to become chicken pâté or part of sausage.

CHICKEN BREEDS

Gallus gallus domesticus . . . or, your friendly domestic chicken. It's everywhere. There are more chickens out there than any other bird. Billions of them, clucking away in twos and threes in backyards, in tens or twenties in farmyards,

Maple Syrup Mustard Chicken

Coauthor Karen says: I made this up one evening. Now it's a fixture.

4–10 deboned or bone-in chicken thighs (each person will eat two thighs, so calculate accordingly; you can also use the drumsticks)

Salt and pepper

1–2 tablespoons unsalted butter

French Dijon mustard*

Maple syrup (Vermont, of course; amber is best because its flavor is more intense)

Dash of hot sauce, if you like

A few tablespoons of orange juice, also if you like

1. Skin the thighs (unless you like fat, in which case, don't).
2. Salt and pepper lightly.
3. Melt the butter over low heat.
4. In a small bowl, mix about 2 to 3 tablespoons of mustard with about 2 to 3 tablespoons of maple syrup (if you want more bite, use more mustard). Also mix in the hot sauce and orange juice, if you're using either or both.
5. Add the melted butter and stir. Pour the marinade over the chicken thighs.
6. Marinate the thighs in the mustard/syrup/butter mixture for about 2 hours in your fridge.
7. Place in an ovenproof, low casserole dish and into a 350°F (177°C) oven until done (if you're uncertain, cut into one), turning once and basting occasionally. This normally takes from 30 to 40 minutes when I make it.
8. Serve over white rice or couscous with the liquid from the pan.

* The mustard I use is wonderful and is called Moutarde Forte de Dijon (not expensive). Do *not* use regular ballpark mustard; this should have a bite. The success of the recipe depends entirely upon the quality of the mustard.

in thousands upon thousands in poultry-raising barns, and lined up neatly in plastic packages on supermarket shelves.

For some aficionados, they're collectibles—available in all colors, sizes, and featherings. A few people even keep them as pets. Hey, if parrots can be pets, why not chickens? And you've probably heard about the latest trend sweeping America: urban chickens. Some cities have passed zoning laws allowing people to keep a few chickens in their backyards (other cities won't hear of it!). I don't see the problem. Which is louder? Sirens, pneumatic drills, and honking cars, or three small chickens?

But where did they come from?

Most scientists think that domestic chickens originated from Southeast Asian wild junglefowl (still out there scratching around) about 10,000 years ago.

About 3.6 million years ago, red junglefowl diverged from early green junglefowl, then much, much later (vaguely dated between 58,000 to 12,000 years ago), the domestic chicken diverged from the red junglefowl.

And when you take a look at this picture of a red junglefowl living happily in California today, you can easily see the resemblance. In fact, I can't tell the difference.

When Did It All Happen?

That's a little hazy. But chicken bones found at the Neolithic site of Cishan in northern China were dated to 8,000 years ago recently, much earlier than the oldest domesticated chicken bones found in India (those were only 4,000 years old).

The Chinese bone fragments were slightly larger than those of wild junglefowl but smaller than those of modern chickens. Most of the bones were from cocks (roosters), indicating that people were killing and eating the male animals and saving the hens for breeding and eggs.

Since the red junglefowl isn't native to the region where the bones were found, it's thought that

Can you tell that this isn't a domesticated chicken? It's a Red Jungle Fowl. *Courtesy Kong Vang, http://redjunglefowl.webs.com.*

domestication took place much earlier, perhaps in the region of Vietnam and Thailand, where the junglefowl is endemic.

Comin' to America

When the Spanish conquistadores "discovered" the Incas in 1532, guess what they found, besides lots of gold? Chickens. Lots of them, raised by the Inca people. How did chickens get from Asia to America?

The answer lies—as it so often does—in archaeology. Chicken bones from a site called El Arenal in Chile were recently radiocarbon-dated to earlier than about 1425 c.e.—well before Columbus's first trip to our shores. The scientists compared DNA from the Chilean chicken bones with Polynesian chicken bones discovered on early sites across the Pacific, and found that the two were very closely related.

Which proves that the first American chickens were the descendants of chickens brought to the shores of the continent by early seafaring Polynesians.

You wouldn't believe some of the shapes, sizes, feathers, and colors of chickens. There are more than 60 recognized breeds, with names that are a delight to say aloud—such as the Rumpless Araucana, the Frizzle, and the Sabelpoot (say that one three times fast)—as well as the ones we all know, like the Bantam, the Plymouth Rock, the Rhode Island Red, and the White Leghorn.

Show Chickens

There are ornamental breeds—look but don't eat—like the Phoenix (long long tails), the Chinese Silky (fluffy, not feathery), and on and on. Take a gander at some of them below . . . but don't eat them. They're show birds!

Appenzeller Spitzhauben.

Sumatra rooster.

Buff Orpington.

Modern Game.

Courtesy Tim Daniels, poultrykeeper.com

Meat Breeds

It surprised me to learn that there aren't very many meat chicken breeds out there.

The main reason there are fewer meat breeds than egg-producing breeds is that—according to Shannon Burasco, president of CWT Farms International (a sister company of Aviagen)—"Most meat chickens out there are heritage breeds and there has never been a true broiler among heritage breeds; in other words, no heritage bird has ever been designed as a meat producer."

CWT raises broiler hatching eggs for the meat chicken sector, and offers three meat breeds: the Red Ranger™, the Naked Neck, and a dual-purpose (meat and egg) bird called the Rainbow.

Photo by Jeanette Beranger, The Livestock Conservancy.

Brahma

The Brahma chicken is considered a superb meat bird, as well as a great egg layer. It's large, with hens reaching 13 to 14 pounds and roosters up to 18 pounds, and comes in three color varieties—light, dark, and buff. From the mid-1850s to about 1930, this was the leading American meat breed, possibly because one large Brahma could feed the entire family.

Brahmas do best in colder climates, and are easy to raise—they don't range widely and are happy to stay in one place.

Photo by Jeanette Beranger, The Livestock Conservancy.

Cochin

The Cochin is a huge bird with a calm nature, so easily tamed that many become "house chickens." Developed in China for large size, Cochin excel as capons, when slaughtered at about 16 months old. Cochins are heavily feathered, so they do well in cold climates. They're hardy and adaptable.

Photo by Jeanette Beranger, The Livestock Conservancy.

Cornish Rock Cross/Cornish x Rock

This ubiquitous breed is the base of the commercial meat industry. Home flock owners who have tried them sometimes complain that they're "top-heavy" and often have difficulty walking. This is because most Cornish Rocks have been selectively bred for extremely large breasts.

Photo by Jeanette Beranger, The Livestock Conservancy.

Jersey Giant

The Jersey Giant chicken is—as you can tell from its name—a large bird. Roosters can weigh up to 13 pounds and hens, 10 pounds. This is a dual-purpose breed, but is best known for its excellent meat—perfect for roasting. It will take about eight to nine months for a Jersey Giant to reach good slaughtering weight.

Courtesy S&G Poultry.

Naked Neck

The Naked Neck developed in very hot climates. Since chickens can't sweat like us, a naked neck is an efficient cooling mechanism. So no, they weren't plucked to look this way. Naked Necks are often considered the tastiest eating bird; their meat texture is superb, and their thin skin crisps perfectly when the bird is roasted.

Photo by Jeanette Beranger, The Livestock Conservancy.

Orpington

The Orpington is a hardy, fast-growing chicken that produces both eggs and meat. Orpington chickens are excellent broilers that will weigh about 2 to 2½ pounds at 8 to 10 weeks of age. They can also be grown on to about five months as roasters.

Courtesy S&G Poultry.

Red Ranger™

The Red Ranger™ (trademarked by CWT Farms) is a meat chicken that delivers a different taste and texture from the commercial Cornish Rock; it's described by many as richer and chewier. Its breast is in proportion to the rest of its body (not overlarge); it has yellow skin, shanks, and beak, strong legs (no problems walking), and dark feathers. The bird is an excellent forager, is usually fed an all-vegetable diet, and does well in organic programs.

Photo by Jeanette Beranger, The Livestock Conservancy.

Rhode Island Red

Rhode Island Red is a name that's probably familiar to you. That's because it's very popular as a farm chicken. These birds are very good layers of brown eggs, perhaps the best layers of all the dual-purpose breeds.

Chapter Eleven

Making Sausage and Value-Added Products

In essence, there are three reasons to make sausage. (1) It tastes good. (2) It tastes good. (3) It tastes good. There are as many kinds of sausage as you have the imagination—or daring—to create.

If you have dipped your toes into butchery, you'll have discovered that during the cutting process you created a certain amount of trimmings. Don't throw them away! These are the glorious essence of so many delectable fresh sausages. You could use the trimmings to make ground meat, but using them for sausage opens up a new world of cooking and eating pleasure.

Ready to make some sausages?

LET ME COUNT THE SAUSAGES

You've got breakfast sausage, dinner sausage, and sausage for the tailgate grill. Sausage is a fun, convenient food . . . like the hot dog, which (you guessed it) is a sausage. Usually sausage is stuffed into a casing, making it easy to put on a bun or eat as it is with Dijon mustard or another condiment. But sausage can also be made into patties, which fit on a hamburger bun or can be served without one on a plate. Sausage can be grilled, broiled, fried, or chopped up to flavor other recipes.

There are British bangers (a traditional dinner sausage), Italian hot and sweet, Greek or Italian luganega, Polish, American breakfast, and chorizo. There are sausages incorporating cheese, fruit, wine, or brandy and other liquors. There are sausages with pistachio and other nuts such as pine nuts. There are lamb sausages, beef sausages, pork sausages, and wild game and poultry sausages.

Some folks buy a whole pig and turn the entire thing into sausage. Getting a bunch of friends together for a sausage-making party is also a lot of fun, especially when it comes time to cook and taste your creations.

Some see scrap—I see sausages!

For this chapter, we'll be focusing on pork sausage. Once you see how easy this is, you'll be off to explore all the other meat options.

WHY IT'S GOOD FOR YOU

It's pretty obvious that if the pork you use for your sausage comes from a local, humanely raised pig and your other ingredients are fresh, the sausage you'll be making will be much healthier than anything you'd buy from a supermarket.

I like my sausage to contain about 25 to 35 percent of fat. Keep in mind that most store-bought sausage can contain up to 50 percent fat. Making your own ensures that you know what's in it.

WHAT YOU'LL NEED

Sausage making is pretty simple, but you will require some basic tools:

- A meat grinder (can't make 'em without one). If you're making only 4 to 5 pounds at a time, you can work with a sausage grinder attachment for your food processor, rather than investing in a meat grinder.
- A sausage-stuffing attachment for the meat grinder (which is merely a funnel). This is very inexpensive, and probably came with your grinder.
- Plastic food-grade gloves.
- Two or three good-sized plastic food-grade bins (luggers) to hold your sausage trimmings, for mixing (which you'll do by hand), for the finished product, for waste, and so on. You bought these for butchering (see the CD), so you already have them.

Suspicious Sausage

Be cautious about sausage sold in your local store labeled STORE-MADE (unless you know the store staff and owners well and trust them). Most often these are made with a commercial mix of pre-measured seasonings and are definitely not the store's own recipe. And many stores use questionable product in their sausage, such as chops or roasts that are on the edge of spoiling.

Before a sausage class. (Your meat grinder won't look like this one!)

- A scale to weigh your trimmings. This isn't a "nice to have"—it's very important to use the exact weight your recipe calls for. If you change the proportions of ingredients in a sausage recipe, you change the flavor of the sausage. So, for example, if a recipe calls for 5 pounds of pork, I would not allow more than a 4-ounce fluctuation from that amount.
- Casings—ah, the nub of the thing. *Always* use natural casings (casings made from animal intestines that have been cleaned and washed thoroughly). Some grocers carry natural hog or sheep casings or can order them for you. Buy salted casings rather than pre-flushed casings. Yes, you'll have to flush out the salted casings yourself, but salted casings have almost an indefinite shelf life when refrigerated, and flushing is easy. And don't think all sausage must be in a casing . . . if you don't want to get into stuffing, make sausage patties instead.

Casing Choices

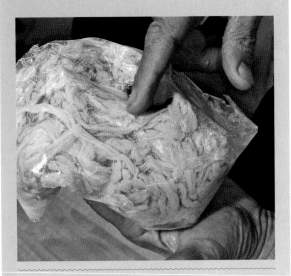

A bag of hog casings.

Hog casings will serve almost any purpose, although there are other casings. Sheep is the smallest in diameter of the natural casings—in the ¾- to 1-inch range. Hog casings are available in a few different sizes, from the smallest 1¼-inch range for things like Italian sausage to a medium size of 1½ to 2 or more inches for blood sausage. You may prefer smaller or larger sausages; the choice is yours. I use hog casings for everything except hot dogs, for which I use sheep casings.

SAUSAGE MAKING STEP-BY-STEP

For the beginner, I recommend starting with a small batch—5 pounds or so. I generally grind the meat once through the medium-hole plate of my grinder (although for a meatier texture and certain recipes, you may wish to use a coarse plate).

Step 1: Cubing the Meat

No matter what meat you're using (pork, veal, beef, poultry, Vermont crocodile), you start the same way—by cutting it into 1- to 1½-inch cubes. You can estimate its fat content by eye (go for it!), trying to keep total fat content in the 25 to 35 percent range.

Step 2: Chilling

After you've cut your meat into cubes, it's a good idea to chill it in the freezer for 30 minutes or so before grinding. This will make it easier to handle. If you're using a sausage grinder attachment for a food processor, put the attachment in the refrigerator, and keep it there until you're ready to start grinding.

Step 3: Grinding

Using the medium plate on your grinder, grind the meat once into a food-grade plastic tub or pan large enough to give you room to work easily.

Step 4: Add Dry Ingredients

Next, add the dry ingredients. Your recipe may call for herbs, spices, dried fruits, or even nuts: The photo shows dried apricots and cranberries. Make sure everything is pre-measured and ready to go.

Step 5: Hand-Mix Lightly

Loosely mix the dry ingredients into the ground meat by just lifting the meat and turning it over. Always mix gently, but thoroughly: No squeezing or packing!

Step 6: Add Wet Ingredients

Once the mixture is well mixed, add any wet ingredients that your recipe calls for: wine, juice, or liquor.

Step 7: Final Mix

Continue to mix thoroughly by hand until everything is well blended. Gently, gently, please.

Step 3.

You can make sausage using a combination of meats. This bowl holds ground pork and beef ready for mixing.

Step 4.

Step 5.

If you plan to sell any of your sausage, wear gloves while mixing!

Step 6.

Casings or Not?

Your road now splits. Do you want traditional sausages? Next stop—casings. Would you rather not bother? Form the mixture either into patties and cook, or into a log for chilling and cutting into uniform slices. Or just use the mixture in bulk for stuffing other cuts of meat. How about turkey stuffing? Mmmm.

A final note: Your sausage will taste best if it's refrigerated overnight, giving all the different flavors time to intermingle.

Now let's return to the step-by-step instructions for stuffing sausage. First comes casing preparation. I buy my casings from Syracuse Casing Company. Not only is this a great company that will deliver to individual customers, it's the *only* casing company left in the United States that doesn't source its casings from China. Yes, you read that right. Their casings come shirred onto a plastic sleeve so that they don't tangle and can be easily removed for flushing or insertion onto the stuffing funnel of the meat grinder. This makes them much easier to use for beginners. Note that the correct term for this is *tubed casing*. We've included the website for Syracuse Casing Company in the resources section.

Step 8

Take the casing—still on its sleeve—and rinse the salt off by holding it under a small stream of cold running water. Then slip the casing off the plastic sleeve and find one end. (Obviously, there are two ends!)

Step 9

Hold that end open while the rest of the casing sits in a clean dish or bowl. Let a gentle stream of cold water run into the casing. Lift the rest of the casing a bit at a time out of the pan so that the water runs through the entire casing. This is called flushing.

Step 10

Repeat the flushing process two to three times, then drop the spotlessly clean flushed casing back into the dish. Leave the end of the casing hanging over the edge of the dish (so you can find it when you're ready to start stuffing).

Try to flush out the exact amount of casing you'll need for your batch. Usually this works out to 1 foot of medium hog casing to 1 pound of sausage. So if you're making 5 pounds of sausage you'll need 5 feet of casing; 8 pounds of sausage, 8 feet of casing; and so on.

Step 11

Let your flushed casings sit in cold water for 5 to 10 minutes. This will help them achieve exactly the right consistency: easy to slip onto the sausage funnel and the perfect "slidiness" for stuffing.

Step 12

With your meat grinder set up for sausage making, gently slip the casing onto the meat grinder's stuffing funnel.

Step 13

Let the machine do the work; just guide the meat mixture with your hand. Don't squeeze!

Step 14

You will end up with one *very* long sausage tube.

Step 15

Decide what length you want your sausage link to be, then pinch the tube at each end and twirl until the sausage forms a link (it will happen naturally, and is much easier than it looks). Trust me on this.

Step 16

Once you've twisted the whole sausage into links, use your small boning knife to cut between each link. Voilà! Sausages!

Step 17

Unfortunately, the final step in the process is cleanup! Take a look back at chapter 6 and re-read the cleanup section. Remember that you're working with organic products, so sanitize and clean thoroughly.

Step 7.

Casings before flushing.

Step 10.

Step 12.

It's easy. With my right hand, I'm gently pushing the ground meat into the machine with a plastic "pusher" (unseen), letting the machine do the work of stuffing. It will happen quickly, but it's very easy to do.

This alternative view is from a sausage class at Sterling College; I've included it to give you a better sense of what the sausage casing will look like as it's being filled. In this case, the students are using an industrial sausage stuffer.

Just-stuffed, but not yet twisted into links.

Step 15.

Step 16.

MIXING IT UP

There are zillions of sausage recipes out there (well, at least hundreds). You can pick one that tempts you or make up your own. Once you know the basic procedure, you're ready to go.

One sausage that's popular in my state is breakfast maple sausage. I like to grind apples, cranberries, and dried apricots into the sausage meat. I've included the recipe, which features pure Vermont maple syrup.

Everyone's taste buds are different; you can easily alter a recipe to suit your taste. Some folks like less salt in their sausage. With sausages like British bangers that call for bread crumbs, I choose to use fewer bread crumbs than the recipe calls for, or even make gluten-free bangers—eliminating bread crumbs altogether; something I was always willing to do at a customer's request.

Adding your favorite wines or liquors to sausage also gives you a variety of unique flavors. I make a special sausage at Christmas that I call Porketta with fresh garlic, Romano cheese, white wine, fresh parsley, and roasted red peppers. It has a great Christmas color with its red and green ingredients, plus it's absolutely delicious. So don't limit yourself. Let your imagination guide you.

Recipes

Cole's Porketta Sausage

9 pounds pork shoulder or boneless bottom butt (23–35% fat)

3 tablespoons kosher salt

12 cloves garlic, minced

1½ cups grated Parmesan or Romano cheese

2 tablespoons lemon zest

6 tablespoons coarsely ground black pepper

1 cup dry white wine

3 large roasted red peppers, chopped
½ cup chopped fresh parsley

1. Grind the meat once through the medium plate of your grinder into a large food-grade plastic tub. Mix all the dry ingredients together loosely, including the grated cheese.
2. Next add the wet ingredients, roasted red peppers, and parsley, and mix thoroughly again.
3. Stuff the mixture into medium hog casings, pinching and twisting into your desired lengths.
4. Cut the links and chill overnight in the fridge, allowing the flavors to intermingle. Reminder—casings are not necessary; this sausage can be used in bulk or made into patties. Grill, broil, or pan-fry—delicious! This is my Christmas sausage; its red peppers and green parsley give it a nice Christmas touch.
5. Will keep for 2 to 3 months in the freezer.

Cole's Maple Breakfast Sausage

6 pounds pork shoulder or boneless pork butt (23–35% fat)
2½ tablespoons kosher salt
1 tablespoon ground sage
1½ teaspoons McCormick Poultry Seasoning
2 teaspoons finely ground black pepper
1½ teaspoons sugar, white or brown
¼ teaspoon red pepper flakes
⅓ cup dark pure Vermont maple syrup (don't use anything but the real deal, and remember that the best maple flavor comes from a darker grade of syrup)

Grind the meat once through the medium plate of your grinder into a large food-grade plastic tub. Add all dry ingredients and mix together loosely. Add the maple syrup and thoroughly mix again. Stuff into casings, pinch and twist, cut, and chill.

Making Sausage Patties

I prefer my maple breakfast sausage as patties. Whatever recipe you try, you may want to make patties rather than links, so you don't have to fuss with casings.

Prepare for making patties by cutting a large sheet of freezer paper and laying it on a counter or other flat surface—waxed- or shiny-side up. Place the sausage mixture on the freezer paper and form it into a log, leaving about 5 inches of paper on each end of the log. Try to make the log as uniform as possible and approximately 4 to 5 inches in diameter.

If stuffing sausages into casing isn't for you, don't despair. Just fry up some patties.

RED BANK, NEW JERSEY: CITARELLA'S MEATS AND DELI

Citarella's Meats and Deli's website says it all: "From Our Family to Yours: Family Owned and Operated Since 1901." This is a wonderful story of an immigrant family that prospered doing what they love.

Andrew and Carmela Citarella came to the United States in 1900. Soon Andrew was selling meat from his front porch, eventually opening a store in Red Bank, New Jersey. Carmela took meat-cutting classes. A few years later, their son Ralph joined the business. He remembered people riding to the store on horseback after local fox hunts. Meat deliveries were made by horse and wagon, then by a Model T truck.

Years rolled by, as Ralph and his son Andrew continued the tradition. When a new location became available one block from where Andrew had grown up, he bought the building. And that's where we found *his*

son Ralph—the fourth generation. Ralph notes, "In my family the first son is named after his grandfather—it's a sign of respect."

Citarella's covers all the bases, offering every kind of meat in a variety of specialty presentations, along with homemade sausages, Italian specialties like eggplant Parmesan, meatballs, spaghetti sauces, soups, potpies, and much more. People drive miles for their sandwiches, and they also cater. Ralph recalls:

My great-grandfather and my grandfather used to go to the meat houses and pick out their own meat. By the time my father took over, the meat houses knew that if what they sent him wasn't up to Citarellas' standard it would come right back . . . plus they'd get a very unpleasant phone call! I have seen him do it myself, and I've done it, too.

Ralph and his mother, Maria Citarella. *Courtesy Ralph Citarella.*

Our standards are simple: We want meat that is the least processed, preferably without growth promotants or antibiotics. Our pork and turkeys are farm-raised, vegetable-fed without growth promotants or antibiotics, and are minimally processed. We only carry American lamb. Today they call Australian and New Zealand lamb "imported lamb." Sounds better, like imported cheese or imported wine. But in fact this is just a creative way of marketing what I believe is an inferior product.

When it comes to beef we look for the best marbling. Notice I didn't say the most *marbling. We buy right where the fine line of prime grade and high-choice grade meet. You want a good degree of marbling, but not too much or the meat becomes fatty. Grass-fed beef has no marbling. Today a lot of stores are selling beef as grass-fed, claiming it's healthier. If it's marbled, chances are it's not grass-fed.*

People think butchers are expensive. But we're very competitive with the supermarkets and deliver better quality. I had a customer last winter come into my store and tell me she went to a supermarket to buy a beef tenderloin. When they told her the price, she said for that much she could go to Citarella's and get a better one. Well, she was wrong, because although she did get a better one from me, it cost $30 less than the supermarket!*

As supermarkets and discount chains try to pawn themselves off as high-end butcher shops, it makes it more challenging for small butcher shops to survive. Today both men and women work, and [their] time is limited—you could say that convenience has become our biggest competitor.

Supermarket meat-cutters pretty much use the band saw to cut everything up as fast as they can. Butchers mostly use knives and cut with more detail and precision, which makes for a better presentation. It could be as simple as cutting steaks or as fancy as preparing a crown of lamb (or pork) or a prime rib roast giving your holiday meals a real gourmet look.

Personal service and better quality is what keeps our customers coming back. It has for over 100 years. As long as there are people that know the difference between quality and commercial, there will be butcher shops.

Apple-stuffed loin of pork. *Courtesy Ralph Citarella.*

When you have the log formed as tightly and uniformly as you can, fold the ends of the freezer paper up over the ends of the sausage log and start rolling it up in the freezer paper. Once you have it rolled, tape it so it stays together and place it into the freezer for an hour.

Remove the sausage from the freezer, unroll it from the paper, slice the log to your desired thickness, and package as desired and freeze. This is good for two to three months in the freezer. Pan-fry for a delicious breakfast sausage.

Storage, Wrapping, and Freezing Tips

Fresh sausage can be kept in the fridge for about three days. Most will keep quite well in the freezer for two to three months if properly wrapped or packaged. If you're freezing sausage, there are a few options.

Vacuum-packing is probably the best, but I've had very good luck with freezer storage bags if you squeeze out as much air as possible, then wrap the sausage-filled storage bag in freezer paper as well. This really does work nicely.

Thaw frozen sausage slowly, in the refrigerator; it will retain more of its natural juices this way. So get your friends, neighbors, and family together and plan a sausage party. Make teams and challenge each to invent a different sausage. The really fun part will be the sampling. Don't forget the wine!

VALUE-ADDED PRODUCTS

Value-added products are simply cuts of meat prepared in a way that adds value, thus increasing their attractiveness (as well as the price if you buy them at the store). For example, boneless skinless chicken breast might retail for $9.95 per pound. So say you take one breast, two slices of prosciutto, two slices of Cabot slicing cheddar, and a small bunch of asparagus. Then roll the asparagus tips in the prosciutto and cheddar cheese, butterfly the boneless skinless chicken breast, and tuck the rolled-up asparagus, prosciutto, and cheddar cheese inside the butterflied breast. Then dip the whole shebang into buttermilk, roll it in bread crumbs, and garnish with a little paprika and parsley. Ta da! Now you have a value-added product called chicken roulades. These will sell for $10.99 per pound. Learning to make this kind of toothsome delight is essential if you aspire to become a gourmet butcher someday. But even if you don't, it's great to learn how to make this type of value-added meat product for yourself to save some money and to enjoy the fullest range of possibilities for enjoying the meat you eat.

MORE EXAMPLES, PLEASE

Flank steak is another good candidate for the creation of a value-added product. My recipe for Cole's Pig in a Flanket is easy, tasty, and eminently salable. Once again, you have added value to an

Don't they look delectable? *Public domain*

$8- or $9-per-pound flank steak, increasing its price to $12 or $13 a pound.

Ground meat is another great value-added candidate. Think of the many ways it can be pre-prepared: Jalapeño cheddar, bacon cheddar, mushroom and Swiss, pickle and cheese, and many other tasty ingredients can be mixed right into gourmet burgers, for instance. The list goes on. Many butcher shops have kitchen facilities just for their line of value-added products. Much can be produced in a small kitchen: meat sauces, shepherd's pie, meat pasties or empanadas, ready-to-go meals, pâtés, stuffed burgers, whatever the mind can dream up.

Displaying these items in a meat case decorated with a bit of blanched kale will entice any customer to buy. Blanching the kale, then running it under cold water, will give it a nice bright green color. Remember that customers buy with their eyes. And of course it's crucial to use only the freshest product and freshest ingredients as well as the highest quality. You only have one time to make that first impression on the customer. If you want it to be a good one, you had better do it right the first time.

The following are just some of the products and cuts to which I've added value, thus fetching a higher price:

- Flank steak: Pinwheels, Pig in a Flanket
- Chicken breast: Roulades, stuffed breast
- Flat iron steak: Pinwheels
- Ground meat: Gourmet burgers, eight varieties
- Ground meat: Spaghetti sauce
- Ground meat: Meatballs
- Ground pork: Orange meatballs
- Ground lamb: Lamb sausages
- Pork chops: Stuffed with homemade quality stuffing
- Pork loin: Pork pinwheels with fruit and jam
- Pork butt: 80 different varieties of sausages

The list goes on and on. Again, value-added, like sausage, is limited only by your imagination.

Cole's Apple Cranberry Chicken Roulades

Take half a chicken breast and butterfly it open.

On one side of the butterflied breast, lay down two slices of Vermont Cabot cheddar cheese.

Sliver about half a red apple, skin and all (you want an apple that's both sweet and tart).

Take some juice-sweetened cranberries or craisins, wad them up in your hand so they form a tight little ball, and set those on top of the apples. (How much exactly? Enough to fit into the palm of your hand.)

Put another slice of cheddar on top of everything, then fold the breast over and squeeze everything together so it's nice and tight.

Dip into buttermilk, then roll in unflavored bread crumbs (I like to make my own, but you don't have to).

Bake in a 350°F (177°C) oven for about half an hour.

Option: Spread some orange marmalade over the breasts just before baking.

Empanadas are made by folding dough around a meat or vegetable stuffing. *Public domain*

The Gourmet Butcher's Pig in a Flanket

1 flank steak, butterflied by you (if you've taken my
  course) or your butcher
1 tablespoon Montreal Steak Seasoning
  (available everywhere)
5–8 ounces fresh spinach leaves
4–5 rectangular slices (⅛ inch thick) Cabot cheddar
  cheese (Cabot, because it's one of Vermont's best
  products—so use it!)
5–6 *roasted* red peppers, sliced into manageable pieces
1 pork tenderloin

1. Lay out the butterflied flank steak and season with
   the steak seasoning.

2. Layer evenly with one-third of the fresh spinach.
3. Then layer with cheddar cheese slices.
4. Then layer with the second third of the spinach.
5. Press down with both hands (important).
6. Lay out the roasted red peppers evenly.
7. Finish with the final third of the spinach.
8. Place the pork tenderloin at the edge of the flank
   steak nearest you, and tightly roll everything up with
   both hands, tucking in loose spinach as you go.
9. Tie the center of the roll with string, then tie string
   around each end.
10. Finish by adding string around the rest of the roll,
    making sure that each (future) slice (which will be
    1½ to 2 inches thick) has at least two strings around it
    to keep it together.
11. Slice and place into a baking pan.
12. Bake in a 350°F (177°C) oven for approximately
    30 minutes.
13. Let rest for 5 minutes and serve.

Spread the flank steak with an initial layer of spinach.

Add slices of cheddar.

Top with red peppers.

Add the final layer of spinach.

Add the pork tenderloin . . .

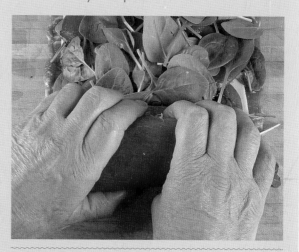

. . . and roll up.

Here's the tied Pig in a Flanket.

Bake in slices.

Chapter Twelve
A Better Way of Thinking About Meat

In North America, the predominant and by far the largest system of meat production is built on the premise that producing enough meat to feed a huge and increasing population requires a highly industrialized approach. This system—the conventional meat production sector—is a multibillion-dollar business where profit is often measured in pennies per pound. It is optimized for high production and economy of scale.

The other meat production system involves a patchwork of thousands of small to medium-sized farms and ranches representing a diverse range of production systems and variety of animal breeds. Can this system feed millions? Probably not, but this begs the question: *Do we need to eat meat every day?*

If I had my druthers, we'd all be eating humanely raised and slaughtered meat from animals that had spent their lives outdoors. But this is perhaps naive, and takes us into a discussion of land-use practices, demographics, and economic factors that are beyond the scope of this book.

When I set out to write this book, I knew I'd have to explore all facets of the meat business, including the most industrialized methods of raising and slaughtering animals. I expected to be disturbed by what I found, and to some degree I was. Yet my exploration was not the horror show I expected. All the farmers, ranchers, managers of large meat production plants, and feedlot operators I spoke with were demonstrably committed to the welfare of their animals; nobody tried to evade my questions or obfuscate.

Is there room for improvement in conventional production systems? Yes. This was not disputed. Every one of the ranchers and processors I spoke with was as disgusted as I am by certain unethical and cruel animal handling practices out there (the ones you've read about in the media, illustrated by distressing photos). But these—as they pointed out—represent the "bad apples" in their industry. I'm certain we could find similar distressing images and cruelty in the "local farm" sector (in fact, I know of some).

As for the killing floor, change is happening there, too. As Dr. Temple Grandin's humane slaughtering practices are increasingly adopted by big processors like Cargill, slaughter itself is becoming much more humane.

I'm not an apologist for the commercial meat sector or for any meat production system. They

A quartet of baby lambs explore their world.

Maple the pig snoozes away her day at Sterling College farm, Vermont.

each exist for a different reason and are aimed at different customers. What I've attempted to do in this book is to present my industry as accurately and dispassionately as possible. To do this, I talked to a wide range of experts: meat scientists, animal welfare organizations, grass advocates, feedlot operators, marketing boards, breeders, historians, nutritionists, veterinarians, doctors, federal and state meat inspectors, and many others (see a list of their names at the end of this book). More important, whenever I came across a claim or statement of opinion about elements of meat production, I verified it via at least two sources. My aim has been to present as close to the truth in as verified a way as possible.

TAKE-AWAY MESSAGES?

I discovered that the impact of hormone implants is not as significant as I had previously believed (this was confirmed by almost every one of the scientists, farmers, and even some activists I spoke with), compared with our daily exposure to hormones or hormone-like substances in other foods, plastics, and many other materials that surround us. However, the use of sub-therapeutic doses of antibiotics is most definitely concerning. I agree with many of those I spoke with that ongoing research is confirming a strong link between the use of antibiotics in meat animals *for purposes other than the necessary treatment of a disease* and the development of drug resistance in microorganisms that can infect humans.

I remain uncomfortable with certain practices in some (I hope very few) retail meat stores and supermarket meat departments that expose customers to unnecessary risk. We all need to become more wary and more informed.

I find it strange that the subject of meat eating has become so politicized; a phenomenon I attribute to humanity's predilection for believing unverified claims and parroting what the other guy says without checking it out for ourselves. There's a lot of disinformation out there—who knows why? Self-interest? Unease about forces we perceive as beyond our control? How about focusing on facts for a change? My take-away lesson from this is to verify everything for yourself. With facts that can be checked.

The strongest conclusion I reached after working on this book for a year has to do with animal welfare. The more I learned about meat raising methods, the more passionate I've become about the way we treat meat animals. So may I ask you for a personal favor? Do some investigation on your own about the ways meat animals are raised. If you'd like some suggestions, check the resources list at the back of this book. Or visit local farms. Or both. That's all I ask.

Keeping Honesty and Integrity in the Local Movement

I'm all for keeping my food dollars in my local community. I'm all for eating locally raised healthy meats. What I am *not* for is paying three times more for local meat and getting the exact same thing as conventionally raised meats!

When a farm advertises "Vermont Raised" meat, I expect its animals to have been born on that Vermont farm and to have spent their entire lives on that farm. If the farmer buys cattle from somewhere else, brings them back to the farm for finishing, and advertises the meat as "local," is it really local? Don't think so. And don't assume that this doesn't happen; it does, much more often than you'd believe. But more to the point—*how would you know?*

Farmers who cheat in this way are doing exactly what the local movement is trying to turn away from. "Locally raised" means nothing if the farmer is buying culled cattle from another state and feeding them for a few months, without knowing (or caring) what the animals ate or how they were treated before they got to him or her.

Local doesn't necessarily mean healthy. It has about as much meaning as "all natural" or "free range." It's a claim intended to give people a sense of security about what they're eating. The local movement is a great movement, but increasingly I see the term prostituted by cheats trying to cash in on it.

Another practice that bothers me has to do with farm names. You'll see a name like Joe's Black Angus Farm, which implies that they only raise and sell true Black Angus beef cattle, when in fact it's just a name. Picked on purpose.

Then there are claims such as "grass-fed." Well, was it grass-fed and grass-*finished*? Did it eat just grass all its life right to slaughter or did it graze on grass and spend the last few months of its life eating additive-laced feeds? *How would you know?*

Another thing I've seen and find appalling is one or two of my state's larger beef farms bringing frozen meats to some retail stores and allowing those stores to thaw it and sell it as fresh local meat. *How would you know?*

We also have farms selling vacuum-packed pre-ground beef to stores, which then open the packs, re-grind the meat, and sell it as "Fresh Store Ground." *I'm getting redundant, but how would you know?*

If the local movement (which perhaps ought to be called the healthy movement) is to survive and deliver high-quality, healthy, humanely treated meat to local tables, then something needs to change. Honesty and integrity don't automatically happen, because there will always be cheats waiting for an opportunity to jump on the bandwagon of a good cause for their own gain.

North America has many thousands of ethical farmers with integrity and passion about what they do. These farmers truly practice sustainable, humane, and healthy farming. You've met some of them throughout these pages. However, there are some other farmers out there who just see dollar signs in the local food movement and are only trying to make a buck. I know a few of them and don't like what they stand for. I see everything from deceptive labeling to misrepresentation of products. I see them doing everything that the local movement is trying to turn away from. I see it with individual farmers, with retailers, and with CSAs.

So to answer the implied question in the title of this section, keeping honesty and integrity in the local food movement will depend on the active participation of people who are prepared to defend best practices and speak out against deceptive ones. I believe it would be a good idea to create a set of guidelines to be monitored by an organization similar to those that monitor and guarantee organic production. Until something like this happens, many consumers will continue to pay three times more for local meat with no guarantee that they're getting anything essentially different from commercial meat.

Cole's Notes

CSA, or community-supported agriculture, is a locally based method of agriculture in which a network of individuals pledge to support one or more local farms. CSA members pay for a share of the anticipated harvest; once harvesting begins, they receive weekly shares of that harvest. There are meat CSAs as well as those offering vegetables or fruits.

VALUING OUR FOOD

There was a time in my youth when it was common to hear my parents, aunts, and uncles talk of how hard things were during the Great Depression. If anyone would know, they would: My mother was one of 17 children, and my father one of 13.

My mother and father divorced when I was eight years old. Things were pretty tough for a single mother of 10 with 7 children still at home. It was the late 1950s and money was scarce even though Mother worked full-time.

Wages for women at that time left a lot to be desired. The food budget was very limited, and waste of any kind was not even thinkable. We used every scrap of food. Dessert was a luxury reserved for holidays. Often for days at a time a meal consisted of bread and gravy. Mealtime was a time for family and a time to give thanks for what we did have to eat.

The labor my mother went through to make a dozen loaves of bread on a wood cook stove did not go unappreciated. Today my family honors her memory for all of the effort, hard work, and love that she put into providing for her children.

Valuing the Animals We Eat

In today's world we seem to take the things we have for granted. Most of us have no idea where the food we eat comes from or how our meat is raised and processed.

Mother's milk is delicious. *Courtesy Patrea L. Pabst, Beaver Creek Farm.*

THE FUTURE

Despite the dearth of butchers and disappearance of traditional butcher shops from our streets, the craft is not dead. Not yet. Increasingly, young people are testing the waters via either internships, courses, or seminars.

Sterling College is a progressive liberal arts college in northern Vermont, in the beautiful old community of Craftsbury. Its 100-plus students come here because of Sterling's commitment to the environment and grassroots sustainability. It's one of only seven Work–Learning–Service colleges in the nation; all Sterling students work on campus, earning a portion of their tuition fees while serving the community.

Anne Obelniki is the director of sustainable food studies, part of the Sustainable Agriculture team. One of the program's initiatives is called Farm Scale Production of Value-Added Products, part of a larger five-week course called Vermont's Table, now in its second year.

And for the second year in a row, I was recently invited to teach a one-day seminar in hog butchering. The students were from all over . . . talking to them gave me hope for the future.

Anne Obelniki.

We butchered one of the college farm's own pigs, and then the students invented their own sausages. Lots of fun—but for me, it was more interesting to talk to some of them. Why were they here?

Nadine Nelson

I grew up in Toronto and now live in New Haven, Connecticut. I'm a chef and owner of Global Local Gourmet, a "gastonomical oasis" for people who like food, art, and care about diversity. I also teach culinary classes.

My family is Jamaican, so I've seen goats and chickens butchered, but never pig. Growing up in Jamaica, we all knew the difference between supermarket and natural. Lots of people don't understand that the choices they make are unhealthy. Yet we can have some control over our health. I want to show people how to eat better and help them understand that doing it this way is actually much easier. My goal is to have my own culinary studio with a farm link, where I can teach cooking classes. Most of our meal experiences are passive—I want food to be more interactive.

Matthew Ehrlich

I live in New York, but I'm originally from Los Angeles. I have a BA in Spanish history, so you're probably wondering what I'm doing here. What really interests me are the eating and culinary habits of different cultures. I'm looking to go into the academic side of culinary history, to teach as a professor and to write.

In the last 10 to 15 years, I think food studies has come into its own as a legitimate subject for serious study. People have taken food for granted, so it wasn't considered "worthy" of serious study.

I'm here to learn about the physical aspects of food production, to gain a better understanding of what it takes. I do think that awareness of healthy eating is spreading fast. Both Boston University and New York University now have food studies programs, so things are changing.

Corrie Weikle

I'm from Minnesota. I'm an environmental studies major at St. Catherine and Macalester colleges. I'm very interested in the connection people experience with their food—especially in urban environments. I believe that when somebody has an "Aha!" moment, they'll change. I want to help people experience the "Aha!" I'm here to learn more about our food systems.

My ultimate goal is to create a nonprofit that connects people not only to their food, but also to their community, and helps spur the "rescaling" of food and community infrastructure.

I grew up with a father who hunted and fished. Humans are omnivores, and should eat meat, but in an ethical manner. Anything that feels deserves ethical treatment.

Nadine Nelson.

Matthew Ehrlich.

Corrie Weikle.

A cheer at the end of the day!

Every piece of meat we eat represents the life of an animal. Isn't it our responsibility to understand how it was raised and slaughtered? I think so. Today's highly industrialized meat industry should be understood by all consumers, to ensure that the animals we eat are not subjected to miserable lives and miserable deaths. And no animal—for any reason—should be abused or unnecessarily stressed. They are not "walking meat packages"; they are living creatures that are much more intelligent than they've been made out to be. If you refuse to face facts and choose to pretend that the unpleasantness of certain aspects of the conventional meat sector do not exist, you're a coward.

A MORE HOLISTIC APPROACH

But you know I'm not advocating vegetarianism. Humans have always eaten meat; it's a delicious food that's good for us. What I am suggesting is that we choose more ethical (and often tastier) options. I'm astonished how passive people are when it comes to their food—especially meat. I doubt you'd be so accepting if it involved, say, a financial investment.

An ewe and her two lambs in a pasture.

You can source local and ethically raised meats. You can opt for organic. You can insist on antibiotic-free meats, or choose grass-fed and -finished, or a grain-finished product that has been pasture-raised by an ethical rancher. If your budget is limited and some of these options seem expensive, you can select cheaper cuts (which are just as delicious as prime rib if cooked properly), or simply eat a little less meat. You can find a meat market or butcher shop whose staff are informed and will talk to you about their products. You can connect with caring farmers or look for farms certified by one of the animal welfare associations. You have a world of choice.

Meat animals have no choice—they give their lives to feed us. It's our responsibility to make certain that the life they live is good and as nature intended it to be. This is one part of honoring the animal.

Honoring the Animal

The other aspect of honoring the animal is to use it all. You've heard the phrase *nose-to-tail eating*—that's what this means. Each and every part of the animal has a use—from the hide to the hooves. If you're going to do your own butchery, I urge you to explore this. You'll get much more value from

This young steer is surrounded by the glories of autumn in a Vermont pasture.

Courtesy Patrea L. Pabst, Beaver Creek Farm.

your carcass and be honoring the animal by using all of it.

I used to hunt with my son—now I just walk in the woods. Often I've come across the carcass of deer shot out of season and found just the hindquarter missing. To kill an animal and take only the choicest cut, leaving the rest to rot, is an insult not only to the animal but to nature itself. Even though it is illegal to shoot wild game out of season, I'd be a bit more forgiving if I knew that the hunter had taken all the meat.

The same applies to the domestic animals we raise for meat. The greatest honor you can bestow on your animal is to make use of all of it. Be an upstanding human being and think about how you eat.

Creating Ties with the Land

I have a bumper sticker on my truck that says NO FARMERS; NO FOOD. I believe that seeking out and supporting local agriculture has terrific benefits. Consumers become more knowledgeable about the food they eat and develop links with the land—a

spiritual as well as practical benefit. Farmers benefit through exposure to larger markets and often better prices for their animals. Small local shops that carry local meats develop stronger ties with customers and have a wider range of exciting products to advertise and sell.

Insisting on Excellence

Food is the bedrock of our lives, and those who produce it deserve to be honored—if they are approaching their work in the right spirit and with a commitment to excellence. In my early days, respect was bestowed on the art of butchery—not only by customers but by fellow butchers as well. There was a certain etiquette among butchers that seemed to vanish during the 1970s, except in truly high-end butcher shops, where butchery was often a work of art.

Now what I see are "Oh, that's good enough" owners who take no pride in what they offer their customers. This is the sort of thing that disgusts a butcher of my generation. If I put time, energy, and expense into raising an animal for meat, I want the best that animal has to offer. I want to honor it by masterfully cutting it to maximize every ounce and use every edible part. I want my cuts to be beautiful and uniform as possible—with a perfect trim—laying the meat out in table-ready cuts as though I were painting the *Mona Lisa*.

So my conclusion? It's time for a change.

How about this? Let's not accept second or third best. Let's not condone cruel and unethical practices. Let's learn about where our food comes from and teach our kids. Let's support the farmers who feed us and stop pretending that what happens outside cities doesn't matter. Let's try sitting down as a family around the dinner table to give thanks for the food and its producers. Maybe thank the animal it came from, too.

And let's bring back traditional butchers. Hey, have I got a steak for you.

GLOSSARY

A

Abbatoir: Another word for a slaughtering facility. Used often in Canada.

Abomasum: The fourth stomach compartment of cattle; the one that is most like the "true stomach" of nonruminant animals (like us).

Abscess: Pus in a cavity formed by tissue disintegration.

Starting with Something Completely Irrelevant . . .

How many of these names for animal groups do you know?

- A herd or drove of cattle
- A gaggle of geese
- An army of caterpillars
- A zeal of zebras
- A pod of whales
- A kettle of vultures
- A mutation of thrush
- A muster of storks
- An (I love this one) escargatoire of snails
- An exultation of larks
- A wreck of seabirds
- An unkindness of ravens
- A prickle of porcupines

Aging: A process beginning at slaughter and continuing over time, during which the connective tissue in an animal slowly breaks down, making the meat more tender.

American Meat Institute: A trade association of meatpacking and -processing companies.

Angle trim: A way of trimming one edge of a steak or piece of meat on the diagonal, so that from the top face, the meat looks nicely trimmed. But when it's turned over, the other side may have excess fat. A nice scam.

Antibiotic: A product that destroys or inhibits the growth of organisms such as bacteria.

Aurochs (*Bos primigenius*): Ancestor of today's cattle. Hunted to extinction in the mid-17th century.

B

Backfat: The amount of fat over an animal's back, usually measured at the 12th to 13th rib.

Backgrounding: A growing program for cattle (whether they're grazing in a pasture or being fed harvested feed) whose duration is generally from the time calves are weaning until they're being finished in a feedyard.

Baconer: A pig reared for bacon only.

Banger: British name for a type of sausage; the name comes from the fact that they used to be made with water and would explode on the grill.

Bantam: A miniature chicken.

Barrow: A neutered male pig.

Beef: Meat from cattle (bovine species) other than calves. Meat from calves is called veal.

Beef Belt: A US region that includes Texas, Oklahoma, Kansas, Nebraska, South Dakota, North Dakota, and the eastern parts of New Mexico, Colorado, Wyoming, and Montana.

Beef Breeds Council: US national organization of beef breed associations.

Bloat: An abnormal condition in cattle in which the rumen is distended because of an accumulation of gas.

Blood sausage: Also called black pudding (*boudin noir* in France), this is a type of sausage made by cooking fresh or dried blood with a filler such as grain, onions, ground meats, rice, or suet until it is stiff enough to stuff into a casing. A variety of animal bloods are used. The final product is usually fried. And it tastes much better than it sounds!

Bloom: The chemical process in which the dark red or purple myoglobin protein in freshly slaughtered meat becomes bright cherry red or pink when exposed to light and oxygen.

Boar: An intact (uncastrated) adult male pig that can be used for breeding.

Boar taint: A strong smell and taste in pork, generally because the meat is from a boar that has not been castrated.

Bolus: A small rounded mass (like chewed food), at the moment of swallowing.

***Bos indicus*:** The scientific name for zebu (humped) cattle, including the Brahman breed in the United States.

***Bos taurus*:** The scientific name for most breeds of cattle in North America and Europe.

Bovine: Term for the family group of cattle.

Bovine spongiform encephalopathy (BSE): A degenerative disease that affects the central nervous system of cattle, causing degeneration of the brain; also called mad cow disease.

Boxed beef: Beef that has been cut up at the slaughtering plant, put into vacuum packages, and shipped out in cardboard boxes to distributors and end users. Almost all North American beef is sold this way.

Branded beef: A specifically labeled beef product differentiated by its brand name, such as (for example) Certified Angus Beef, Laura's Lean, Cattlemen's Collection, and so on. There are many branded beef programs in North America.

Break joint: A cartilaginous area of a sheep's cannon bone that has not yet turned into bone.

Breed: Animals that share a common origin and characteristics that distinguish them from other groups within the same species.

Broiler: A young chicken bred for its meat; generally slaughtered at about 9 to 10 weeks old and usually weighing between 2½ and 3½ pounds.

Broken mouth: A term for an animal with some teeth missing or broken.

Bull: An intact (uncastrated) adult male cattle animal. The term is also applied to other animals—bull elk, bull moose, bull elephant, et cetera.

Bullock: A young bull generally less than 20 months old.

By-product: A product of less value than the major product. The hide and offal are by-products; the meat is the major product. There are two categories of by-products: non-edible (teeth, hair, bones, skin) and edible (liver, kidney, tripe).

C

Calf: A male or female bovine animal under one year of age.

Canadian Food Inspection Agency: The Canadian equivalent of the Food and Drug Administration (FDA) in the United States, its role is to enforce Canadian food safety and nutritional quality standards and carry out inspection and enforcement of animal health and plant protection.

Cannon bone: The leg bone in hoofed animals between the hock (a joint in the hind leg corresponding to our ankle) and the fetlock (the projection just above the animal's hoof).

Capon: A castrated rooster.

Captive bolt: A method of slaughter that uses a gun-like mechanism to fire a bolt or shaft into the animal's head. There are two types of captive bolt:

- **Penetrating**, in which the bolt penetrates the skull; unconsciousness is caused by physical brain damage.
- **Nonpenetrating**, in which the bolt does not penetrate the skull and unconsciousness is achieved by impact.

Casing: The membrane that encloses the filling of a sausage. There are two types of casing: natural and artificial. Natural casings are made from the intestines (very well cleaned) of meat animals. Artificial casings are made from materials such as collagen, cellulose, or plastic, and may be edible or non-edible.

Cattle: Domesticated mammals of the genus *Bos*, including cows, steers, bulls, and oxen, often raised for meat and dairy products. The two most prominent groups are *B. taurus* (or European-type) and *B. indicus* (or the Asian/African-type).

Cattle pack: An entire cattle carcass that has been cut up (probably into primals) and vacuum-packaged into several cardboard boxes as boxed beef.

Cell grazing: See *Rotational grazing*.

Cephalization: Having a head. Scientifically, an evolutionary process in which nervous tissue becomes concentrated at one end of a creature, producing a head.

Certified Angus Beef: A branded beef product supplied by Angus or Angus-crossbred cattle that meets very specific carcass specifications.

Chop mark: A mark on a bone that is short and broad.

Chorizo: Spanish type of pork sausage, usually involving red peppers. Tasty and hot.

Cleaver: A heavy, broad-bladed knife used by butchers for chopping meat.

Close-dated product: A product shipped from the warehouse or wholesaler to the retailer with an imminent use-by or sell-by date. In other words, once opened, it will have little or no shelf life and must be used immediately.

Closed herd: A herd in which no outside breeding stock are introduced.

Cockerel: A male chicken less than one year old.

Composite breed: A breed that has been formed by crossing two or more breeds.

Conjugated linoleic acids (CLA): A family of unsaturated fatty acids especially in the meat and dairy products of ruminants such as cattle.

Continuous grazing: A method of grazing where animals have unrestricted access to an entire grazing unit for most or all of a grazing season.

Controlled grazing: A system of grazing designed to improve the efficiency of the forage production cycle by either (1) dividing pasture into subunits or paddocks and allowing animals to graze in each paddock for a given number of days (generally less than five) or (2) varying stocking rate (number of animals per paddock) to match forage growth rate and availability.

Cooler: A room in a packing plant where carcasses are chilled after slaughter but before processing or cutting up.

Cow: An adult female cattle animal that has had a calf.

Cow–calf operation: An animal-raising farm or ranch that maintains a breeding herd and produces weaned calves.

Creep/creep feeding: Maintaining an enclosure where calves can enter to feed but cows cannot.

Crossbred: Produced by crossing two or more breeds.

Crossbreeding: A controlled process of animal mating from different breeds, usually to gain the advantage of hybrid vigor (*heterosis*).

CSA: An acronym that stands for "community supported agriculture," a locally based agricultural system built around a network of customers who pledge to support local farmers by purchasing meat and produce directly from them.

Cud: The bolus (compressed mouthful) of feed that cattle regurgitate for further chewing.

Cull: To eliminate one or more animals from the breeding herd or flock.

Custom feeding: A system by which a farmer or other entity provide facilities, labor, feed, and care but does not own the cattle.

Cutability: Refers to the fat, lean, and bone composition of the beef carcass. Used interchangeably with *yield grade*. More cutability equals more meat yield.

Cutter: A pig between pork and bacon weight, raised to produce larger joints.

D

Dam: Female parent.

Dark cutter: Term for beef that is a dark purplish red, much darker than the color of fresh beef, and often sticky in texture. Dark cutter meat is caused by stress at slaughter.

Denaturation: The breakdown of the molecular bonds that maintain the three-dimensional structure of proteins (such as meat).

Dhabihah: Muslim law regarding animal slaughter.

Diethylstilbestrol (DES): One of the first hormones used in feedlots, with a nasty history. From 1940 to 1970 it was given to pregnant women to prevent complications, but was shown to cause vaginal tumors in girls exposed to the drug when their mothers were pregnant. It was then banned by the USDA, but further studies have shown that its effects continue to cause significant medical problems through the lives of those exposed to it. A cautionary tale for those who make assumptions about the "safety" of drugs.

Disarticulate: To cut up a carcass into pieces, generally along joint lines.

Domesticate: To tame an animal over generations of breeding to live in close association with people. The animal usually loses its ability to live in the wild.

Dressed beef: Carcasses from cattle.

Dressing percentage: The percentage of the live animal's weight that becomes carcass weight at slaughter. Dressing percentage is determined by dividing the carcass weight by the live weight, then multiplying by 100. Also referred to as yield.

Dry-aging: A process whereby beef carcasses or parts of carcasses are stored without protective packaging at refrigeration temperatures for one to five weeks. This creates a unique flavor and tenderness, but is more costly than wet-aging. Generally, dry-aged beef can only be found in specialty butcher shops or high-end steak houses.

E

Ear marking: A method of permanent identification by which slits or notches are cut into an animal's ear.

Ear tagging: A method of identification that uses a numbered, lettered, and/or colored tag in the ear.

Empanadas: Small stuffed pastries, often with meat.

Encephalization: Quantity of brain mass relative to body size. Generally, the bigger the brain, the smarter the animal.

Environmental Protection Agency (EPA): An independent agency of the US federal government established to protect the nation's environment from pollution. Environment Canada, a federal department, handles this role in Canada.

***E. coli* (*Escherichia coli*):** A common bacteria in the lower intestines of humans and warm-blooded animals such as cattle. *E. coli* helps the body break down and digest food. Most *E. coli* bacteria are harmless, but some strains can cause food poisoning. Symptoms include stomach cramps, vomiting, and diarrhea.

Essential fatty acids: Fats that the human body cannot make itself, but which we need to maintain good health.

Estradiol: A natural hormone normally administered as an ear implant in cattle. Estradiol increases growth rate by 10 to 20 percent, lean meat content by 1 to 3 percent, and feed efficiency by 5 to 8 percent.

European breed: A breed originating in European countries other than England (in which case, they are called British breeds).

Ewe: Female sheep.

F

Farm flock sheep: A flock of sheep kept in smaller pastures than a range band flock of sheep (see *Range band flock*).

Farrowing: Applied to pig raising, the period from birth to weaning.

Farrow-to-finish operation: A swine-raising system that includes all production phases, from breeding to gestation, farrowing, nursery, grow-finishing, and market.

Fat thickness: The amount of fat that covers muscles; typically measured at the 12th and 13th rib as inches of fat over the longissimus dorsi muscle (rib eye).

Feed additive: An antibiotic or hormone-like substance added to an animal's diet.

Feed efficiency: The amount of feed required to produce a unit of weight gain; or, the amount of gain made per unit of feed.

Feeder lamb: A lamb sold to someone for further feeding to a heavier weight.

Feeder pig operation: When the pig breeder sells pigs to a finishing operation, where they will grow to market weight. The pigs sold are then referred to as finisher pigs.

Feedlot/feedyard: A feeding facility in which cattle are fed grain and other concentrates for usually 90 to 100 days. Feedyards are made up of paddocks or pens in which the animals live and range in size from less than 100-head capacity to many thousands.

Feral: A term generally applied to a once-domesticated animal that has returned to the wild.

Fillet: To cut meat into boneless strips.

Finished cattle: Fed cattle whose time in the feedlot is completed; they are now ready for slaughter.

Finishing ration: The feed given to cattle during the latter part of the feeding period. It's usually high in energy so that the animals will put on weight and fat.

Food and Drug Administration (FDA): The US government agency responsible for protecting the public against unsafe food, drug, veterinary, and other products. In Canada, this is handled by the Canadian Food Inspection Agency.

Forage: Herbaceous plants eaten by animals such as horses, pigs, sheep, cattle, and so forth. Most forage is a mixture of plants like grasses, forbs, and legumes.

Forage production: The total amount of dry matter produced per unit of area on an annual basis (pounds per acre per year).

Forager: An animal that wanders in search of its food.

Forb: A nonwoody, broad-leafed plant that is not a grass. Dandelions are forbs, as is alfalfa.

Freemartin: A female born as the twin to a bull (approximately 90 percent of such heifers will never conceive).

Free range: Defined by the USDA to mean that a poultry flock was provided shelter in a building, room, or area with unlimited access to food, fresh water, and continuous access to the outdoors during their production cycle. Use of this term is regulated by the USDA.

Full mouth (solid mouth): A sheep with two central incisors, two middle incisors, two lateral incisors, and two corner incisors (an adult sheep over four years old).

Genetic engineering: Changing the characteristics of an animal by altering or rearranging its DNA. Cloning is genetic engineering, as is gene manipulation.

Genetics: The branch of biology that studies heredity and the patterns of inheritance of specific traits.

Genotype: The genetic makeup—*not* the physical appearance—of an individual. Genotype determines the animal's hereditary potential.

Genus: A group of very similar species. The term is used in the science of taxonomy, which identifies, describes, classifies, and names living beings.

Gestation: The time from conception until the female gives birth.

Gestation period: The length of a pregnancy.

Gilt: A young female pig that has not yet had piglets.

Glatt kosher: A subset of the Jewish dietary laws of Kosher in which an animal's lungs after slaughter must be free from adhesions or other defects.

Glycogen: A white substance that makes up the primary carbohydrate storage material in animal muscle and is converted into lactic acid.

Grain: A cereal grass or the seeds of a cereal food grass such as wheat, corn, rye, oats, rice, or millet. Grain can also refer to the collective—"We harvested grain."

Grass: Any plant of the botanical family *Gramineae*, with jointed stems, slender sheathing leaves, and seed-like grains.

Grass-fed: A term defined by the USDA to refer to animals that receive a majority of their nutrients from grass throughout their lives. A GRASS-FED label does not limit the use of antibiotics, hormones, or pesticides.

Green chop: Freshly harvested forage fed directly to livestock (in other words, the forage is brought to the animal).

Green meat: A term for meat that has not aged enough to be acceptable.

Gross weight: The sum of net weight (product alone) plus tare weight (the weight of the container).

Ground beef log: A document that records information about every shipment of ground meat, including the name of the firm from which the meat was purchased, the fat content, the lot number, and the date(s) on which it was further ground. Ground beef logs must be kept by every retail meat store that sells ground beef by fat content, and serve to provide "trackability" in case of contamination or food poisoning. Without an accurate log, it would be impossible to identify the source of contamination.

Guild: An organization formed to protect a specific craft and its practitioners.

Gummer: An ewe that has lost all her teeth.

Haggis: Scotland's national dish, make by mixing a sheep's lungs, heart, and liver with suet and oatmeal, stuffing it into the sheep's stomach, and boiling for three hours. Often served with neeps (turnips) and tatties (potatoes). It's delicious.

Halal: A specific term applied to foods allowable under the dietary laws of the Muslim faith.

Hardware disease: Infection of the heart sac, lungs, or abdominal cavity in cattle caused by ingestion of sharp objects that are captured in the rumen and pushed into the reticulum, where contractions force the object into the peritoneal cavity.

Hay: The fodder harvest of a pasture, cut and dried for animal feed.

Heifer: A female cattle animal that has not had a calf; generally under three years old.

Hen: A female chicken over one year old.

Herbivore: An animal that eats only plants.

Herdbook (herd register, herd record): The official record of individual animals and pedigrees of a recognized breed of livestock, often dating back to the 1800s. In today's livestock business, the ancestral record of an individual animal is a major determinant of its market value.

Heterosis: Vigor or strong growth of offspring that perform better than the average of both parents. Also called hybrid vigor.

Hog: Another term for a pig.

Holocene: Relating to the geological time span that began at the end of the last Ice Age. We are living in the Holocene period.

Hominid: The scientific term for humans and any member of the species of animal we are most closely evolved from.

Hospital pen: An isolated pen in a feedlot where sick cattle are treated.

Hot carcass weight: The total weight of a carcass just prior to chilling.

Humane: A label that suggests that the animals were treated humanely during the production cycle. Since verification is not regulated under a single USDA definition, it is important to seek out an ethical humane-certification program.

I

In-pig: When a gilt or sow is pregnant—"with child" for pig mamas.

Intensive grazing management (IGM): A grazing system in which a pasture is subdivided into sub-units or paddocks and the animals moved from one to another, generally remaining in each paddock for just a few days. See *Rotational grazing.*

Intermuscular fat: Fat located between muscle systems; also called seam fat.

Intramuscular fat: Fat within the muscle; known commonly as marbling.

Ionophore: An antibiotic that enhances feed efficiency by changing the microbial fermentation in the rumen of an animal.

K

Kashrut: Jewish dietary laws.

Kosher: A specific term applied to foods allowable under the dietary laws of the Jewish faith.

Kosher meat: Meat from ruminant animals (with split hooves) slaughtered according to Jewish law.

L

Lamb: Young sheep, generally up to a year old. In sheep operations, the term refers to an animal with eight milk teeth and no permanent teeth. Lamb is also the meat of a young sheep.

Lard: Rendered pig fat. Some claim the best piecrust is made with lard. I wouldn't know; I simply cannot make a pie.

Leaker: A package of vacuum-packed meat that has been punctured.

Legume: A flowering plant that produces seeds in pods, such as beans, peas, alfalfa, and clover. Legume crops enrich the soil where they are planted; legume seeds are rich in protein.

Linebreeding: A form of inbreeding whose intention is to concentrate a particular bull's genes within a herd.

Litter: The offspring of a single farrowing, in pigs.

Locker plant: A refrigeration establishment with rentable lockers for food storage.

Loss leader: A product offered at an artificially low price (often at a loss) in order to attract customers into the store.

Luganega: A traditional Italian pork sausage, seasoned with spices.

M

Macellum: Latin word for a butcher's stall or meat market; root of the modern Italian word *macelleria*, which means "butcher shop."

Management intensive grazing: See *Rotational grazing.*

Marbling (intramuscular fat): White fat flecks or streaks visible within lean pieces of meat. Marbling is usually evaluated in the rib eye between the 12th and 13th ribs.

Mast: The term for the nuts or fruit of forest trees, which are eaten by animals as they forage.

Meat by-products: Animal-derived products such as hides, feathers, dried blood, fat, and protein meals like bonemeal.

Meatpacker/meat processor: A plant that slaughters, cuts up, processes, packages, and distributes meats for human consumption and meat by-products for other uses.

Melengesterol acetate: A growth-promoting animal feed additive.

Middle meats: The rib and loin of a beef carcass. These primals generally yield the highest-priced beef cuts.

Modified atmosphere packaging (MAP): A way of lengthening the shelf life of a fresh food by sealing it in a package containing a mixture of gases in proportions designed to retard decay by slowing oxidation and microbial growth. The gas mixtures vary according to the food involved.

Muscling: The amount of lean meat in a slaughter animal or carcass.

Mutton: The meat of a sheep older than 24 months.

N

N6/N3 ratio: The number of grams of omega-6 fatty acids versus the number of grams of omega-3 fatty acids in a food.

Natural: This term is narrowly defined by the USDA to refer to meat, poultry, and egg products that are minimally processed and contain no artificial ingredients. It does not define standards regarding farm practices and applies only to the *processing* of meat and egg products. Thus, a NATURAL label can be misleading, sometimes deliberately.

Net weight: The weight of the product alone (without its packaging).

No added hormones (raised without hormones): Note that federal regulations have never permitted hormones or steroids in poultry, pork, or goat.

Nutrient: A substance that nourishes the metabolic processes of the body.

Nutrient density: Amount of essential nutrients relative to the number of calories in a given amount of food.

O

Offal: Organs removed from inside the animal during the slaughtering process.

Omasum: One of four stomach compartments of a ruminant animal.

Omnivore: An animal (like me, unfortunately) that eats everything.

Organic: Defined by the USDA to mean that the food or other agricultural product has been produced through approved methods that foster cycling of resources, promote ecological balance, and conserve biodiversity. Synthetic fertilizers, sewage sludge, irradiation, and genetic engineering may not be used. To be granted a USDA ORGANIC seal, the product must be certified by a USDA-approved third party and must contain 95 percent or more organic content.

Outbreeding: Continuously mating females of a herd to unrelated males of the same breed. Sometimes called outcrossing.

Ovis: Scientific name for sheep.

Ox: An ox is a cattle animal used for draft purposes. Oxen (the plural) are usually castrated adult male cattle.

P

Packing plant: A facility where animals are slaughtered and processed.

Paddock: A pasture subdivision.

Paleolithic: The period in earth's history that began about two million years ago. Also called the Stone Age. It's divided into the Lower Paleolithic (until about 200,000 years ago), the Middle Paleolithic (from 200,000 years ago to about 40,000 years ago), and the Upper Paleolithic (from 40,000 years ago to about 10,000 years ago).

Pannage: A method of raising animals in which they are pastured in a forest.

Pasture: Land used for grazing animals. Pasture is generally enclosed by a fence and is managed by the farmer. Pasture is a mixture of plants such as grasses, forbs, and legumes.

Pasture-raised: A term used to denote that an animal was raised in a pasture. There are many variables involved in pasturing animals, so the term is vague and there is no USDA definition. A pasture-raised animal may have been fed growth hormones or not, may have been finished on grain or not, and so on.

Pathogen: A biologic agent (bacterium, virus, protozoan, nematode) that can cause disease.

Phenotype: The characteristics of an animal that can be seen or measured (for example, its color, weight, horns/no horns, and so on).

Piglet: A young pig.

Polled animal: An animal that is naturally hornless.

Portion-cut beef: Generally, 10-pound boxes of identical cuts (such as rib eyes), each vacuum-packed for convenience.

Power scavenging: The process of driving away a predator that has just killed an animal in order to steal the carcass.

Preconditioning: The preparation of feeder calves for marketing and shipment. The process may include vaccinations, castration, and training calves to eat and drink in pens.

Price lookup code (PLU): An identification number affixed to food products in stores to make checkout and inventory control easier and faster. PLUs are programmed into the checkout cash registers.

Primal cut: A large section of a carcass. A whole carcass is first cut into a side; then each side of the carcass is broken down into large portions called primals, from which subprimals and then retail cuts can be made.

Progesterone: A hormone secreted by the female reproductive system that causes the uterus to change so that it provides a suitable environment for a fertilized egg.

PSE meat: A term used in pork production for "pale, soft, exudative" meat. PSE occurs in pork and poultry and is characterized by a pale off-color and poor consistency. Simply stated, it's mushy. The condition is related to stress.

Pullet: A female chicken under one year old.

Purebred: An animal eligible for registry with a recognized breed association.

Q

Quality grade: Meat grades such as prime, choice, and select group carcasses into value- and palatability-based categories, generally according to the degree of marbling and the age of the animal.

R

Ram: A male sheep.

Range band flock: Generally, a flock of over 1,000 sheep grazing in large fenced or open-range pasture, eating only what they can find in the natural environment. The flocks are managed by shepherds and sheepdogs.

Rangeland: Open land that may or may not be grazed. It is not managed.

Red meat: Meat from cattle, sheep, pigs, and goats.

Registered: Referring to an animal that is recorded in the herdbook of a breed.

Retail cuts: Meat in sizes for purchase by the consumer.

Reticulum: A stomach compartment of a ruminant animal lined with small compartments, giving a honeycomb appearance.

Rib-eye area (REA): The area of the longissimus dorsi muscle, measured in square inches, between the 12th and 13th ribs. Also referred to as the loin-eye area.

Righting reflex: The physical actions by an animal to move itself into a normal lying, sitting, or standing position.

Ritual slaughter: A slaughtering method particular to a specific religion.

Rooster: A male chicken more than one year old (also called a cock).

Rotational grazing: An important component of sustainable agriculture, rotational grazing involves dividing a pasture into smaller paddocks ("cells") and moving animals from paddock to paddock. Animals are usually kept in one paddock for several days. Also termed cell grazing or management intensive grazing.

Roughage: Feed that's high in fiber, low in digestible nutrients, and low in energy.

Rumen: A compartment of the ruminant stomach similar to a large fermentation pouch where bacteria and protozoans break down fibrous plant material swallowed by the animal.

Ruminant: A mammal with a four-part stomach made up of the rumen, reticulum, omasum, and abomasum. Cattle, sheep, goats, deer, and elk are ruminants.

Rumination: Regurgitating undigested food to be chewed and then swallowed again. Also, me, gazing off into the distance.

Runt: The smallest piglet in a litter.

S

Salumi: Italian cured meat products; salami is a specific type of salumi.

Self-service meat case: The most common way of presenting meat, this is generally a long shelf-like case where pre-cut meat packaged in clear plastic film is laid out for customers to pick up themselves.

Service meat case: Generally, a glass-fronted meat case where products are displayed and a service clerk or butcher personally serves customers. Service meat cases operate on a higher profit margin than self-service meat cases.

Shechita: Jewish ritual slaughter.

Shoat: A growing pig. This term has been largely replaced with the terms *nursery pig* or *grow-finish pig*.

Shochet: Person who performs Jewish ritual slaughter.

Silage: Fresh, high-moisture fodder that has been fermented and stored in a silo or plastic-covered bale that keeps air out.

Slap mark: A form of identification where the pig has been tattooed with an identification number.

Slaughter lamb: A lamb ready to slaughter—generally from 100 to 135 pounds.

Slice mark: A mark on a bone that is narrow and linear.

Sounder: Term for a group of wild boar.

Soundness: Degree of freedom from injury or defect.

Sow: An adult female pig that has had piglets.

Spool joint: The ossified area of an animal's cannon bone.

Stag: A boar that is castrated after sexual maturity.

Steer: A castrated male cattle animal.

Stocker: Weaned cattle that eat high-roughage diets (including grazing) before going into the feedlot.

Stocking density: The number of animals or total animal live weight in a defined area (such as a paddock) at a particular point in time. Stocking density is usually defined on a per-acre basis.

Straw: The dry stems and dead leaves of cereal plants like wheat, oats, and barley after the grain has been removed.

Stun/kill: A slaughtering method in which the animal is stunned before killing.

Subcutaneous fat: External (visible) fat on the carcass.

Subprimal cuts: Cuts that are smaller than primal cuts: top round, bottom round, eye of the round, and sirloin tip, for example.

Sus: The scientific name for the genus of animals that includes wild and domestic pigs.

Sus scrofa: The scientific name for the domestic pig.

Swine: A term used to describe all pigs.

T

Tare weight: The weight of an empty container (without its product).

Terminal sire: A sire used in a crossbreeding system in which all its offspring (male and female) are marketed.

Testosterone: A natural hormone often combined with estradiol in ear implants.

Tourtière: A French Canadian meat pie, often of pork, although other meats may be used.

Trenbolone: A synthetic veterinary drug and steroidal hormone used in cattle ear implants to increase muscle growth and appetite. Trenbolone is sometimes used illegally by human athletes such as bodybuilders to increase body mass.

U

Ungulate: A hoofed mammal, including goats, cattle, deer, pigs, and so on.

V

Vaccine: A biological preparation used as a preventive inoculation to confer immunity against a specific disease, usually employing an innocuous form of the disease agent, such as killed or weakened bacteria or viruses to stimulate antibody production.

Value-added product: Generally, a product that has been further processed or pre-prepared to offer customers almost-ready-to-eat meals. Examples are stuffed chicken breasts or roulades, fresh (in-store-prepared) sausage, and more. Value-added meat products command higher prices than standard cuts of meat.

Variety meats: Edible organ meats including liver, heart, tongue, tripe, and so on. More generally known as offal.

Veal: Meat from young cattle often under three months of age. Veal typically comes from dairy bull calves.

W

Weaner: An animal separated from its mother and eating only solid food.

Weaning: Separating young animals from their mothers so that they can no longer suckle.

Weaning weight: Weight of the young animal when it's removed from its mother.

Wet-aging: After meat has been vacuum-packed and placed into boxes for shipping, it continues to age in its own juices—a process known as wet-aging.

Wether: A castrated male sheep.

White meat: Meat from poultry.

WOGS: A butcher's term for a small (2½- to 3½-pound) chicken sold without giblets.

Y

Yardage: The per-head daily fee charged by a feedlot to the owner of the cattle. The fee is usually in addition to the cost of medicine and the feed markup.

Yearling: An animal approximately one year old.

Yearling weight: Weight when approximately 365 days old.

Yield grade: Measurement of the amount of meat that is left after all bone and superfluous fat have been removed; also called cutability.

Z

Zeranol: A synthetic hormone used in cattle ear implants. It is approved in the United States for livestock, and in Canada just for cattle; it is not approved in the European Union.

RESOURCES

Animal Handling and Slaughter

Cargill: Slaughterhouse video tour narrated and hosted by Dr. Temple Grandin; definitely worth viewing.
www.cargill.com/connections/beef-processing-tour-dr-temple-grandin/index.jsp

Dr. Temple Grandin: Comprehensive discussion of animal handling and behavior, humane slaughter procedures, and more.
www.grandin.com

Associations and Industry Groups

American Grassfed Association: A comprehensive site for those seeking information about the growing grass-fed movement.
www.americangrassfed.org/about-us

American Lamb Board: An excellent consumer-oriented site with lamb information, recipes, and more.
www.americanlamb.com

American Meat Institute: A national trade association representing companies that process 95 percent of the red meat and 70 percent of the turkey in the United States.
www.meatami.com

American Meat Science Association: Community and professional development for those who apply science to provide safe, high-quality meat.
www.meatscience.org

American Sheep Industry Association: A national organization representing more than 80,000 US sheep producers.
www.sheepusa.org

American Society of Animal Science: Supports scientists and animal producers by fostering scientific knowledge concerning the responsible use of animals.
www.asas.org

Animal Welfare Institute: Seeks to alleviate suffering inflicted on animals by people.
http://awionline.org

Canadian Cattlemens' Association: A mega-site for the Canadian beef sector.
www.cattle.ca

Canadian Food Inspection Agency: Works with industry and consumers to protect Canadians from preventable health risks related to food and zoonotic diseases.
www.inspection.gc.ca/eng/1297964599443/1297965645317

Farm Forward: Implements strategies to promote conscientious food choices, reduce farm animal suffering, and advance sustainable agriculture.
www.farmforward.com

The Livestock Conservancy: A nonprofit working to protect nearly 200 breeds of livestock and poultry from extinction.
www.livestockconservancy.org

National Cattlemen's Beef Association: A mega-site for US beef producers and consumers.
www.beef.org

National Institute for Animal Agriculture:
A source for information, education, and solutions regarding challenges facing animal agriculture.
www.animalagriculture.org

National Livestock and Meat Board (NLSMB): A US organization providing nutrition, research, education, and promotional information on beef, pork, and lamb.

National Pork Board: A pork producers' mega-site.
www.pork.org/home.aspx

National Pork Producers Council: Conducts public-policy outreach on behalf of 43 affiliated state associations of US pork producers and industry stakeholders.
www.nppc.org

Pork. Be Inspired: Recipes, cuts, and cooking.
www.porkbeinspired.com/index.aspx

Prairie Swine Centre: A nonprofit conducting research on behalf of western Canadian pork producers (the website features a live webcam of a modern swine facility).
www.prairieswine.com

USDA (United States Department of Agriculture): A huge and helpful site for everything ag-related, although you'll need to dig to find what you're looking for.
www.usda.gov/wps/portal/usda/usdahome

USDA Food Safety and Inspection Service (FSIS): The public health agency responsible for ensuring that the nation's commercial supply of meat, poultry, and egg products is safe, wholesome, and correctly labeled and packaged.
www.fsis.usda.gov

USDA FSIS Meat, Poultry, and Egg Product Inspection Directory: List of all slaughtering facilities in the United States, with their codes so that you can identify where your packaged meat came from.
http://www.fsis.usda.gov/wps/portal/fsis/topics/inspection/mpi-directory

USDA US Meat Animal Research Center:
www.ars.usda.gov/main/docs.htm?docid=2340

USDA Meat and Poultry Labeling Terms:
www.fsis.usda.gov/factsheets/meat_&_poultry_labeling_terms

USDA National Organic Program: A 2012 list of certified USDA organic operations.
http://apps.ams.usda.gov/nop

Butchers Featured in This Book

Antica Macelleria Falorni, Greve, Italy: Will ship anywhere in Europe (sorry, North America!).
www.falorni.it/en/macelleria/home

Boucherie Viandal, Montréal, Canada: No website, but what a place!
550 rue de l'église, Verdun, Québec, H4G 2M4; 514-766-9906 (they speak English)

Citarella's Meats & Deli, Red Bank, New Jersey: A wide selection and very tempting, but you'll need to pick it up yourself.
http://citarellasmarket.com

Don and Joe's Meats, Seattle, Washington:
www.donandjoesmeats.com

Lindy & Grundy's Meats, Los Angeles, California:
http://lindyandgrundy.com

MacDonald Bros. Butchers, Pitlochry, Scotland: Mail order only in the UK, but oh, what a selection!
www.macdonald-bros.co.uk

R. J. Balson & Son, UK and US: This is the American website for the oldest UK butcher shop, featuring mail-order British meat products like back bacon and bangers, recipes, and much more. They'll ship to you!
www.balsonbutchers.com

Savenor's Market, Boston, Massachusetts: A wonderful website, with events and classes, store locations, profile of supplier farms, recipes, and the like.
www.savenorsmarket.com/web

Equipment (Knives, Scrapers, Sausage Equipment, Et Cetera)

Butcher & Packer: Sausage-making and butcher supplies.
> www.butcher-packer.com

Hobart: Commercial butcher equipment— maker of Cole's favorite power meat saws.
> www.hobartcorp.com

Hubert USA: A comprehensive array of chef and butcher equipment. www.hubert.com

Kasco Sharptech: Professional meat room and butcher supplies.
> www.kascosharptech.com/butcher
> -supplies.html

Syracuse Casing Company: Cole's recommended hog and sheep casing supplier (they will sell direct to individuals; we checked!).
> www.makincasing.com/x/home.php

Victorinox: Maker of Cole's favorite meat-cutting knives.
> www.victorinox.com/stories

Farmers Featured in This Book

Applecheek Farm, Vermont: John Clark's certified-organic beef, veal, pork, poultry, guinea fowl, duck, turkey, and eggs, available on the farm or in Boston.
> www.applecheekfarm.com

Border Springs Farm: Craig Rogers's award-winning natural lamb, recipes, cooking tips, and more.
> www.borderspringsfarm.com

Dickinson Cattle Company, Ohio: Darol Dickinson's branded Longhorn Head to Tail free-range lean, low-calorie beef products, including ground beef, sausages, and jerky.
> www.head2tail.com/beef

Grass Roots Farm, Vermont: Paul List's grass-fed and -finished Lowline Angus beef.
> http://grassrootsfarmvt.com

Morning Star Ranch, Texas: Pam Malcuit sells Dexter cattle and herd sires, miniature donkeys, and goat fiber.
> www.morningstarranch.net

North Hollow Farm, Vermont: Mike and Julie Bowen's grass-fed and -finished beef, natural pork, chicken, and goat, plus "the real thing" maple syrup.
> http://vermontgrassfedbeef.weebly.com

Pinnacle Pastures Farm, Wisconsin: Wendy and Cody's grass-fed and -finished beef and lamb; also pastured pork, plus ducks and geese.
> www.pinnaclepastures.com

Sugar Mountain Farm, Vermont: Walter Jeffries's superb pork. The online store has many products, including tusks and T-shirts.
> http://sugarmtnfarm.com/home

History

Cattle Raisers Museum:
> www.cattleraisersmuseum.org

Pointe-à-Callière Museum, Montréal: Re-creation of an 18th-century public market.
> http://pacmusee.qc.ca/en/calendar-of
> -activities/cultural-activities/18th
> -century-public-market

Smithfield Market: Website for the market, with photos and history.
> www.smithfieldmarket.com

TaurOs Programme: Returning aurochs to Europe's wilderness.
> www.taurosproject.com

The Good Wife's Guide (*Le Ménagier de Paris*): Can be ordered online, and is available as an ebook.
> www.cornellpress.cornell.edu/book/?GCOI
> =80140100274850

The Market Assistant by Thomas F. De Voe: Now an ebook; conduct a Google search for "The Market Assistant."

Sourcing Meat

Eat Wild: A terrific site about sustainable food, including grass-fed and -finished meat. Extensive state-by-state lists of farmers/ranchers who sell to the public.
www.eatwild.com

Eat Wild Canada:
www.eatwild.com/products/canada.html

Local Harvest: Similar to Eat Wild, only exclusively organic foods, including meats. State-by-state lists.
www.localharvest.org

Local Harvest Canada:
www.localharvest.ca

Murray McMurray Hatchery: Meat, poultry, and other fowl. Sells directly to individuals.
www.mcmurrayhatchery.com

THANKS

Many people helped us by graciously sharing their expertise, reading chapters, and verifying scientific or factual information. We want to thank them for the time they invested with us that they probably could have spent doing something else (like eating a nice, juicy steak).

They are:

Terry Ackerman, Canadian Lamb Producers Cooperative

Brian Arnold, vice president of member outreach and youth development, National Swine Registry

Denise Beaulieu, PhD, research scientist, nutrition, Prairie Swine Centre

Jeanette Beranger, research and technical programs manager, The Livestock Conservancy

Mark Boggess, PhD, national program leader, Food Animal Production 101; animal production and protection national program leader, Rangeland Systems 215; Natural Resources and Sustainable Agricultural Systems USDA-ARS

Imam Bukhari, Islamic Centre of Quebec

Shannon Burasco, President of CWT Farms International

Jason Care, manager/auditor, Manitoba Hog Grading

Marty Carpenter, executive director, market development North America, Canada Beef Inc.

Steven M. Carr, PhD, professor of biology, Memorial University of Newfoundland

Stephen Cave, deputy director, Grading and Verification Division, USDA Agricultural Marketing Service

Jenn Colby, Pasture Program coordinator, University of Vermont Center for Sustainable Agriculture

Cindy Delaloye, manager, Canadian Beef Grading Agency

Paul Ebner, PhD, associate professor, Department of Animal Sciences, Purdue University

Mike Engler, CEO, Cactus Feeders, Amarillo, Texas

Peter R. Ferket, PhD, WNR Distinguished Professor of Nutrition, Prestage Department of Poultry Science, North Carolina State University

Stephen E. Gerike, director of food service marketing, National Pork Board

Ronald Goderie, director, TaurOs Programme, the Netherlands

Ben Goldsmith, executive director, Farm Forward, Portland, Oregon

Temple Grandin, PhD, professor of animal science, Colorado State University

John B. Hall, PhD, professor, extension beef specialist and superintendent, Nancy M. Cummings Research Extension and Education Center, University of Idaho

Dr. Gail Hansen, senior officer, Pew Charitable Trusts

John Hardiman, chief scientific officer, Cobb-Vantress Inc.

Marcia E. Herman-Giddens, PA, DrPH, adjunct professor, Department of Maternal and Child Health, School of Public Health, University of North Carolina

Susan Holtz, senior policy analyst, Canadian Institute for Environmental Law and Policy

Robin Horel, manager, Chicken Sector, Canadian Poultry and Egg Processors Association

Chris Hostetler, PhD, PAS, director of animal science, National Pork Board

Salima Ikram, PhD, chair SAPE Department, professor of Egyptology, American University in Cairo, Egypt

Tabatha Jeter, marketing manager, PIC (Pig Improvement Company), Hendersonville, Tennessee

Dena M. Jones, Farm Animal Program manager, Animal Welfare Institute, Washington, DC

Greger Larson, PhD, reader, Department of Archaeology, Member Centre for the Coevolution of Biology and Culture, Durham University, UK

Aaron Lavallee, deputy assistant administrator, Office of Public Affairs and Consumer Education, Food Safety and Inspection Service, USDA

Rick Machen, PhD, professor and specialist, animal and natural resource management, Texas Agrilife Research and Extension Center

Mark A. McCann, Extension animal scientist, Animal and Poultry Sciences Department, Virginia Tech

Madeline McCurry-Schmidt, scientific communications associate, American Society of Animal Science

Richard J. McIntire, public affairs specialist, Food Safety and Inspection Service, USDA

Bob Oros, CSP, CMC, Center of the Plate consultant

Thomas Powell, PhD, CAE, executive director, American Meat Science Association

Randy J. Quenneville, Meat Programs Section chief, Vermont Meat Inspection Service

Gil J. Stein, PhD, director of the Oriental Institute, professor of Near Eastern archaeology, Oriental Institute, University of Chicago

Emily Teeter, PhD, Oriental Institute, University of Chicago

David R. Thompson, DVM, public health veterinarian and deputy district manager, USDA FSIS OFO, Jackson District

Gwen Venable, vice president of communications, US Poultry and Egg Association

Kimberly Vonnahme, PhD, associate professor in animal sciences, Department of Animal Sciences, North Dakota State University

Bridget Wasser, senior director, meat science and technology, National Cattlemen's Beef Association

Sydney Weisman, WHPR, Los Angeles

Tommy L. Wheeler, PhD, research leader, Meat Safety and Quality Research Unit, US Meat Animal Research Center, USDA Agricultural Research Service

Bud Wood, owner, Murray McMurray Hatchery, Iowa

REFERENCES AND SOURCES

Chapter 2:
A Long Tradition

Thomas F. De Voe, *The Market Assistant* (New York: Hurd and Houghton, 1867).

Jimmy Dunn, "Cattle: The Most Useful Animal of Ancient Egypt," www.touregypt.net /featurestories/cattle.htm.

Jimmy Dunn (writing as Taylor Ray Ellison), "The Origins of Egyptian Religion," www.touregypt .net/featurestories/religiousorigin.htm.

Ann Gibbons, "The First Butchers," *Science Now*, 2010, http://news.sciencemag.org/sciencenow /2010/08/the-first-butchers.html.

Haskel J. Greenfield, Ehud Galili, et al., "The Butchered Animal Bones from New Yam, a Submerged Pottery Neolithic Site off the Carmel Coast," *Journal of the Israel Prehistoric Society* 36 (2006): 173–200.

Salima Ikram, *Choice Cuts: Meat Production in Ancient Egypt* (Leuven, Peeters Press and Department of Oriental Studies, 1995).

Dwayne James, "The Nineteenth Century Farmer in Upper-Canada: A Comparative Butchering Analysis of Four Historical Sites in Ontario," (dissertation, Trent University, Ontario, Canada, 1997).

Maison Saint-Gabriel Museum, Québec, www.maisonsaint-gabriel.qc.ca.

Mark Maltby, "Chop and Change: Specialist Cattle Carcass Processing in Roman Britain," *TRAC 2006: Proceedings of the 12th Annual Theoretical Roman Archaeology Conference* (2006): 59–76, http://eprints.bournemouth.ac.uk/11697/1 /Maltby_revised_TRAC_paper.pdf.

Shannon P. McPherron, Zeresenay Alemseged, et al., "Evidence for Stone-Tool-Assisted Consumption of Animal Tissues Before 3.39 Million Years Ago at Dikika, Ethiopia," *Nature* 466 (2010): 857–60.

Elia Psouni, Axel Janke, et al., "Impact of Carnivory on Human Development and Evolution Revealed by a New Unifying Model of Weaning in Mammals," *PLoS ONE* (2010): www.plosone.org/article/info%3Adoi %2F10.1371%2Fjournal.pone.0032452.

Caroline Seawright, "History of Egypt: Upper Paleolithic: 30,000–10,000 BC," www.touregypt.net/ebph3.htm.

Krish Seetah, *Meat in History: The Butchery Trade in the Romano-British Period* (UK: University of Cambridge, 2004).

Pat Shipman, "A Worm's View of Human Evolution," *American Scientist*, www.american scientist.org/issues/pub/a-worms -view-of-human-evolution/1.

Craig B. Stanford, *The Hunting Apes: Meat Eating and the Origins of Human Behaviour* (Princeton, NJ: Princeton University Press, 1999).

Maguelonne Toussaint-Samat, *A History of Food* (Chichester, UK: Blackwell Publishing & John Wiley and Sons, 2009).

J. B. Wheat, *A Paleo-Indian Bison Kill: New World Archeology* (San Francisco: W. H. Freeman, 1974), 213–21.

Chapter 4:
Getting Meat to the Table

Jack Kyle, "Rotational Grazing," Ontario Ministry of Agriculture, Food, and Rural Affairs, www.omafra.gov.on.ca/english/crops/field/news/croptalk/2009/ct-0909a5.htm.

Chapter 7:
A Side of Beef

"Antibiotic Resistance," *Science Daily*, www.sciencedaily.com/articles/a/antibiotic_resistance.htm.

Antibiotic Use in Food Animals: A Dialogue for a Common Purpose. Information synthesized from an October 26–27, 2011, symposium in Chicago, IL. Institute for Animal Agriculture.

"Backgrounding Beef Cattle in Saskatchewan," Saskatchewan Agriculture, Food and Rural Revitalization, August 2003, 3MISBNo-88656-689-8 LB0113.

Ruth Bollongino, Joachim Burger, et al., "Modern Taurine Cattle Descended from Small Number of Near-Eastern Founders," *Molecular Biology and Evolution* 29(9) (March 11, 2010): 2101–2104.

Paul D. Ebner, Jiayi Zhang, et al., "Contamination Rates and Antimicrobial Resistance in Bacteria Isolated from 'Grass-Fed' Labeled Beef Products," *Foodborne Pathogens and Disease* 7(11) (November 2010): 1331–36, http://online.liebertpub.com/doi/abs/10.1089/fpd.2010.0562.

Dan S. Hale, Kyla Goodson, et al., "Beef Quality and Yield Grades," Department of Animal Science Texas AgriLife Extension Service College.

John B. Hall and Susan Silver, "Nutrition and Feeding of the Cow–Calf Herd: Digestive System of the Cow," Virginia Cooperative Extension, Virginia Tech, May 2009, http://pubs.ext.vt.edu/400/400-010/400-010.html.

Ray V. Herren, *The Science of Agriculture: A Biological Approach*, second edition (Clifton Park, NY: Delmar, 2004).

Susan Holtz, *Reducing and Phasing Out the Use of Antibiotics and Hormone Growth Promoters in Canadian Agriculture*, (Canadian Institute for Environmental Law and Policy, 2009).

Dan Loy, "Understanding Hormone Use in Beef Cattle," Iowa State University Extension.

National Agricultural Statistics Service (NASS), Agricultural Statistics Board, United States Department of Agriculture (USDA). *Cattle on Feed.* August 17, 2010, OSSM 1948-9080.

Northwest FCS Cattle Knowledge Team, "Industry Perspective: Feedlot 2007," Northwest Farm Credit Services, http://agr.wa.gov/fof/docs/feedlot.pdf.

Christopher Raines, "Hormones in My Organic Food? Yep," comment on Meat Blogger posted on December 5, 2009, http://meatblogger.org/2009/10/05/hormones-in-my-organic-food-yep.

Frederick K. Ray, "Meat Inspection and Grading," Oklahoma Cooperative Extension Service, ANSI-3972, http://pods.dasnr.okstate.edu/docushare/dsweb/Get/Document-1950/ANSI-3972web.pdf.

Jerry Speir, Marie-Ann Bowden, et al., "Comparative Standards for Intensive Livestock Operations in Canada, Mexico and the United States," (Alberta Agriculture and Rural Development, 2003).

Chapter 8:
A Side of Pork

"Designing Feeding Programs for Natural and Organic Pork Production," University of Minnesota Extension, BU-07736 2002.

Daniel W. Gade, "Hogs," in *The Cambridge World History of Food*, ed. Kenneth F. Kibble and Kriemhild Conee Ornelas (Cambridge: Cambridge University Press, 2001): 536–542.

Michael Hosenberg, Clark Nesbitt, et al., "Hallan Cemi Tepesi: Some Preliminary Observations Concerning the Fauna1 Assemblage," *Anatolica* 21 (1995).

R. L. J. M. van Laack, et al., "Evaluating Pork Carcasses for Quality," National Swine Improvement Federation Annual Meeting, December 1, 1995.

Greger Larson, Thomas Cucchi, et al., "Genetic Aspects of Pig Domestication," in *Genetic Aspects of the Pig*, ed. Max F. Rothschild and Anatoly Ruvinsky (CAB International, 2011): 14–37.

Mick Vann, "A History of Pigs in America," *Austin Chronicle*, April 10, 2009, www.austinchronicle.com/food/2009-04-10/764573.

John Noble Wilford, "First Settlers Domesticated Pigs Before Crops," *New York Times*, May 31, 1994, www.nytimes.com/1994/05/31/science/first-settlers-domesticated-pigs-before-crops.html?pagewanted=all&src=pm.

Chapter 9:
A Side of Lamb

Lloyd Arthur, Brett Kessler, et al., "Lamb Carcass Evaluation," Purdue Animal Sciences, http://ag.ansc.purdue.edu/sheep/ansc442/semprojs/carcass/442.htm.

C. Dwyer, "The Behaviour of Sheep and Goats," VetMed Resource, www.cabi.org/vetmedresource.

Richard Gray, "Sheep are Far Smarter than Previously Thought," *Telegraph*, February 20, 2011, www.telegraph.co.uk/science/science-news/8335465/Sheep-are-far-smarter-than-previously-thought.html.

K. Kris Hirst, "Domestication History of Sheep," About.com, http://archaeology.about.com/od/shthroughsiterms/qt/Sheep-History.htm

"Yield Grades and Quality Grades for Lamb Carcasses," Virginia Department of Agriculture and Consumer Services, www.vdacs.virginia.gov/livestock/lambcarcass.shtml.

Chapter 10:
Cluck—I'm Your Domestic Chicken

Hiromi Sawai, Hie Lim Kim, et al., "The Origin and Genetic Variation of Domestic Chickens with Special Reference to Junglefowls *Gallus g. gallus* and *G. varius*," *PLoS ONE*, May 19, 2010, www.plosone.org/article/info:doi/10.1371/journal.pone.0010639.

Alice A. Storey, José Miguel Ramírez, et al., "Radiocarbon and DNA Evidence for a Pre-Columbian Introduction of Polynesian Chickens to Chile," *Proceedings of the National Academy of Sciences USA* 104 (2007): 10335–39.

INDEX

Note: Page numbers in *italics* refer to figures and photographs. Page numbers followed by *t* refer to tables.

green
press
INITIATIVE

ABOUT THE AUTHORS

COLE WARD grew up in the tiny Vermont town of Sheldon Springs. At the age of fourteen he began working part-time for a local butcher, washing meat trays and stuffing sausages for 20 cents an hour. At fifteen, he became an apprentice meat cutter at the local IGA, and in very few years was a master butcher specializing in whole-animal culinary butchery. In his early twenties, wanderlust took him out west to a job at LaFrieda Prime Meats at Los Angeles's celebrated farmers' market. The famous butcher shop was next to CBS studios, and Ward's celebrity clients soon included Billy Crystal, Bernadette Peters, Perry Como, Edith Head, and Raymond Burr.

In 1982, Cole returned to Vermont, where most of his large family lives. He worked in markets and supermarkets around the state, managed meat departments, and eventually began giving workshops and doing on-farm cutting. Today Cole mixes hands-on butchering with teaching, and his encyclopedic knowledge of the meat sector makes him a sought-after lecturer and seminar leader at culinary academies, colleges, and agricultural and sustainable-living conferences. His full butchery course was recently released on the two-DVD set *The Gourmet Butcher*.

KAREN COSHOF began her career as a commercial photographer, shooting campaigns for Air Canada, Sheraton, Clairol, and other clients, as well as fashion spreads and magazine covers. She then joined Stonehaven Productions, a communications company that produces television as well as specialized projects for the Canadian government and international corporations. At Stonehaven, Karen worked as a scriptwriter, proposal writer, print broker, director, producer, and executive producer.

While at Stonehaven, she conceived and produced *The Great Warming*, a three-hour Discovery climate-change series narrated by Keanu Reeves and Alanis Morissette that was broadcast in fifteen countries. She also co-executive produced the one-hour national PBS special *Global Warming: The Signs and the Science*.

She has been a keynote speaker at the Pacific Islands Environmental Conference, the US Fuel Ethanol Workshop, the Canadian Renewable Fuels Association, the World Business Council on Sustainable Development, and the Caribbean Society of Trust and Estate Planning.

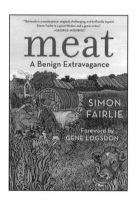
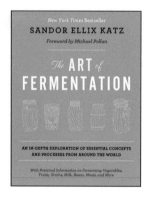
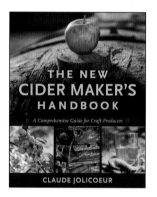